FOR SCIENCE, KING & COUNTRY

With best wishes to
Gordon.

Russell Egdell

Roy Macleod.

FOR SCIENCE, KING & COUNTRY

The Life and Legacy of Henry Moseley

Edited by
Roy MacLeod, Russell G Egdell,
and Elizabeth Bruton

This edition first published by Uniform
an imprint of Unicorn Publishing Group

Unicorn Publishing Group
101 Wardour Street
London W1F 0UG

All rights reserved. No part of this publication may be reproduced, stored in a retrieval system or transmitted, in any form or by any means, electronic, mechanical, photocopying, recording or otherwise, without prior permission in writing from the publisher.

© Unicorn Publishing Group, 2018
www.unicornpublishing.org

A catalogue record for this book is available from the British Library

ISBN 978-1-910500-71-2

Cover design Uniform Press
Typeset by Vivian@Bookscribe

CONTENTS

PREFACE	vii
ACKNOWLEDGEMENTS	viii
INTRODUCTION: *Roy MacLeod and Russell G. Egdell*	1

PART ONE: LIFE

1.	HENRY MOSELEY AT ETON AND OXFORD *Clare Hopkins*	14
2.	MOSELEY IN MANCHESTER *Neil Todd*	45
3.	READING BETWEEN THE LINES: DECONSTRUCTING MOSELEY'S DIAGRAM *Kristen M. Frederick-Frost*	67
4.	ANTONIUS VAN DEN BROEK, MOSELEY AND THE CONCEPT OF ATOMIC NUMBER *Eric Scerri*	90
5.	SACRIFICE OF A GENIUS: MOSELEY THE SIGNALS OFFICER AND SIGNALLING IN GALLIPOLI *Elizabeth Bruton*	107

PART TWO: LEGACY

6.	MOSELEY AND THE POLITICS OF NOBEL EXCELLENCE *Robert Marc Friedman*	136
7.	MOSELEY AND THE MATTEUCCI MEDAL *Russell G. Egdell, Francesco Offi, Giancarlo Panaccione*	155
8.	X-RAY SPECTROSCOPY AND THE DISCOVERY OF NEW ELEMENTS *Russell G. Egdell*	175
9.	X-RAY SPECTROSCOPY 100 YEARS ON *Russell G. Egdell*	199
10.	ACCOUNTS OF MOSELEY AND VERSIONS OF HIS LAWS *John L. Heilbron*	224
11.	ARTEFACTS AND ARCHIVES: PRESENTING MOSELEY IN A MUSEUM CONTEXT *Elizabeth Bruton, Silke Ackermann, Stephen Johnston*	244

APPENDIX I	270
APPENDIX II	275
APPENDIX III	281
BIBLIOGRAPHY	286
CONTRIBUTORS	298
INDEX	302

PREFACE

As the ancients wrote, they whom the gods love die young. Seldom in the history of modern science is there recalled a tragedy as great as that of the early death, in the Great War, of Henry Gwyn Jeffreys Moseley, a Signals Officer in the Royal Engineers, killed in action on 10 August 1915 at Gallipoli, shortly after arriving and near the climax of the Allies' ill-fated Dardanelles campaign. At the age of twenty-seven, young Moseley was regarded by many – both then and since – as among the most promising British physicists of his generation. Before his death, he was nominated for Nobel Prizes in both Physics and Chemistry. Had he survived, it is widely believed that he was destined to win a Prize in 1916 or later.

Moseley's death was not unique, or unprecedented, in a war that claimed the lives of so many talented men and women, amongst the many millions who were lost to history. But it provoked an unusual, international outpouring of grief that echoed across the lines and through the popular press around the world. Inevitably, in Britain, there came many reassessments of the roles that scientists might be asked to play in wartime. But grief at his loss was combined with increasing respect for his work, as the intellectual power of his achievements became apparent. Today, these achievements are widely recognised. Across many disciplines, his legacy stretches from the research laboratory to the limits of outer space. For the historian, his brief life says much about the nature of physical science, and physical scientists, at the turn of the 20th century.

For these reasons, in 2015, the Museum of the History of Science at Oxford took the opportunity presented by the centenary of his death, and the recovery of his experimental apparatus, to hold an exhibition in his honour. This exhibition, *Dear Harry… Henry Moseley: A Scientist Lost to War,* ran from 14 May 2015 to 31 January 2016. Although the story was seen by many – and is recaptured on the web at http://www.mhs.ox.ac.uk/moseley – a more detailed analysis was called for, and this book is the result – a testament to his life and to his legacy, and to the enduring power and excitement of experimental science.

Roy MacLeod
Russell G. Egdell
Elizabeth Bruton

Sydney, Oxford and London
October 2017

ACKNOWLEDGEMENTS

A book of this nature, with such a rich and technical history, and with such a wide diversity of authors, inevitably acquires many debts, which we as editors are grateful to acknowledge. Especially for their contribution to the first chapter, we wish to thank Eleanor Hoare, Archivist, and Sally Jennings, Collections Administrator of Eton College; Laurence Dardenne, Librarian of Summer Fields, Oxford; Sharon Cure, Librarian and Bryan Ward-Perkins, Fellow Archivist, of Trinity College, Oxford; Alysoun Saunders, Archivist of Macmillan Publishers International; and Mr Robert Ingham, a graduate of Trinity, who generously lent his family photographs to the College archives. Throughout our enquiries, the support of Will le Fleming, a great-nephew of Harry Moseley, has been invaluable. For their assistance with Nobel and other Archives, we wish to thank the Centre for History of Science of the Royal Swedish Academy of Science, Stockholm; for assistance with information, we wish to thank the Library of the Royal Society of London, and the Fisher Library of the University of Sydney. We are also indebted to Mrs. Antonella Grandolini, Archivist of the *L'Accademia Nazionale delle Scienze detta dei XL*, for her help in locating material within the Rome archive.

For their timely and valuable help at many points, we would like to offer our warmest thanks to Tony Simcock, Archivist of the Museum of the History of Science, Oxford; Mme. Catherine Kounelis of ESPCI ParisTech; Ms Isabel Holowaty of the University of Oxford; Dr James Peters and Professor Wendy Flavell of the University of Manchester; Mr Adam Perkins of the University of Cambridge; Ms Catlin Crennell of the Archives of the Science Museum, London; Ms Fiona Keates of the Royal Society of London; Jane Harrison of the Royal Institution; and Amanda Nelson of the American Institute of Physics. We also wish to recognize Mr Alan Gall, at the Institute for Science and Technology in Sheffield, for his helpful correspondence regarding the instrument maker Chas. Cook. For sharing his detailed research on the Battle of Chunuk Bair, and the circumstances surrounding Moseley's death, we are grateful to Mr. D. H. Harding.

The testing and analysis of the Matteucci Medal was facilitated by the assistance of Mike Jenkins, Emeritus Fellow in Materials Science, Trinity College, Oxford; George Smith, FRS, Emeritus Professor of Materials Science, Oxford; and Chris Salter and Alison Crossley of the Department of Materials, University of Oxford. Our thanks go also to Professor Justin Wark, Dr Jim Bennett, both of Oxford, and to Professor Victor Henrich, of the Department of Applied Physics, Yale University. We should also like to extend thanks to Professor John Heilbron, for his encouragement, and for his pioneering contributions to the study of Henry Moseley and his work. For her patience and help in the preparation of this book, Roy MacLeod is indebted to his wife, Dr Kimberley Webber. We owe a special debt to Douglas Matthews for producing a particularly comprehensive and useful index, at short notice, and with all his customary attention to deal. We want also to thank our editor, Vivian Head, for taking such care with our text, and for helping us ensure accuracy and consistency.

Finally, we wish to express our appreciation to the President and Fellows of Trinity College, Oxford; the Museum of the History of Science; and the Department of Physics, Oxford for their generous support of this work, and for their commitment to the memory of Henry Moseley.

INTRODUCTION

ROY MACLEOD AND RUSSELL G EGDELL

Henry Gwyn Jeffreys Moseley (1887–1915) was born in Dorset and graduated from Trinity College in Oxford with a degree in Natural Science. Between September 1910 and November 1913, he worked in the Manchester laboratories of Nobel laureate Ernest Rutherford, and then moved back to Oxford to continue with his experiments until the outbreak of the Great War. In a period of less than two years between October 1912 and June 1914 he laid the foundations of what became the field of X-ray spectroscopy, which led in turn to the development of ideas about the structure of atoms that underpin most areas of contemporary chemistry and physics. He also predicted the existence of four 'missing' elements – those now known as technetium, promethium, hafnium and rhenium. Moseley was killed on 10 August 1915, shortly after landing in Anzac Cove on the Gallipoli peninsula during the ill-fated Dardanelles campaign. Before he died at the age of twenty-seven he had been nominated for Nobel Prizes in both Chemistry and Physics.

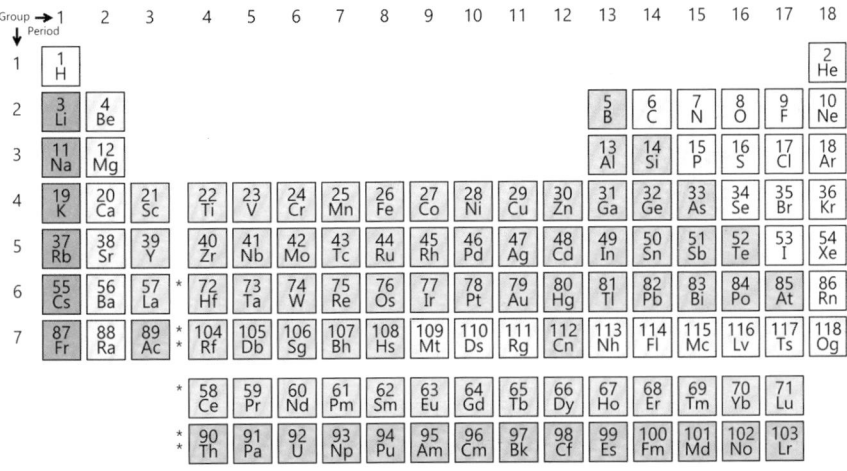

Fig. 1 The periodic table of the chemical elements, with each element identified by its atomic number. The four 'missing' elements predicted by Moseley are those with atomic numbers 43, 61, 72 and 75. (Source: https://commons.wikimedia.org/wiki/File:Simple_Periodic_Table_Chart-en.svg#file, accessed 27 November 2017, by courtesy of Offnfopt via Wikimedia Commons).

The periodic table of the chemical elements ordered on the basis of the charge on their nuclei – their atomic number – is recognised as one of the most enduring achievements of twentieth-century science (Fig. 1). Moreover, the iconic image of the periodic table has escaped the narrow confines of chemistry text-books and lecture theatres to find a place in the wider cultural landscape of the twenty-first century (Fig. 2). Elements within the same 'group' of the table have similar properties, but in moving across 'rows', chemical properties change in a regular way within each row. The organisation of the elements in this way now seems self-evident, but the anguished history of attempts to systematise chemical knowledge in a simple table extends back to the early days of the nineteenth century,[1] whilst the final element appearing in Fig. 1 (Og, oganneson, element 118) was only formally ratified and named in November 2016.[2]

Fig. 2 A taxi outside Oxford railway station decorated in the livery of the periodic table. As shown in the inset for element 66 (Dy, dysprosium), each element is identified with its name, chemical symbol, atomic number and atomic weight (which is obscured by the rear door handle in the inset). The energies of the strongest X-ray lines are also given. Element 61 (Pm, promethium) is one of Moseley's 'missing' elements. (Source: Photograph taken by R. G. Egdell, 2009, by courtesy of Oxford Science Park).

Those who have studied chemistry or physics may know that Moseley is generally credited with establishing the concept of atomic number, based on his measurements of the frequencies of X-rays emitted from different elements; and that Moseley's new ordering of the elements superseded that developed by Mendeleev in the nineteenth century,[3] based on atomic weights. But very few will know in detail the origins and gestation of Moseley's ideas, his debts to others, his differences with authority, and the novelty of his insights. Taking account of these factors, as seen both in his time and in retrospect, the reader is quickly taken beyond easy formulaic generalisations about the nature of scientific discovery, and into the more human heartland of intelligent guesswork and good judgment, if not also good luck

Although Moseley was publicly mourned and widely remembered, for nearly sixty years he lacked a definitive biography, which came only in 1974, with Professor John Heilbron's *H. G. J. Moseley: The Life and Letters of an English Physicist, 1887–1915*.[4] Since then, interest has grown in the man and his subject, and new archival information has surfaced through renewed contact with later generations of the Moseley family via the 'Dear Harry' exhibition, held at the Museum of the History of Science in Oxford (MHS) between May 2015 and February 2016. The centenary of Moseley's death in 1915, combined with national and international commemorations of the Great War, afforded the MHS an opportunity to draw upon the key collections of Moseley's equipment and notebooks, and his mother's diaries, to mount the exhibition. For the first time we are therefore in a position to re-examine key themes in a life that scientists have long taken for granted, and to see the relevance of an object – and instrument – centred museum-led approach to the history and public understanding of science. Both as art, as artefact, and as a very human subject, Moseley quickens our curiosity, claims our attention, and wins our respect.

In the forty years since Heilbron's biography, there has been episodic interest in Moseley, but no new comprehensive account of his life and legacy. Indeed, his achievements, which register today on the borderlines of physics and chemistry, cannot easily be comprehended within a single discipline, and have not been attempted by a single author. What we have, instead, is the need to see Moseley from a comparative, interdisciplinary point of view. This volume does not attempt simply to retrace the ground covered in earlier work, but rather to explore a series of interlinked themes relating to his life and legacy, using much hitherto unpublished material. The chapters that follow assess (1) Moseley's early life and education at Eton and Oxford; (2) his pioneering experiments in X-ray spectroscopy in Manchester and Oxford, and how others influenced his thinking about atomic number; (3) Moseley's service as a Signals Officer in Gallipoli, with accounts of his death; (4) the deeper history of Moseley's nomination for the 1915 Nobel Prizes in Physics and Chemistry and the posthumous award of the Italian Matteucci Medal for Physics in 1919; (5) the discovery of elements missing from the periodic table of 1914, with a focus on the role of X-ray spectroscopy in these discoveries; (6) an appraisal of the development of X-ray spectroscopy in the twentieth and twenty-first centuries; (7) a retrospective view of the interpretations and mis-interpretations of Moseley's Law, and speculation as to what he might have achieved had he survived the Great War; and finally (8) a discussion of the genesis and presentation of the 'Dear Harry' Moseley Exhibition at the Museum for the History of Science in Oxford.

Looking back over a century, what can be said about Henry Moseley's fundamental achievements? Moseley's Law as enunciated in his two key papers published in 1913 and 1914 shows that the square roots of the frequencies of the X-rays emitted from different elemental targets assign an ordinal number to each element that in turn defines its position in the periodic table. Moseley also suggested that this number was the same as the charge on its nucleus.[5]

X-rays were discovered at the end of the nineteenth century as a highly penetrating form of radiation emitted when energetic cathode rays (now known as electrons) fall on a metal target in an evacuated tube. The immediate backdrop to Moseley's experiments was provided by two more recent discoveries. First, the finding by Charles Barkla that so-called secondary X-rays emitted from different elements when irradiated with primary X-rays from a tube have different powers to penetrate through thin metal foils, leading to the idea that X-rays are 'characteristic' of an element. Barkla also established that a given target could emit two different sorts of characteristic X-rays: harder (more penetrating) K-type rays and softer (less penetrating) L rays.[6] The second discovery was the observation of symmetrical spot patterns when X-rays are transmitted through a crystal onto a photographic plate, leading to the idea that crystal diffraction could be used to separate an X-ray beam into components of different wavelength.[7] Closer to home, the unexpected finding by Geiger and Marsden in Manchester that highly energetic alpha particles (now known to be doubly charged helium ions) could be scattered through large angles when incident on ultra-thin metal foils[8] led Rutherford to propose a model of atomic structure where the positive charge on an atom was concentrated in a small 'nucleus', whose size was much smaller than that of the atom.[9] This contrasted with J. J. Thomson's earlier 'plum pudding' model of the atom, where negatively charged electrons were moved within a spongy background of positive charge whose size matched that of the atom.[10]

When Moseley started his pioneering experiments on the high-frequency [X-ray] spectra of the elements in 1913, Rutherford's now-familiar nuclear atom was still in its infancy, and the charge on the nucleus was presumed to be half the atomic weight. However, the idea that the charge on the nucleus should be equated with the order number of the element in the periodic table had already been articulated in the work of an obscure Dutch lawyer, Antonius van den Broek.[11] In a letter to *Nature* published in 1913, Moseley states that 'my work was undertaken with the express purpose of testing Broek's hypothesis'.[12] However, van den Broek remains a little known figure in the history of science. On the other hand, it is widely recognised that the famous Danish scientist Niels Bohr, who was a frequent visitor to Manchester, had an important influence on Moseley's experiments.

Early in 1913, Bohr still adhered to the consensus view in Manchester that the nuclear charge was half the atomic weight.[13] But during an interview conducted in 1962, Bohr implied that just before Moseley had started his experiments on different elements in August 1913, the idea of order number being the same as nuclear charge was common currency in Manchester.

> [Rutherford] even spoke to me in one of his later years and said that he felt that I had had much more to do with the atomic number than van Broek. And I said, 'Of course, I had'…. I got to know Moseley really partly in this discussion about whether the [*sic*] nickel or cobalt should be in the order of their atomic weights. … Moseley came to me and Darwin and he asked what I thought about that. And I said, 'There can be no doubt about it. It has to go according to the atomic number'. And then he said, 'We will try to see'. And that was the beginning of his experiments. And then he did it at tremendous speed, and then, of course, he died, so things were so difficult.[14]

In a letter to *Nature* submitted on 6 December 1913 Rutherford gives due credit to van den Broek, Bohr and Moseley:

> The original suggestion of van den Broek that the charge on the nucleus is equal to the atomic number and not to half the atomic weight seems to me very promising. This idea has already been used by Bohr in his theory of the constitution of atoms. The strongest and most convincing evidence in support of this hypothesis will be found in a paper by Moseley in the *Philosophical Magazine* of this month.[15]

The Bohr model[16] itself was undoubtedly an important stepping stone between the understanding of the structure of atoms as it was before 1913 and our contemporary understanding. Bohr envisaged electrons as circulating around the atomic nucleus in orbits having a well-defined radius. Limitations on the radii and energies of orbits were imposed by the condition that each was constrained to have an angular momentum quantised in units of $h/2\pi$, where h is the constant introduced by Planck in his theory of black-body radiation. The problem that a circulating charged particle should radiate energy was circumvented by the *ad hoc* assumption that electrons in a Bohr orbit occupied a privileged 'stationary state', and energy was only absorbed or emitted when an electron jumped between two different orbits. Bohr was then able to show that the energies of the single electron in hydrogen should take on values E_n where:

$$E_n = -R/n^2$$

The n in this formula can have integer values 1, 2, 3 and so on. R is the ionisation energy of hydrogen (also known as the Rydberg constant), which was shown to be a simple function of h and the charge and mass of the electron.

Bohr's formulae accounted almost perfectly for the atomic spectrum of hydrogen and were readily extended to other 'hydrogenic species' with only one electron, such as He^+, where R was replaced by Z^2R, Z being the charge on the nucleus. Atoms with two or more electrons proved more problematic. The idea of 'rings' of electrons still held but it seemed unlikely that all of the electrons should be in the same ring. The major problems were how to determine the distribution of the total number of electrons between the rings; and how to evaluate the energies of the electrons taking account of inter-electron repulsion as well as electron-nuclear attraction. In broad terms, it was realised that a necessary consequence of repulsion between electrons would be to reduce Z mentioned above to $Z-\sigma$, where σ is what would now be called a shielding constant. Bohr was optimistic that Moseley's result would provide important insights into ring occupancy. However, in the end, the Bohr Model proved incapable of dealing with atoms with more than one electron.

Our current understanding of the electronic structure of atoms and the nature of X-ray spectra emerged only after the introduction of the ideas of wave mechanics and quantum theory, as encapsulated in the Schrödinger and Dirac equations.[17] The pattern of quantum numbers that must be assigned to the electrons in an atom also determines the overall structure of the periodic table. A brief technical account of the electronic structure of atoms is given in Appendix I, which also explains the relationship between the conventional descriptions of electronic structure used today and the arcane system used to label and classify X-rays that was first introduced before the Great War. This labelling system is still used in the contemporary scientific literature, and of necessity appears in many chapters in this book.

The ideas of wave mechanics and quantum theory are radical departures from classical physics. Although Moseley's tangible legacies in pioneering the technique of X-ray spectroscopy and in reshaping the periodic table are widely appreciated, perhaps his most enduring contribution to physical science is in the broader way in which his technique was critical in consolidating general acceptance of the Bohr-Rutherford Model; and in thereby paving the way for the transition from the classical to the quantum world. In Bohr's own words:

> … you see actually the Rutherford work [the nuclear atom] was not taken seriously. We cannot understand today, but it was not taken seriously at all. There was no mention of it any place. The great change came from Moseley.[18]

The collection of chapters in our book brings together specialists in history, science, and the history of science. The first chapter by Clare Hopkins deals with Moseley's early family life, as well as his education at Eton and Oxford. It explores his growing interest in science – particularly chemistry – at Eton; and his progress as an undergraduate in Oxford. The next chapter by Neil Todd looks at his formative time in Manchester, and the ways in which radioactive contamination in the Manchester laboratories influenced his research career. This chapter also explores Moseley's interactions with Rutherford and individuals within his group; and shows how Moseley moved his research away from radioactivity (the dominant area of interest to Rutherford at the time) to the newly emerging area of X-ray physics.

Kristen Frederick-Frost brings the story back to Oxford. Her chapter focuses on how the apparatus available to Moseley – particularly the X-ray tubes – shaped the form of the famous plot which led to exposition of Moseley's Laws. To the modern eye, Moseley's plot of atomic number *against* the square root of the X-ray frequency looks distinctly strange: today one would think of the atomic number as *determining* the X-ray frequency, so the two axes would be reversed. In fact Niels Bohr switched the axes around in this way during his Nobel Prize lecture in 1922.[19] However, it needs to be remembered that in 1914 the concepts of the atomic nucleus and the atomic number as the charge on the nucleus were still controversial: the fixed point was provided by the measured X-ray frequency.

This discussion is followed by Eric Scerri, who explores the little-known story of how Antonius van den Broek arrived at his ideas concerning atomic number. Scerri describes in detail how van den Broek re-analysed data from the Geiger-Marsden experiments conducted in Manchester to reach the conclusion that the charge on the nucleus is the order number of elements in the periodic table, and not half the atomic weight.

Returning to Moseley's wider world, Elizabeth Bruton recounts his life and destiny as a Signals Officer in Kitchener's New Army, and the circumstances surrounding his death at Gallipoli. The task of a Signals Officer was one of the most dangerous on the frontline. That Moseley volunteered for the military instead of seeking a reserved occupation reflects much about the patriotic spirit of the man, evident since his early days at school and university. The chapter also analyses the evidence from primary accounts of Moseley's death, which throws new light on the battle which formed the 'climax' of the Dardanelles campaign.[20]

Moseley's death provoked comment, sympathy and outrage both nationally and internationally. Over the years, there has been widespread speculation as to whether he was destined to win a Nobel Prize had he lived; and whether prizes awarded to Charles Barkla in 1918 and Manne Siegbahn in 1925 should be regarded as proxy prizes for Moseley. Robert Marc Friedman weighs the evidence surrounding Moseley's nomination for both the Chemistry and Physics Prizes in 1915, and the decisions taken by those in Stockholm who controlled the award of the Prize. Despite not becoming a Nobel Laureate, Moseley did win posthumously in 1919 another major international prize in physics – the Italian Matteucci Medal – as discussed by Russell Egdell, Francesco Offi and Giancarlo Panaccione. The next recipient of the Medal in 1921 was Albert Einstein, for his theoretical work on relativity. It appears that the Matteucci Committee was more consistent than their counterparts in Sweden in judging its nominees based purely on the merits of their science: Friedman argues that the Nobel Committees were all too often influenced by political considerations.

In the first of his publications on the 'High-Frequency Spectra of the Elements' Moseley realised that X-ray spectroscopy could be both an invaluable tool for the discovery of new elements and an important new technique for chemical analysis.[21] The next two chapters by Russell Egdell explore these aspects of Moseley's legacy. X-ray spectroscopy did lead to discovery of the elements hafnium and rhenium. However, for the four other elements 'missing' after X-ray spectra had been measured for all accessible elements up to uranium, X-ray spectroscopy proved to be a blind avenue of research –

these elements all transpired to be highly radioactive, with short half-lives and almost vanishingly low natural abundances, which rendered measurement of X-ray spectra impossible. Nonetheless, the technique remained a port of call in the search for new elements well into the 1970s. On the other hand, X-ray spectroscopy did develop to become arguably the most important of all chemical analytical techniques, as well as establishing for itself an ongoing role in exploration of the electronic structure of matter in all its forms. Egdell's next chapter provides an overview of X-ray spectroscopy 100 years after the first experiments with the technique and emphasises the enormity of the field of scientific research spawned by Moseley's two pioneering 'High-Frequency' papers.

Taking a long view, forty-years on from publication of his book, John Heilbron then offers a compelling retrospective, exploring the ways in which Moseley's life and legacy have been celebrated, and appropriated, by science and the scientific community. In many ways, Heilbron and Scerri, in recounting Moseley's interactions with Bohr, expose differences in historical interpretation, which remain to be resolved. In the meantime, Professor Heilbron asks us to see and take account of this and a range of other factors that routinely feature in our understanding and appreciation of scientific research, and suggests competing, counterfactual outcomes for our current appreciation of Moseley, had he survived the war and returned to research. The 'what if' question commonly preoccupies historians as well as scientists; in this case, the answer Heilbron suggests, could well have had bearing on the future of nuclear science, not to mention world history.

As our authors show, despite his short life in research, there is much yet to be said and learned about Moseley, his work, and his methods. Our collection concludes with an assessment of the significance of 'Dear Harry: Henry Moseley, a scientist lost to war', the prize-nominated exhibition of Moseley and his work held at the Museum of the History of Science in Oxford. For the many who visited the Museum, this demonstrated the value of bringing together the objects and instruments of experimental research and the material culture of science. Working from the object to the text, we trust this book will inspire historians and scientists to take greater account of museum media, and to engage a wider public. In this way, we hope it will take the story of Moseley – a life and legacy in science – to a wider audience.

Notes and References

[1] Eric R. Scerri, *The Periodic Table: Its Story and Its Significance* (New York: Oxford University Press, hereafter, Scerri, *Periodic Table*); William H. Brock, *The Fontana History of Chemistry* (London: Fontana Press, 1992).

[2] https://iupac.org/iupac-announces-the-names-of-the-elements-113-115-117-and-118/ (accessed on 27 November 2017)

[3] Ulf Lagerkvist, *The Periodic Table and a Missed Nobel Prize* (Singapore: World Scientific, 2012).

[4] J. L. Heilbron, *H. G. J. Moseley: The Life and Letters of an English Physicist, 1887–1915* (Berkeley: University of California Press, 1974).

[5] H. G. J. Moseley 'The High-Frequency Spectra of the Elements', *Philosophical Magazine Series 6*, 26 (156), (1913), 1024–1034 (hereafter Moseley, 'High-Frequency I'); H. G. J. Moseley, 'The High-Frequency Spectra of the Elements. Part II', *Philosophical Magazine Series 6*, 27 (160), (1914), 703-713.

[6] C. G. Barkla and C. A. Sadler, 'Homogenous Secondary Röntgen Radiations, *Proceedings of the Physical Society of London*, 21 (1907), 336–373; C. G. Barkla, 'The Spectra of the Fluorescent Röntgen Radiations', *Philosophical Magazine Series 6*, 22 (129), (1911), 396–412; C. G. Barkla and J. Nichol, 'Homogenous Fluorescent X-radiations of a Second Series', *Proceedings of the Physical Society of London*, 224 (1911), 9–16.

[7] A. Authier, *Early Days of X-Ray Crystallography* (Oxford: Oxford University Press, 2013).

[8] H. Geiger and E. Marsden, 'On a Diffuse Reflection of the α-Particles', *Proceedings of the Royal Society A*, 82 (557), (1909), 495–500; H. Geiger and E. Marsden, 'The Laws of Deflexion of α-Particles through Large Angles', *Philosophical Magazine Series 6*, 25 (148), (1913), 604–628.

[9] E. Rutherford, 'The Scattering of α and β Particles by Matter and the Structure of the Atom', *Philosophical Magazine Series 6*, 21 (125), (1911), 669–688; J. L. Heilbron, *Ernest Rutherford and the Explosion of Atoms* (Oxford: Oxford University Press, 2003), 57–81; Scerri, *Periodic Table, op. cit.* 164.

[10] J. J. Thomson, 'On the Structure of the Atom: an Investigation of the Stability and Periods of Oscillation of a number of Corpuscles arranged at equal intervals around the Circumference of a Circle; with Application of the Results to the Theory of Atomic Structure', *Philosophical Magazine Series 6*, 7 (39), (1904) 237–265; Scerri, *Periodic Table, op. cit.* 185–186.

[11] Eric R. Scerri, *A Tale of Seven Scientists and A New Philosophy of Science* (New York: Oxford University Press, 2016), 41–62; Scerri, *Periodic Table, op. cit.* 165–169.

[12] H. Moseley, 'Atomic Models and X-Ray Spectra', *Nature*, 92 (2307), (1914), 554. This short letter responds to criticisms of his paper 'High-Frequency I' by Dr. F. A. Lindemann.

[13] N. Bohr, 'On the Theory of the Decrease of Velocity of Moving Electrified Particles on Moving through Matter', *Philosophical Magazine*, 25 (145), (1913), 10–31. This paper was published in the first issue of the *Magazine* for 1913 and states that 'we shall expect the values [for the number of electrons in an atom] equal to about half the atomic weight'.

[14] Interview of Niels Bohr by Thomas S. Kuhn, Leon Rosenfeld, Aage Petersen, and Erik Rudinger on 1962 November 1, Niels Bohr Library & Archives, American Institute of Physics, College Park, MD USA, http://www.aip.org/history-programs/niels-bohr-library/oral-histories/4517-2 (accessed on 17 January 2018). Cobalt has a higher atomic weight than nickel, but chemical properties suggest that cobalt should come before nickel in the periodic table. Moseley's paper 'High-Frequency I' showed clearly that the atomic number of nickel was higher than that of cobalt.

[15] E. Rutherford, 'The Structure of Atoms', *Nature*, 92 (2302), (1913), 423. We believe this paper to be the first to make explicit use of the term 'atomic number'.

[16] N. Bohr, 'On the Constitution of Atoms and Molecules', *Philosophical Magazine Series 6*, 26 (151), (1913), 1–25. Although the symbol Z is now generally used to specify the charge on a nucleus, in Moseley and Bohr's papers published before the Great War the symbol N is often used for atomic number and nuclear charge. In the current volume Z is used in most instances and N is reserved for quotations from primary publications.

[17] For accessible accounts of the quantum theory of atoms, see P. W. Atkins, *Molecular Quantum Mechanics, Parts I and II* (Oxford: Oxford University Press, 1970); and F. L. Pilar, *Elementary Quantum Theory* (New York, McGraw-Hill International, 1990).

[18] Interview of Niels Bohr by Thomas S. Kuhn, Leon Rosenfeld, Aage Petersen, and Erik Rudinger on 1962 November 1, Niels Bohr Library & Archives, American Institute of Physics, College Park, MD USA, http://www.aip.org/history-programs/niels-bohr-library/oral-histories/4517-1 (accessed on 17 January 2018).

[19] http://www.nobelprize.org/nobel_prizes/physics/laureates/1922/bohr-lecture.pdf (accessed on 27 November 2017).

[20] The failure of the August Offensive of the Dardanelles Campaign is discussed in detail in R. Crawley, *Climax at Gallipoli* (Norman: University of Oklahoma Press, 2014).

[21] Moseley, 'High-Frequency I', *op. cit.* 1030.

PART ONE
LIFE

CHAPTER ONE
HENRY MOSELEY AT ETON AND OXFORD

CLARE HOPKINS

Introduction

Between the ages of thirteen and twenty-two, Henry Gwyn Jeffreys Moseley attended two of Britain's most elite institutions – Eton College and Trinity College, Oxford – where he received an education that was to shape both his scientific ambitions and his short military career. His natural ability in physics, mathematics, and chemistry was quickly recognised at Eton, and, more slowly, at Oxford; and his schoolmasters and tutors did their best to nurture his gifts. This chapter explores the different ways he took advantage of, but was sometimes frustrated by, the educational opportunities that were made available to him.[1]

Harry – as he was known to his family – entered Eton on 20 September 1901, aged thirteen years and not quite ten months.[2] Almost half his life was already behind him. A study of his education reveals a youth who relished challenges and thrived on competition, qualities that were strongly fostered by the teaching methods offered to the upper middle classes in Edwardian Britain. Harry's letters home demonstrate that he could endure discomfort with stoicism, and disregard opposition with insouciance. Already in place were attributes that were to be characteristic of, and important

Fig 1. Henry Moseley, born 23 November 1887. This is the earliest known photograph of him, taken in July 1889. (Source: Ludlow-Hewitt family collection, 1889, by courtesy of Will le Fleming).

in, his brief but momentous life. In 1910, Harry left Oxford for Manchester, where his work was to win him a reputation as perhaps the most remarkable scientist of his generation.

Eton was founded in 1440 by Henry VI; it is among the most prominent of England's public schools.[3] These independent foundations have traditionally sent to Oxford (and Cambridge) young men of the upper and upper-middle classes destined to become the nation's leading politicians, churchmen, writers, and administrators. The city of Oxford is home to the most ancient university in the English-speaking world; its origins are obscure, but teaching has taken place in the city since at least the late eleventh century. By the middle of the thirteenth century, the first colleges had been established, and in the second half of the thirteenth century, Oxford was home to the renowned Franciscan Roger Bacon (*c.* 1219/20–*c.* 1292), who pioneered a new natural philosophy grounded in experiment and observation. The second half of the seventeenth century saw a significant flowering of science in Oxford, as founding members of the Royal Society – including John Wilkins (1614–1672), Thomas Willis (1621–1675), Robert Boyle (1627–1691), and Robert Hooke (1635–1703) – met together at Wadham College.[4] Their legacy can be seen for example in the naming of Oxford's first chair in Physics as that of Experimental Philosophy. But these were graduate Fellows undertaking private research, and, for most of its history, the University made absolutely no provision for teaching science to undergraduates.

During the half century before Harry's birth, prompted by rising continental competition, successive British Governments had made strenuous efforts to reform the organisation of Oxford and Cambridge, to equip graduates to run an Empire, and to meet the scientific and technical demands of the modern world.[5] Changes made by the Oxford University Examination Statute of 1800 soon evolved into an Honours degree based almost exclusively on classical languages, philosophy and history. The natural result of such a restricted curriculum was that the public schools concentrated on the classics, as most useful for their pupils. Many potentially able science candidates from non-public schools were deterred by the fact that, until 1920, a reading knowledge of ancient Greek was a compulsory requirement of Oxford entrance.[6]

Oxford's Honour School of Natural Science was established in 1850. However, in the fields of science where Cambridge, London, and the provincial universities were making significant progress, Oxford remained slow to change.[7] In 1860–1864, only 8.3 per cent of honours graduates were taking degrees in the subject. Steadily, the picture changed, but as late as 1910–1914, only 13.7 per cent of honours students read science.[8] At Cambridge, where the Natural Science Tripos had been introduced in 1848, the comparative figure had risen from 2 per cent in 1861 to 18 per cent by 1910.[9]

This was the educational reality that confronted Harry. As we shall see, the Moseley family had a strong connection with Oxford, and his education and natural ability together ensured that he had no difficulty meeting the required standards in Latin and Greek. He also imbibed the tenets of established Anglicanism, attending compulsory chapel services at both Eton and Oxford. There is no evidence that he ever questioned, or showed any particular enthusiasm for, the teachings of Christianity.

In 1901, Britain was in the midst of a tragic, imperial war in South Africa, a conflict which was to foreshadow the horrors to be unleashed in 1914. Educated Edwardians expected that every gentleman should 'play his part', a belief sanctified by insistence upon attendance at chapel and chivalry towards women. Eton had an 'Army Class' that taught the skills and knowledge necessary for a regimental officer, and every boy could enlist in the school's Rifle Volunteers as soon as he reached fifteen. Membership of the Corps was hugely popular. In Harry's first half, the *Eton College Chronicle* carried a thrilling report of a 'field day' in which two 'forces', totalling some 299 boys, battled cross-country to take control of Sunningdale Railway Station.[10] For whatever reason, Harry chose not to join the Corps; one possible explanation is simply that it would have taken up too much of his time.[11]

Eton

Harry entered Eton as a King's Scholar, one of the seventy-strong intellectual elite of the school. He was one of six new scholars from his Oxford preparatory school, Summer Fields. This school specialised in Eton success, and each year's young candidates were thoroughly coached by their formidable headmaster, Charles Williams (1851–1941). Indeed, 1901 was only the second year in the decade since 1891 that a Summer Fields boy had not headed the Eton list; and a tradition arose in the Moseley family that Harry would have been top, had he not been unwell in a mathematics exam.[12]

Each Scholar won his place in a highly competitive examination, and lived a privileged life in the heart of the College precinct, set apart from the thousand 'Oppidans' who were divided between Houses in the town. To succeed in the Eton 'election' required ability and preparation. There were eight examinations over three days, in Latin and Greek translation, grammar, and verse composition; mathematics; French; and a general paper, which included classical and British history, geography, science, and biblical knowledge. Harry was very able at both Classics and Mathematics. On 15 July 1901, he placed fifth of nineteen boys elected.[13]

Fig 2. Harry Moseley aged 13 (back row, centre) with the scholarship class of Summer Fields School, July 1901. According to his mother, Harry disliked being photographed. The direct but slightly frozen expression in this image is characteristic of all known portraits of him. (Source: Summer Fields School, 1899, by courtesy of the Headteacher and Governors of Summer Fields School, Oxford).

On his first day at Eton, Harry signed himself into the Entrance Book with a large and firm hand; he may, however, have been nervous, for he misspelt the second of his middle names, writing Jefferies for Jeffreys.[14] The next day, a Friday, he wrote home to his mother. Of all the influences on Harry as he grew up, Amabel Moseley was surely the strongest. She had a scientific background, being the youngest of the four daughters of the conchologist and mollusc specialist John Gwyn Jeffreys (1809–1885). Jeffreys was articled to a Swansea solicitor at the age of 17, and became a prominent local lawyer, and later practiced in the Chancery Court in London. In 1866, he retired to a country estate, serving for many years as a J.P., and as the High Sheriff of Hertfordshire in 1877. Like many Victorian scientists, Jeffreys was largely self-taught. During his career, he developed his childhood interest in collecting shells and became an active member of various scientific institutions, including the Linnean Society of London. In 1840, he was elected to the Fellowship of the Royal Society of London, and in retirement, gave more time to his scientific work, funding dredging expeditions in Scotland and the Mediterranean, and publishing extensively.[15]

Fig. 3 Amabel Moseley. (Source: Ludlow-Hewitt family collection, undated, by courtesy of Will le Fleming).

It was most likely through her father that Amabel met her future husband, the naturalist Henry Nottidge Moseley (1844–1891). Henry Nottidge was the son of the mathematician and mechanical engineer Henry Moseley (1801–1872). Moseley senior had enjoyed a distinguished scientific career, although it was only after being ordained in the Church of England that he was appointed, in 1831, as the first professor of natural and experimental philosophy and astronomy at King's College, London. He was elected a Fellow of the Royal Society of London in 1839.[16] Despite his efforts to promote scientific education, Henry Moseley would have preferred his son to read classics or mathematics at Oxford. Nonetheless, when Henry Nottidge matriculated at Exeter College in 1864, he persuaded his father to allow him to read Natural Science under the Linacre Professor of Physiology, George Rolleston (1829–1881). He graduated with First Class Honours in 1868, and he went on to study in London and Germany, before embarking on a series of overseas expeditions, the most notable of which was as a naturalist on HMS *Challenger*'s four-year circumnavigation of the globe. Henry Nottidge Moseley also became a Fellow of the Royal Society, in 1879. Amabel married Henry Nottidge in February 1881, just a few months before he succeeded Rolleston as Linacre Professor (from 1882 the chair was in Human and Comparative Anatomy).[17]

Amabel shared her husband's enthusiasm for the natural world and his delight in travel. They settled in a large house on St Giles, Oxford, where four children were born to them within five years: Elizabeth (Betty) in 1883; Margery, in 1884; Amabel, in 1886; and Henry Gwyn Jeffreys – Harry – in 1887. The *Eton Chronicle* describes Harry's 'hereditary gift for mathematics and science'.[18] He never met either of his grandfathers, however, and if he retained any memories of his father, they would not have been of the tireless worker, brilliant conversationalist and keenly observant naturalist beloved of his many friends, but of a voiceless invalid. Before his son's birth, Henry had fallen victim to what was diagnosed as cerebral sclerosis. In the words of his friend, Sir E. Ray Lankester (1847–1929) who was to succeed him as Linacre Professor, Henry's final illness 'began with violent headaches and fits of an epileptiform character and rendered him mentally imbecile. It showed itself quite suddenly and for four years he was cheerful and without pain, but his mind was, so to speak, gone'.[19]

By 1891, seemingly on account of Henry's illness, the family had withdrawn from Oxford to a large house at Clevedon, Somerset, with five servants, including a male nurse and a nanny.[20] On 28 May, Harry's five-year old sister Amabel died of tubercular meningitis, and on 10 November, their father succumbed to bronchitis, four days before his forty-sixth birthday, and thirteen days before Harry's fourth.

Harry had lost both his father and his closest sibling at a very vulnerable age. His approach to life as a schoolboy and young adult owed much to the example and nurture of his mother who, with characteristic resilience, devoted herself to the upbringing of her three surviving children. When, tragically, Betty too contracted tuberculosis, Amabel moved her household

Fig 4. Harry Moseley aged 6, with sisters Margery and Betty (aged 10 and 11); taken in April 1894. (Source: Ludlow-Hewitt family collection, 1894, by courtesy of Will le Fleming).

from Somerset to a new home in the wholesome air of the North Downs, near Guildford in Surrey. Betty died aged sixteen, in May 1899; and Amabel once again packed up and sought a fresh start, returning to a large house in Oxford at 48 Woodstock Road. It was to be conveniently near when Harry embarked on his undergraduate career at the University.

Harry's letters to his mother reflect her understandable concern for his health, but also the unfailingly positive outlook they both shared. There was much in Harry's first letter from Eton to reassure and encourage his 'dear mum'. He was settling in well. The first page does not survive, but the second begins with the words 'sausages and potato *mashé*' – a succinct summary of the first evening's meal – and the sixth and last is signed proudly, 'Your Loving Son H. Moseley KS' [King's Scholar]. Harry's untidy scrawl shows considerable self-assurance, his spirits seemingly undaunted by a gruelling day of tests. On that first morning, he reported, he 'got up at 6.50 and had roll, butter, and tea. Then we did an exam Latin prose for 1.30 [90 minutes]. Then we had breakfast…'. Later, 'I did a Latin Trans. Paper for 1 hour and Latin grammar for ½ hr. All the exams are very easy'. Indeed, 'the grammar paper was just a

Fig 5. Harry Moseley in Eton collar, aged 14, taken in his second half at Eton, January 1902. (Source: Ludlow-Hewitt family collection, 1894, by courtesy of Will le Fleming).

race who could write fastest. I wrote the last word as the bell struck'.[21]

Harry's first impressions of Eton's masters were favourable. 'I like Mr Alington and Mr Goodheart very much,' he wrote. Cyril Alington (1872–1955) was his tutor, and Arthur Goodheart, the Master in College, or housemaster to the King's Scholars. A graduate of Trinity College, Oxford, Alington had arrived at Eton two years before Harry in 1899, newly ordained as a deacon and with three years' experience as Sixth Form Master at his old school, Marlborough. In 1904, he married Hester Lyttleton, and thus equipped, took over the role of Master in College (a wife being essential for this role *in loco parentis*).

Cyril Alington was a notable preacher, calm and kind, who exerted a strong moral influence on the boys in his care. Hester was a woman of great charm, who threw herself into her duties. On one occasion in his final year, Harry described dinner with the Alingtons as 'a very boring occupation, as Mrs Alington does all the talking and never talks interestingly', although she was, he conceded, 'very nice and very amiable'. Hester's opinion of Harry's conversation can be gauged in a comment, most likely by her husband, in an obituary in the *Eton Chronicle*: he was 'one of the most cheerful and good-tempered of mortals, frequently expounding scientific theories to a slightly unsympathetic audience'.[22] The minutes of the College Debating Society, to which Harry was elected in 1905, give a flavour of his speech. Members wrote up their own part in a debate, and Harry's are notably succinct, very different from the flowery prose of his peers. Indeed, as Horace Fletcher noted in February 1906, he combined 'the convincing manner of Pericles with the epigrammatic terseness of Pitt'.[23]

As a new boy, Harry seems to have been unruffled by the intimidating environment of a large public school. The youngest scholars were assigned

Fig. 6 The Reverend Cyril Alington and the King's Scholars, Eton 1905. Harry Moseley, aged 17, sits in the third row back, with his arm around the railing. His friend Julian Huxley stands directly behind, four rows above. (Source: Eton College archives, 1905, by courtesy of the Provost and Fellows of Eton College).

to one of fifteen curtained cubicles in the 'Long Chamber'. Harry summed up his new home simply: 'My bed is very uncomfortable and my pillow very low, but I shall soon get over that. The chief objection to the long chamber is that one can hear snores from one end of it to the other'. He noted that he was conveniently situated, should he be interrupted by a summons to run an errand for a member of the Sixth Form: 'They have the absurd fag plan here I find. The boy calls "Here" and the last arrival is made useful. I am very near the 6th form rooms and have very little way to run so I am lucky'.[24]

Throughout his five years at Eton, Harry relished his academic work. Removes to higher divisions were achieved by success in examinations known as 'trials'. In his first half, Harry was placed in the twenty-first division within the lower fifth form, and progressed to the middle fifth in the summer of 1902, and to the upper fifth, a year later. In the summer of 1904, he reached the 'First Hundred' of the School (a distinction based on success in examinations). In addition, Harry regularly went in for, and won, school prizes. At the end of his second half, he was awarded the Assistant Masters' Latin Prose Prize, and in July 1902, he won the same Prize for Mathematics. In 1905, he took both the Chemistry and Physics Prizes, and the College Prize in Geology in 1906.[25] Harry was selective. In his first year, he eschewed the essay topic of 'Birds of Eton & neighbourhood', on which he was already something of an expert; and in his final term, he magnanimously declined to enter for Chemistry, on the grounds that, 'it would be rather grabbing to take it, while if I am not in there will be quite a good competition'.[26]

Eton's masters had to earn Harry's respect. Early on, he commented to Margery on the lower school Maths master: 'I got 95 for the last problem paper. The one I have just shown up I expect to get 90–100 for, but the next one is one of Mr Hunt's, and is absolutely impossible. He is absolutely trying to show off his knowledge, like last time'.[27] Excellent work would be copied out neatly and 'sent up' to be praised by the Headmaster. Harry achieved this distinction eleven times, and regularly shared his delight in his studies with his sister Margery. 'I am astonished to hear that Helium is three or four times as heavy as Hydrogen, though it has a lower boiling point,' he wrote. 'Another sum of the [Problem Paper] which I have succeeded in doing after much labour is to prove that 16, 1, 156, 111, 556, 11, 115, 556 etc. [are] all squares'.[28] In May 1905, he sent Margery some fish scales to look at under a microscope.[29]

Science was a minority subject at Eton, although the school did have laboratories, owing largely to the efforts of the pioneering scientist and educationalist T. H. Huxley (1825–1894), who had been appointed as the

Royal Society's representative on Eton's Governing Body under the Public Schools Act, 1868.[30] In October 1901, the roll of Eton masters enumerated twenty-nine teaching classics, twelve, maths, and just four, science.[31] But working things out by experiment was fundamental to Harry's style, and his favourite master seems to have been the chemist Dr Thomas Porter (1860–1933); Harry's closest friend at Eton was Julian Huxley, grandson of T. H. Huxley and King's Scholar, 1900, who remembered Porter as 'a great showman. He had a high-vaulted classroom, and his special "turn" was to blow bubbles of pure hydrogen and explode them with the aid of a long bamboo pole, smouldering at the tip, just before they reached the ceiling'.[32]

In his last year at Eton, Harry was fully engaged with chemistry, writing to Margery, 'My work with Porter is interesting, This half I made some solution of Potassium Permanganate, found its strength, and thence determined the purity of some oxalic acids crystals. I found finally that 99.6 per cent was pure, a great triumph, as the result was probably very nearly exact, 0.4 per cent impurity being just what we would expect, and the possibility of accumulated error being great. Now I am to find the molecular weight of something by the lowering of the freezing, and raising of the boiling point, a very long and difficult business'.[33]

Fig. 7 Dr Porter's lecture room at Eton. Here Harry first studied the chemical elements, and chemistry was to be his favourite subject at school. (Source: Eton College archives, undated, by courtesy of the Provost and Fellows of Eton College).

Harry's introduction to physics was less inspiring. In May 1905, Harry informed Amabel, 'My timetable is being enormously changed, so that I will do no physics this half except Book-work. Mr de Havilland has arranged for me to do practical physics with Mr Eggar next half. The physics which I was doing with Mr de H last week was too elementary & gets on too slowly as the division which I was doing it with is rather dense'. Harry's talent was recognised, and he was allowed a considerable degree of flexibility. He continued, 'I will be doing lots of Chemistry: 13 hrs practical, 1 hour Chemistry private, and 3 hours prep in school, besides some out of school'.[34]

Harry's relations with the physics master William Eggar (1865–1945), were not at first any more congenial than with de Havilland. Eggar was most likely nervous; Julian Huxley recalled that 'his experiments always seemed to go wrong'.[35] Unlike Porter, he favoured teaching from books, and there was to be nothing experimental as pupils worked through the syllabus in their lessons. With his practical instincts seemingly crushed, Harry felt stifled. In November 1905, there was a row, and Mr Alington wrote to Amabel about it. Harry gave his side in an indignant letter:

> It springs from a quarrel which Mr Eggar thinks he had with me, tempered I expect with a feeling that I get my own way more often at Eton than is good for me… However in the case of Eggar a molehill has been made into a mountain. I have no respect for the man because he either cannot or will not explain anything, and refers you to books. The consequence is I don't do my work his way, but annoy him by introducing unnecessary complications, simply to prove that they don't matter, instead of taking everything for granted. Evidently the 'quarrel' referred to arose from my drawing a diagram of the apparatus I used myself, instead of beaming up the diagram used in the text book, which I told him even if I could conceivably remember it, I had not personally found to work, as mine had. This enfuriated [sic] him, but I think my principle is right as the differences were far from vital.[36]

Harry was to be similarly frustrated in maths. In February 1906, he went to see the Headmaster and 'represented that I could not work at mathematics in a room in which Scoones was teaching. I now therefore work in my own room, or in Reading room, and only write my name, in a special book kept for me at school office, at the end of each school'.[37] Fortunately, the senior maths master, John Dyer, took up the challenge presented by a pupil who preferred working on his own. A week later, Harry wrote to Amabel, 'I am taking a fresh interest in Problems, which are this half just hard enough to be interesting, just not too hard to prevent me from doing them all. Dyer, who had, I fancy, considered that doing science had rendered me hopelessly bad at Mathematics, is mildly astonished, and wonderingly announces that I am far stronger than anyone except Marsden and Carter… a fact that I have never doubted for a moment'.[38]

It would be wrong to read too much into either Harry's boasts or complaints, and his teachers may well have thought his arrogance justified. In 1922, William Eggar commemorated his brilliant pupil in a touching four-line stanza, engraved on a memorial tablet on the wall of the school's physics laboratory. It reads:

> The rays whose path here first he saw
> Were his to range in ordered law
> A nobler law made straight the way
> That leads him neath a nobler ray.

The masters and senior boys were generally close, as evidenced by the Eton College Scientific Society. Perhaps surprisingly, given his defensive teaching style, this club was founded by Eggar in 1902, with a membership of six masters and twelve boys. The members all took turns to present papers at fortnightly meetings. The Society's minutes reveal great excitement about recent discoveries. Harry attended his first meeting in September 1903. One speaker that half was the future chemist, Alfred (later Sir Alfred) Egerton (1886–1959), FRS, who found that he had so much to say on 'the most scarce and wonderful chemical element', radium, that he needed two weeks. It was during Egerton's second talk, 'the mystery of radium', that Harry may first have heard the name of Professor Ernest Rutherford, who, 'a year or so ago made some remarkable experiments on thorium…'.[39] Harry's minutes of his only paper, given in February 1904 on 'Deep sea fauna', are peppered with exclamation marks, as he expounded on the marvels and immensity of his subject. He showed slides, for which 'Mr Eggar worked the lantern'.[40]

Every December, the Scientific Society hosted a *conversazione*. In 1904, Harry set up 'ozone apparatus' that he had painstakingly constructed, as testified by a 'sent up' essay of the same date.[41] The following year, he demonstrated a dramatic chemical experiment, which he described gleefully to Margery: 'A most harassing time yesterday evening… I showed phosphine… [PH_3, used as a fumigant and insecticide]. It worked beautifully. Each bubble as it bubbles up through water explodes as it reaches the air with a little flash of light, making a lovely vortex ring of P_2O_5. The smell is of course appalling, but I had a good big draught cupboard. The chief difficulty was gas, which had to be fed through yards and yards of bits of tubing joined together from another part of the room'.[42]

Juvenal's *maxim mens sana in corpore sano* (a sound mind in a healthy body) was very much a part of the public school ethos. Besides academic work and regular attendance at Chapel, Etonians were required to play sport: in winter, the Field Game (an ancient cousin of soccer) and the Wall Game (akin to rugby football), and in summer, either cricket or rowing. Harry was not an outstanding athlete, but he participated well enough. Records of the College Wall reveal that, although not a first choice player, Harry was 'energetic' and 'useful'.[43] In 1903, he played regularly for the Lower College Field team, and was again praised for his energy, although criticised for his lack of speed.[44] In November 1905, he reported to Amabel that 'my time is chockfull, a game every day', and to Margery, that he was obliged to play two matches in one day.[45] Rowing was undoubtedly his preferred sport, which he was to continue at Oxford, although again records suggest competence rather than talent, and enjoyment of exercise, not of competition. He minded, though, about the status of getting 'colours' as a member of a School House crew, and mentioned the internal politics of the situation several times in letters home before, finally, reassuring his mother in May 1906 that, 'I have at last been given my Boats, but do not know which Boat I will row in. I care not at all, that makes no difference'.[46]

Eton College and its playing fields are surrounded on three sides by the River Thames, its pools and tributaries, and it was in the wild and rural environs of the school that Harry spent his spare time. He was an enthusiastic 'birds' nester', a passion he had shared at home with Margery, with friends at Summer Fields, and now at Eton with Julian Huxley. In his first year, Harry described the school trade in eggs to Margery, remarking that Julian was 'very good indeed on birds, knowing them all, by song or form', although, as perhaps befitted the future naturalist, 'he does not collect eggs'.[47]

The boys' shared interest in birds was a strong basis for friendship as they made their way up the school. On one occasion in their final term, they went out in 'old flannel trousers under our Eton trousers, and an extra pair of socks in the tail coat pockets so that we could wade without fear of leaches [*sic*]. Afterwards we carried the wet trousers back in a little bundle'. A week or so later, Harry described how they forced their way 'into the woodpecker hole in one of the king's finest oaks in the park in full sight of the Ascot road, with a saw which I hid under my waistcoat'.[48] Harry's last letter from Eton to Margery comprised a long account of walking with Julian to Ditton, where he collected a swallow's egg, but cracked it 'running back late', after which Julian climbed a tree to inspect a kestrel's nest. They then explored a hollow oak,

Fig. 8 Harry Moseley aged 18, in his final half at Eton, July 1906. (Source: Ludlow-Hewitt family collection, 1906, by courtesy of Will le Fleming).

of which Harry provided a sketch, 'one of several, all a delight to climb', and encountered 'two owlets, different sizes with white fluff thick and woolly'.[49]

This was a happy ending to what could have been an awkward six months. Both young men were on their way to Oxford, where both had aspired to the prestigious Brackenbury Scholarship in Natural Science at Balliol College. Although it was generally accepted that Cambridge was the better university for science, Julian was the son of a Balliol graduate, and Amabel Moseley had long since decided that her son would attend his father's university, and had entered his name at the college which owed its outstanding reputation to Henry Moseley's friend, the renowned Master of Balliol, Benjamin Jowett (1817–1893).[50]

Competition for an Oxford scholarship was as intense as that for Eton election, particularly in the sciences, for which only a few places were available. Each year, over three days in early December, Balliol oversaw a group examination with Trinity and Christ Church. Five awards were advertised for the year beginning the following October: one scholarship at each college worth £80 p.a., and two lesser exhibitions at Balliol and Christ Church, each worth £40. Trinity's scholarship, the Millard, had been established in 1880, and a chemist was preferred. Certainly, in the decade 1901–1910, Harry was the only Millard scholar not to read Chemistry, although the successful candidate of 1900, Robert Lattey, was, by 1906, working in Oxford as a Physics Demonstrator.[51] Candidates took papers and practical examinations in any two of the three natural sciences (physics, chemistry, and biology). In addition, they had to write an English essay and tackle a general paper that included Latin and Greek. A further paper in mathematics was optional.[52]

The scholarship winners were announced a week after the exam: Julian had got the Brackenbury, while Harry was offered the more humble Millard, in a result not without controversy.[53] For Julian had sat the examination in 1904, then returned to Eton for a year to 'broaden [his] mind', studying Greek in the sixth form. It seems that when Harry wrote to the Master of Balliol to put his name down as a candidate for the 1905 examination, he learned that he could not compete for the top prize: the Brackenbury for 1906 had been unofficially promised to Julian ahead of the competition.

In November 1905, Harry forwarded to Amabel a letter (now lost) from the College's young chemistry don, Harold (later Sir Harold Hartley, FRS, 1878–1972), 'in answer to which I said clearly that any scholarship was better than none, but that I was undecided about the [Balliol] Exhibition question, or whether to try later for other scholarships'. Dr Porter intimated to Harry that he was 'rather disappointed… as he evidently thinks me better than Huxley'. Harold Hartley allegedly 'wished' that Harry had been given the place.[54]

Harry's biographer John Heilbron sees the 1905 Balliol scholarship examination as evidence of the need for university reform: 'so much for open competition', he remarked.[55] Harry, however, accepted the outcome with equanimity, a quality which the relentlessly competitive atmosphere of Eton had honed. He informed Amabel that he was 'of course delighted with the news, and very content to go to Trinity', and that Cyril Alington was also 'very pleased, as it is his old college, and he naturally thinks well of it'.[56] In Balliol's defence, a high value was placed by society on a classical education and evidence of all-round ability. Julian Huxley was to follow his First in Zoology with a long and distinguished career as a naturalist and popular proponent of evolution. Of the examination, he later recalled, 'the set essay was, "What would you do if you had a million pounds?" I suggested buying up as much as possible of the unspoilt coastline of Britain. I was later told that this unusual answer helped in getting me the award'.[57]

Julian's return to Eton in January 1906 was not comfortable. There was a falling out with two of the King's Scholars of 1901, George Henderson and Frank Walters, who refused to allow him to stay in a breakfast 'mess' with them and Harry. 'I was so sorry for him, as he was very anxious not to mess alone', wrote Harry to Amabel.[58] Despite the birds' nesting expeditions, Julian felt that his final two halves were 'sadly marred' by this 'bitter blow'; and the detailed Eton chapter in his autobiography does not mention the name of Moseley even once.[59]

Oxford

Oxford marked a fresh start for them both. Harry arrived on 13 October 1906. He did not write home with his first impressions, for his mother was less than ten minutes' walk away, and he could visit her whenever he wished. Her son was now an adult, and Amabel consulted him increasingly about practical matters; but she still worried. As he teased from Eton, 'with me in Oxford itself [there will be] no more anxieties, as a messenger will be despatched post-haste to bring mum and cinnamon when I sneeze'.[60]

Ironically perhaps, Trinity and Balliol stand adjacent on Broad Street. Balliol is older – medieval – and larger: in 1906, it had thirteen fellows who were tutors or lecturers, thirty-two scholars, and twenty-nine exhibitioners; Trinity – Tudor – had nine, seventeen, and fourteen.[61] Balliol's reputation was for academic brilliance; Trinity's for squirearchical heartiness. There was a famous rivalry between the undergraduates, who would abuse (and amuse) each other, hurling ribald songs and missiles over the boundary wall.[62] But many had close friends across the divide, and in the same wall there was a large wooden door: for Balliol and Trinity shared not only their science scholarship exam, but also their chemical laboratories and their science tuition as well.

Throughout the nineteenth century, Oxford's laboratory provision was limited. The Clarendon Laboratory was not opened until 1873. Towards the end of the century, the University's hastily constructed additional laboratories in the grounds of the Oxford University Museum on Parks Road were augmented by the more far-sighted colleges, including Balliol, where two cellars were converted for chemistry work in 1853. These were replaced in 1879 by an underground laboratory in the north-east corner of the college precinct, adjacent to the boundary of Trinity's Dolphin Yard. Six years earlier, Trinity had been awakened to the possibilities of the new Science School by a large bequest from a stranger to the college, Thomas Millard. A science lecturer was appointed, and an engineering workshop constructed. As chemistry grew more popular, an agreement to share facilities with Balliol was formalised in 1890 by the appointment of Trinity's first science Fellow, the chemist David Nagel, who became demonstrator and assistant to Balliol's chemistry tutor, Sir John Conroy. In 1897 Trinity opened its own laboratory, complete with a door into Balliol, and in 1901, Nagel succeeded Conroy as the head of the shared laboratory.[63]

Such was the situation in 1906, when Harry found himself the only student in his college reading physics. Indeed, he was Trinity's sole Honours graduate in the subject in the decade 1901–1910.[64] Examinations taken at Eton had exempted him from Oxford's first compulsory exam, 'Responsions', and so he set to work for his part one examination. Instead of natural science 'Prelims', which merely had to be passed, he chose to read mathematics, which offered the possibility of First, Second or Third Class Honours. As the only Trinity man taking maths that year, Harry was taught by a college lecturer, John W. Russell (whose other pupils were at Balliol), and in the summer of 1907, he took a First in Mathematical Moderations.

In his first year, Harry was assigned rooms on the top floor of Trinity's staircase 14, situated at the corner of the Garden Quadrangle with convenient access to the laboratory (and lavatories) in the Dolphin Yard. This staircase afforded the setting for most college photographs, both of sports teams and, every summer, whole college groups. Harry did not appear in 1907, but the following year, he posed, looking somewhat awkward and aloof, on a

Fig. 9 Harry Moseley, aged 20, in the Trinity College Group Photograph, June 1908. Harry is at the right of the window frame; Henry Jervis at the left. (Source: Trinity College, Oxford, archive, 1908, by courtesy of the President and Fellows of Trinity College, Oxford).

windowsill. Harry had one close friend at Trinity, Henry Jervis, who can be seen two places along in the same group.

Henry Jervis was eight years older than Harry, and from a very different social background. The son of a Birmingham schoolmaster, he was educated at a city training college and matriculated as an unattached mature student in 1905. A year later, he entered Trinity, presumably in order to use the laboratory, but he never had college rooms. It was a shared commitment to the study of science and a pleasure in laboratory work that brought them together, and their friendship endured. In 1914, Harry was Best Man at Jervis's wedding.[65]

During his four years at Trinity, Harry formed a high regard for his tutor, David Nagel, a kindly man, beloved by his students. Maurice Davidson, five years senior to Harry, wrote: 'The atmosphere here [in the laboratory] was one of the greatest friendliness. The place was in fact rather like a very select social club in which all the members knew each other, and over the activities of which he [Nagel] used to preside in an informal way, almost like an old secretary whom everyone knew and loved, and to whom everyone went for advice and help… From four o'clock onwards tea and cake were available'.[66] Nagel did no research of his own, but was renowned as 'a highly effective teacher of both physics and chemistry, and a trusted adviser on the research of others'.[67] By his death in 1920, Nagel (aged 58), was regarded as 'almost an institution' whose pupils 'all came to him for help and counsel'. On one occasion, indeed, Harry made a round trip from Manchester to Oxford to

Fig. 10 Harry's tutor, David Nagel, in the Trinity/Balliol Laboratory, 1910. (Source: Trinity College, Oxford, archive, 1910, by courtesy of the President and Fellows of Trinity College, Oxford).

consult his old tutor, whereupon 'Nagel selected the cleavage fragment of potassium ferrocyanide which Moseley used in his classic investigation of X-ray spectra'.[68]

Physics teaching for the Honours School in Natural Science was provided centrally, and at this date, Oxford had two professors in the subject. Appointed as the first Dr Lee's Professor of Experimental Philosophy in 1865, Robert Bellamy Clifton (1836–1921), presided over the Clarendon Laboratory, which was built to his own design and opened in 1873. The laboratory boasted generously-sized offices, considerable work space for students, and a fine lecture theatre; but physics was not entirely successful. Clifton disliked university politics, and felt pressure from an increasing, although still low, number of students. He considered his budget inadequate, and over the years, developed what might be called a bunker mentality. He has been criticised for his obsession with buildings and an antipathy towards research, although his carefully demonstrated experiments and solidly traditional lectures provided a suitable education for most of his undergraduates, destined as they were to be schoolmasters.[69]

In 1900 an attempt had been made to address the shortcomings of Oxford's physics teaching by the creation of a second chair, and John Sealy Townsend (1868–1957) arrived from Cambridge to become the Wykeham Professor in Experimental Physics. Townsend was more than thirty years younger than Clifton, and the two men did not work together well, or, indeed, at all. No space was found for him at the Clarendon Laboratory, and for ten years Townsend taught in temporary accommodation behind the University Museum. By the time Harry arrived in Oxford, a division had been established whereby Clifton lectured on optics at the Clarendon on Monday, Wednesday and Friday mornings, while on Tuesdays, Thursdays and Saturdays, Townsend taught electricity and magnetism at the Museum. Both facilities opened on the same afternoons for the two professors and their demonstrators to give practical instruction to Honours students. No attendance register was kept at university lectures, but the termly fees of £1 for each Professor were paid regularly on Harry's behalf.[70]

In addition, a personal tutor was found for him, to offer additional support and mentoring: Idwal Griffith, Fellow of St John's College and assistant demonstrator in the Clarendon. Known for his 'unfailing kindness', Griffith was an excellent teacher, whose 'pupils owed much to [his] encouragement and to his insistence that they should give of their best'. Between Hilary Term 1908 and Trinity Term 1910, Griffith was paid £5 a term to tutor Harry, and in his final term, his pupil sought his advice when considering his future career options.[71]

At Trinity, each undergraduate's progress was monitored by internal examinations, known as Collections, held twice yearly at the beginning of Michaelmas and Hilary terms. The college archives suggest that Harry's abilities were not at first fully appreciated. Having, as he did, the innate talent of the naturally curious experimenter, but also needing the methodical grounding of the syllabus, he did, perhaps, fall between two stools. The Senior Tutor's succinct notes read as follows:

Michaelmas Term 1907: 'Good, but flighty'.
Hilary Term 1908: 'Works erratically, but covers a lot of ground. Bad practical work at present'.
Michaelmas Term 1908: 'very erratic & rather untidy but works very hard'.
Michaelmas Term 1909: 'much more methodical; improving steadily'.
Hilary Term 1910: 'improving in definiteness; able & very industrious'.[72]

Alongside his work in science, Harry continued to row as he had at Eton, and was welcomed into the Trinity Boat Club as an experienced oarsman. In his second term, Harry joined in training walks on Sunday afternoons, and was selected for the College Torpid.[73] In his first summer, he and a 'scratch' partner won the College Pairs, Harry's only sporting trophy. But for whatever reason – perhaps a reluctance to train obsessively – Harry never made it into a First Eight, and by his penultimate term, he was back stroking the Second Torpid, a notably unsuccessful crew which the Boat Club Secretary recorded 'were distinctly disappointing and seemed quite unable to race although their stroke gave them a good example'. Despite Harry's efforts, the boat went down six places in six days.[74]

Harry belonged to a number of societies, both in Trinity and in the wider university. In his very first term he was elected to the Junior Scientific Club, a

Fig. 11 Harry aged 19 (far right) on a Sunday training walk in Hilary Term 1907.
(Source: Collection of Wilfrid Ingham in the Trinity College, Oxford, archive, 1907, by courtesy of Mr Robert Ingham).

paper-reading society open to dons and undergraduates, and in Hilary Term 1910 he served as the Club's president. One speaker that term was Robert Lattey, who on 21 January gave a paper on 'the charges of gaseous ions'.[75] Harry was also active in the Alembic Club, which met weekly on Monday evenings, and restricted membership to twenty-one junior members. In May 1914, when living again in Oxford, Harry 'gave a very interesting paper on the Chemistry of the Radio Elements', although the club President, Henry (later Sir Henry) Tizard 'criticised the optimistic character of the paper from the point of view of the ordinary chemist'.[76]

In Trinity, there was the Debating Society, and the smaller Impromptu, which also debated, and for which Harry acted as treasurer in his third year. Debates were often political and international; in December 1908, for example, the Society presciently 'declared the Kaiser dangerous'.[77] One part of Britain, at least, was actively preparing for a European war. In May 1908, the Secretary of State for War, Richard Haldane, visited Oxford to announce the creation of new Officer Training Corps in the public schools and universities. These were to replace existing organisations such as Eton's Rifle Volunteers, and would improve the supply of trained officers for the regular and reserve forces. The Oxford University OTC delegacy opened premises on Alfred Street with professional staff, and offered rigorous training and qualifications, arranged commissions, and introduced Military History as a degree option within the Modern History School. In a single term, over 600 Oxford undergraduates joined up.[78] Harry waited a little, but by December 1909, had absorbed the arguments rehearsed in numerous debates, and shouldered what he saw to be his duty. He wrote to Margery, 'I am an anti-militarist for myself at any rate, by conviction, a soldier by necessity. I can find no sound argument with which to confute the advocate of universal service… We march about the fields – two men to a skipping rope make a section – imagining ourselves a large army; then the section commanders find themselves behind their "men", and advance hastily over the skipping rope to the front – a manoeuvre which would be worth watching if on parade…. Then we skirmish, open into extended order, and fire rapidly wooden cartridges at the trees'.[79]

And finally – Finals (or 'Schools'), which Harry sat in the summer of 1910. He had worked hard; perhaps, he was even over-prepared. But he was distracted by thoughts of the future. In particular, there was the question as to whether he could, or should, seek a position under Professor Ernest Rutherford at Manchester University. A demonstratorship had been advertised in *Nature*, with an application deadline of 13 June. Writing to Amabel in late

UNIVERSITY OF MANCHESTER.

DEPARTMENT OF PHYSICS.

A JOHN HARLING RESEARCH FELLOWSHIP in PURE or APPLIED PHYSICS of the value of £125 per annum will be offered for award in July next. Candidates must give evidence of being able to conduct an independent research, and should state, if possible, the nature of the research they intend to pursue under the direction of the Professor of Physics (Dr. Ernest Rutherford), and give particulars of their previous training and education. Applications should be sent not later than June 20 to the REGISTRAR, from whom further particulars may be obtained.

UNIVERSITY OF MANCHESTER.

The Council of the University are about to make the following appointments :—

DEPARTMENT OF PHYSICS.

An ASSISTANT LECTURER and DEMONSTRATOR, in consequence of the appointment of Mr. S. Russ, D.Sc., to a Beit Research Fellowship.

Fig. 12 Ernest Rutherford's advertisement in *Nature* on 19 May 1910. (Source: Macmillan archives, 1910, by courtesy of the University of Manchester).

May 1910, Harry recounted inconclusive discussions of the options with Griffith – 'very much in favour'; Nagel – 'advises it'; Professor Clifton – 'will… write privately to Rutherford'; and Professor Townsend's demonstrator, Paul Kirkby – 'who very strongly urges me to have nothing to do with it'. There were the foreseeable advantages of a paid position in science, but it was not good pay, and the work would be at the cost of time for study. On the one hand, Harry sensed that 'some teaching would clear my brain'; but on the other, he suspected there would be 'little opportunity left for research'.[80]

Having a secure job in prospect might have reduced Harry's anxiety about 'Schools'. The first two papers were 'easy', but Harry found himself unable to concentrate. The weather was particularly hot, and as he wrote to Amabel on 17 June, he was greatly disturbed by a noisy bird. 'I was fighting inefficiently against time. The heat is overpowering, and an owl squawked all night in the garden, and kept me awake'. Four days later, the last exam was done. He wrote again, 'The heat has been unsupportable and Saturday was the worst of all. I have therefore spoiled all my chances… The two special subjects about which I knew something came on the Friday and Saturday afternoons; on Friday on Heat I did disgracefully, and had no brain at all left for Electricity on the afternoon of Saturday. It was extremely sultry, and the paper on Light in the morning had used up every thought I possessed. In the afternoon I

Fig. 13 Harry Moseley, aged 22, in the Trinity College teaching laboratory, Summer 1910. (Source: Museum of the History of Science, Oxford, 1910, by courtesy of the Museum of the History of Science, Oxford).

could not even write straight…'. His fears were well founded. Harry missed a First, and took a Second Class Honours degree.

The photograph of Harry in the Trinity teaching laboratory was taken in the summer of 1910. Harry's expression is inscrutable. Was it taken before or after he knew his Finals result? Before or after his appointment at Manchester was confirmed? Either way, the end of Harry's education was a watershed in his life. At Eton, he had developed the natural talents that he had inherited from his scientific ancestors, and had tested his instinct and ability to work things out by experiment. His eyes were opened to the exciting work being done in physics and chemistry. At Oxford, he was grounded in traditional physics, and had achieved an acceptable Honours degree. Nurtured by Amabel, he had made the best of the opportunities that had so far come his way, and he had grown into a thoughtful young man who took an interest in current affairs. Harry was a son of whom she could rightly be proud.

Notes and References

1. Throughout this chapter, I am greatly indebted to John L. Heilbron for his comprehensive and perceptive biography, *H. G. J. Moseley: The Life and Letters of an English Physicist, 1887–1915* (Berkeley: University of California Press, 1974). This work includes invaluable memories of Harry from people who knew him, and an edition of his surviving letters. In 1973 these letters were microfilmed by Professor Heilbron for the Archive for the History of Quantum Physics at the University of California, Berkeley. Copies of the microfilm are held at various institutions including the Museum of the History of Science in Oxford. The numbering of the letters in this volume is that given in Heilbron, *Moseley* (although some quotations about bird-nesting and rowing are from letters that are abridged by Heilbron).

2. Eton College Archive (hereafter ECA), Entrance Book, SCH/HM.1/10, f.62.

3. Eton was one of the seven 'Clarendon Schools' that were given independence from state and church under the Public Schools Act, 1868.

4. During the 1860s, proponents of reform had even considered Wadham as a science-specialist college: see M. Curthoys, 'The Colleges in the New Era', in M. G. Brock and M. C. Curthoys (eds.), *Nineteenth-Century Oxford, Part 2, The History of the University of Oxford* (Oxford: Oxford University Press, 2000), vol. VII, 117.

5. The Oxford and Cambridge Act, 1854 removed traditional religious restrictions to university entrance, although many fellowships were still 'closed', i.e. restricted to Anglican clergymen. The Royal Commission on Scientific Instruction (Devonshire Commission) of 1870–75 led to the Universities of Oxford and Cambridge Act of 1877. This resulted in new university and college statutes in 1882, and eventually opened the way to the creation of more college fellowships and tutorships in science. See Roy MacLeod, 'Resources of Science in Victorian England: The Endowment of Science Movement, 1868–1900', in Peter Mathias (ed.), *Science and Society* (Cambridge: Cambridge University Press, 1972), 111–166.

6. Although maths and physics were originally intended to be an element of the Honours degree, this quickly became optional. See L. W. B. Brockliss, *The University of Oxford: A History* (Oxford: Oxford University Press, 2016), 236–238, 391.

7. Roy MacLeod and Russell Moseley, 'Breadth, Depth and Excellence: Sources and Problems in the History of University Science Education in England, 1850–1914', *Studies in Science Education*, 5 (1), (1978), 85–106.

8. Janet Howarth, '"Oxford for Arts": The Natural Sciences, 1880–1914', in Brock and Curthoys, *Nineteenth-Century Oxford, Part 2, op. cit.* 459 (hereafter, Howarth, 'Oxford for Arts').

9. Roy MacLeod and Russell Moseley, 'The "Naturals" and Victorian Cambridge: Reflections on the Anatomy of an Elite', 1851–1914,' *Oxford Review of Education*, 6 (2), (1980), 177–195.

[10] *Eton College Chronicle* (hereafter ECC) no. 937 (10 October 1901), 1089. There were, and are, three 'halves', i.e. terms, in the Eton year.

[11] Harry complained to his mother and sister about the excessive demands of compulsory sport; see, for example Letters 14 and 15, [November 1905], in Heilbron, *Moseley, op. cit.* 149–151.

[12] *Summer Fields Register, 1864–1929*, 130; Heilbron, *Moseley op. cit.* 13, note 30. He may have had hay fever, which afflicted him regularly in the summer time.

[13] No copies of the 1901 papers survive, but see, for example, ECA, 'Election 1902'; ECA 60/10/2/1/2, *Nominata in Collegia Regalia*, Tom. II (unpag.). Only fourteen of the nineteen listed were to start as scholars; they had to wait for places on the foundation to become available.

[14] ECA, SCH/HM.1/10, Entrance Book, f.62.

[15] W. J. Harrison, 'Jeffreys, John Gwyn (1809–1885)', rev. Eric L. Mills, *Oxford Dictionary of National Biography* (Oxford: Oxford University Press, 2004); http:// www.oxforddnb.com/view/article/14705 (accessed 15 December 2016).

[16] B. B. Woodward, 'Moseley, Henry (1801–1872)', rev. R. C. Cox, *Oxford Dictionary of National Biography*, (Oxford: Oxford University Press, 2004); online edn, May 2007; http://www.oxforddnb.com/view/artic19388 (accessed 15 December 2016).

[17] B. B. Woodward, 'Moseley, Henry Nottidge (1844–1891)', rev. Terrie M. Romano, *Oxford Dictionary of National Biography* [http://www.oxforddnb.com/view/article/19389, accessed 15 December 2016]; G. C. Bourne, 'Memoir of Henry Nottidge Moseley', in H. N. Moseley *Notes by a Naturalist on the 'Challenger'* (London: John Murray, 2nd ed., 1892) v-xvi.

[18] ECC, no. 1543 (7 October 1915), 892.

[19] Quoted in George Sarton, 'Moseley – The Numbering of the Elements', *Isis*, 9 (1), (1927), 99.

[20] The National Archives, 1891 Census RG12/1954: Somerset, Clevedon, 19.

[21] Moseley to Margery, [September1900], Letter 6 in Heilbron, *Moseley, op. cit.* 144.

[22] Moseley to Margery, [7 February 1906], Letter 21 in Heilbron, *Moseley, op. cit.* 157–158; ECC, no. 1543 (7 October 1915), 892.

[23] ECA, College Debating Society, vol. 25, 6 and 92.

[24] Moseley to Margery, [September 1900], Letter 6 in Heilbron, *Moseley, op. cit.* 144.

[25] Harry's achievements were listed in the 'Eton School Record': ECA, SCH/SC/1/1, f.242.

[26] Moseley to Margery, [early Eton], Letter 9 in Heilbron, *Moseley, op. cit.* 145; Moseley to

Amabel, [6 June 1906], Letter 31 in Heilbron, *Moseley, op. cit.* 163, 31. In 1921, Amabel donated all of Harry's Eton prizes – 'twenty-one finely bound volumes, mostly of poetry and essays' – to the new War Memorial Library at Trinity College, Oxford; see *Trinity College Report* (1920–1), 5. By the centenary of Harry's death, only one remained in the College: an edition of Darwin's *Descent of Man*, inscribed, 'Problem Prize, Spring 1904'.

27 Moseley to Margery, [early Eton], Letter 9 in Heilbron, *Moseley, op. cit.* 145.

28 Moseley to Margery, undated, Letter 11 in Heilbron, *Moseley, op. cit.* 147–148. These are the squares of 4, 34, 334, 3334; I am grateful to John Heilbron for his beautifully worked solution. See note 3 at the end of this letter, 148.

29 Moseley to his mother, [May 1905], Letter 13 in Heilbron, *Moseley, op. cit.* 149.

30 The Act was concerned only with the seven major public schools, which had been investigated by the Clarendon Commission of 1861–1864.

31 *ECC*, no. 935 (15 October 1901), 1084.

32 Julian Huxley, *Memories* (London: Allen & Unwin, 1970), 49 (hereafter Huxley, *Memories*).

33 Moseley to Margery, [17 February 1906], Letter 23 in Heilbron, *Moseley op. cit.* 159.

34 Moseley to his mother, [May 1905], Letter 13 in Heilbron, *Moseley, op. cit.* 149.

35 Huxley, *Memories, op. cit.* 50.

36 Moseley to his mother, [20 November 1905], Letter 16 in Heilbron, *Moseley, op. cit.* 152–153.

37 Moseley to Margery, [7 February 1906], Letter 21 in Heilbron, *Moseley, op. cit.* 157–158.

38 Moseley to his mother, [16 February 1906], Letter 22 in Heilbron, *Moseley, op. cit.* 158–159.

39 *ECA*, SCH/SOC/SC/1/1, 87, 102–122.

40 *Ibid*. 168–174.

41 *Ibid*. 233–234; ECA, 'Sent up Copies MDCCCIV', 'The Preparation of Ozone' (unpag). Work is arranged alphabetically by pupil, then chronologically.

42 Moseley to Margery, [16 December 1905], Letter 19 in Heilbron, *Moseley, op. cit.* 155.

43 ECA, 'College Wall, 1901–7', 255–256, 259, 262.

44 ECA, 'Lower College Football (1889-1904), 'Characteristics of the L. C. Field XI 1903' (unpag.).

45 Moseley to his mother; Moseley to Margery, [November 1905], Letters 14 and 15 in Heilbron, *Moseley op. cit.* 149–151.

[46] Moseley to his mother; Moseley to Margery, [27 February 1906] and [18 May 1906], Letters 24 and 26 in Heilbron, *Moseley op. cit.* 159–160 and 161.

[47] Moseley to Margery, [early Eton], Letter 8 in Heilbron, *Moseley, op. cit.* 145. Illegal in the UK since the Protection of Birds Act, 1954, collecting eggs was a popular and respectable pastime in Edwardian Britain.

[48] Moseley to his mother, [23 May 1906] and [4 June 1906], Letters 28 and 30 in Heilbron, *Moseley op. cit.* 162 and 163.

[49] Moseley to Margery, [22 June 1906], Letter 32 in Heilbron, *Moseley, op. cit.* 163.

[50] Balliol College Archives, 'Admissions to Balliol College' Tutors Book 1895–1905 (unpag.).

[51] The preference was presumably for convenience: Trinity's only Fellow in science was a chemist. Towards the end of his life, Lattey reminisced about advising Harry on changing his intended degree from Chemistry to Physics. See Bernard Jaffe, *Moseley and the Numbering of the Elements* (London: Heineman, 1972), 13, vi.

[52] Trinity College Archive (hereafter TCA), Notices 10 (unpag.); 11, 1906–1907.

[53] For example, in *The Times* (12 December 1905) 6 and 14 December 1905, 10.

[54] Huxley, *Memories, op. cit.* 56–57; Heilbron, *Moseley, op. cit.* 25–26; Moseley to his mother, [15 December 1905], Letter 18 in Heilbron, *Moseley, op. cit.* 154–155. Elected to the Brackenbury Scholarship in 1897, and to a Balliol tutorial fellowship in 1901, Harold Hartley did much to promote the study of physical chemistry in Oxford before and after the War. Under his guidance, the Baliol–Trinity laboratory became a centre for the new subject, and produced twelve Fellows of the Royal Society of London, of whom five were his pupils. Hartley's war service was distinguished: he became chemical adviser to the Third Army in France in 1915, and eventually led the Chemical Warfare Department of the Ministry of Munitions. He was knighted in 1928, made KCVO in 1944, GCVO in 1957, and CH in 1967. See E. J. Bowen's obituary in *Journal of the Chemical Society, Faraday Transactions I,* 68, 1972, X001-X002; also E. J. Bowen, 'Hartley, Sir Harold Brewer (1878–1972)', Rev. K. D. Watson, *Oxford Dictionary of National Biography* (OUP, 2004) [http://www.oxforddnb.com/view/article/31207, accessed 16 December 2016].

[55] Heilbron, *Moseley, op. cit.* 26.

[56] Moseley to his mother, [15 December 1905], Letter 18 in Heilbron, *Moseley, op. cit.* 154–155.

[57] Huxley, *Memories, op. cit.* 52.

[58] Moseley to his mother, [28 January 1906], Letter 20 in Heilbron, *Moseley, op. cit.* 156–157.

[59] Huxley, *Memories, op. cit.* 56–57.

⁶⁰ Moseley to his mother, [early May 1906], Letter 25 in Heilbron, *Moseley, op. cit.* 160.

⁶¹ *Oxford University Calendar* (1907).

⁶² Bryan Ward-Perkins and Clare Hopkins, 'The Trinity-Balliol Feud', *Trinity College Report* (1989–90), 45–61.

⁶³ Edmund J. Bowen, 'The Balliol-Trinity Laboratories Oxford 1853-1940', *Notes and Records of the Royal Society of London*, 25 (2), (1970), 228–230; Hopkins, *Trinity: 450 Years of an Oxford College Community* (Oxford: Oxford University Press, 2005), 280–281, 301, 453.

⁶⁴ In the first decade of the twentieth century, physics student numbers across the university remained low, with an average of seven each year. The colleges with the most men reading physics were: New College (nine took Honours in 1901–10), Jesus (six), and Balliol, Exeter and Merton (five each).

⁶⁵ Heilbron, *Moseley, op. cit.* 112.

⁶⁶ Maurice Davidson, *Memories of a Golden Age* (Oxford: Blackwell, 1958), 21–22.

⁶⁷ Tony Simcock, 'Laboratories and Physics in Oxford Colleges, 1848–1947', in Robert Fox and Graeme Gooday (eds.), *Physics in Oxford, 1839–69: Laboratories, Learning and College Life* (Oxford: Oxford University Press, 2005), 137.

⁶⁸ H. B. Dixon, Obituary of 'D. H. Nagel', *Nature,* 106 (2658), (1920), 186; E. J. Bowen, *Chemistry in Oxford* (Cambridge: Cambridge University Press, 1966), 9.

⁶⁹ Howarth, 'Oxford for Arts', *op. cit.* 459–461; Robert Fox, Graeme Gooday and Tony Simcock, 'Physics in Oxford: Problems and Perspectives', in Robert Fox and Graeme Gooday (eds.), *Physics in Oxford*, 5–18; Robert Fox, 'Context and Practices of Oxford Physics, 1839–79', *idem.* 59; Graeme Gooday, 'Robert Bellamy Clifton and the "Depressing Inheritance" of the Clarendon Laboratory, 1877–1919', *idem.* 80–82.

⁷⁰ The timetables of both professors were announced at the start of each term in the University *Gazette*. TCA, Tutorial Accounts D/1 (unpag.).

⁷¹ TCA Senior Tutor, D/1; T.W. Chaundy, 'Obituary Notice of Idwal Owain Griffith', *Monthly Notices of the Royal Astronomical Society*, 102, (1942), 64–65.

⁷² TCA, Senior Tutor C/2, Collections (unpag.). No records survive for Hilary Term 1909.

⁷³ Owing to the narrowness of the river at Oxford, crews do not race side by side, but chase each other in single file, aiming to 'bump' the boat ahead, while avoiding being bumped by the boat behind. The main races, 'Eights', take place in the summer, whilst 'Torpids', in February, are restricted to oarsmen who have never rowed in a First Summer Eight.

⁷⁴ TCA B/BC/vii/6 vol. V (1906–13) (unpag.).

[75] *Transactions of the Oxford University Junior Scientific Club 1907*, N.S., nos 28–30 (1907); *idem*. N.S. no. 35 (October 1912), 38.

[76] E. J. B. 'The Alembic Club: the first fifty years', TCA, DD41add/ 4.1. The physical chemist Henry (later Sir Henry) Tizard (1885–1959), FRS, matriculated at Magdalen College, Oxford, in 1904, and returned to Oxford as a Fellow of Oriel in 1911. His father had been on the *Challenger* expedition with Henry Nottidge Moseley, and he was to attend, alongside Harry, the fateful annual meeting of the British Association for the Advancement of Science held in Australia in the summer of 1914. In 1920, he was appointed Reader in Chemical Thermodynamics at Oxford, where he developed the 'octane rating' used to classify petrol. In 1927, he became Permanent Secretary of the Department of Scientific and Industrial Research (DSIR), where his achievements included the development of radar, the establishment of the National Chemical Laboratory, and the first serious studies of UFOs.

[77] *Oxford Magazine*, 27 (3 December 1908), 125; (28 January 1909), 158.

[78] *Ibid*. (21 January 1909), 138.

[79] Moseley to Margery [4–5 December 1909], Letter 37 in Heilbron, *Moseley, op. cit.* 166–168.

[80] Harry hoped to have 'cleared up his ideas by writing them down', but as he admitted, his 'brain [was] limp as putty'. Moseley to his mother, [27 May 1910], Letter 39 in Heilbron, *Moseley, op. cit.* 169–171.

CHAPTER TWO
MOSELEY IN MANCHESTER

NEIL TODD

Dear Prof Rutherford,
Thank you for the letter informing me of my appointment. It will be a great pleasure to me to work in your laboratory, and after my failure in 'schools' I consider myself very lucky to have got the opening which I coveted.

... I would like to be guided entirely by you on the subject which I attempt ... since I cannot profitably choose for myself.

... I will spend August in Oxford and will then read up on Radioactivity, in the hope that your suggestion may lie in that direction.[1]

Introduction

In the autumn of 1910, with Ernest Rutherford's invitation in his pocket, Henry Moseley exchanged Oxford for the New Physical Laboratories of University of Manchester. Following the discovery between 1895 and 1898 of X-rays, radioactivity and the electron, as well as the highly active element radium, Manchester had become the beating heart of what has been called the 'second scientific revolution'.[2] Established in 1900, the 'New Laboratories' were first directed by Arthur (later Sir Arthur) Schuster FRS, and from 1907 by Rutherford himself. With its own stock of radium, Rutherford's department had by 1910 become Britain's leading school of radioactivity. That winter, the birthplace of the Dalton chemical atom witnessed the silent birth of the nuclear atom. Three years later in 1913 the full significance of the nucleus became apparent with the emergence of the Bohr-Rutherford quantum atom.

If a place with Rutherford offered Moseley an opportunity, the stimulus of Bohr's theories gave him a motive, and in 1913 he began the experiments that established that chemical elements may be classified and arranged by the number of positive charges in their nuclei – that is the atomic number. Tragically, the coming of War in August 1914 broke up Rutherford's team at the height off its productivity. Many of his students and collaborators enlisted, but Moseley's death at Gallipoli in 1915 came as a stark reminder of the perils of war. For the next three years, Rutherford managed to keep the laboratory going, and demonstrated that the chemical atom could be artificially transmuted, adding a new dimension to Manchester's atomic odyssey. This chapter outlines the role of Manchester in Moseley's story, and in the history of the city and university to which modern physics and chemistry owe such a great debt.

Manchester, Birthplace of the Chemical Atom

Academic visitors to modern Manchester soon become aware of the legacy left by two of England's most outstanding names in the history of the physical sciences – John Dalton (1766–1844) and James Joule (1818–1889). Dalton arrived in Manchester in 1790 to teach mathematics, physics and chemistry at New College. In 1794 he joined the Manchester Literary and Philosophical Society, famous for its role in the industrial revolution. In 1809, he published *A New System of Chemical Philosophy* which established the principle of the chemical atom.[3] Salford-born Joule studied with Dalton at the Literary and Philosophical Society in 1834, before producing his seminal work on the mechanical equivalence of heat in 1845.[4] He was also associated with the development of the principle of conservation of energy, work he would extend with William Thomson (Lord Kelvin).

These early developments were consolidated by the foundation of Owens College in 1851, made possible by a bequest of £96,942 from the wealthy textile merchant John Owens. The College was to emulate Oxford and Cambridge in what was taught, but at the same time it would be non-sectarian and non-residential, in contrast to Oxbridge. The original Owens building on Quay Street still stands today. Early leaders in British science who worked there included the chemist Henry Roscoe (1833–1915), the physicist Robert Clifton (1836–1921), and the engineer Osborne Reynolds (1842–1912). In 1865 Clifton became Professor of Physics and founder of the Clarendon Laboratory at Oxford. A chair in physics in Manchester was endowed by a gift of £10,000 from E. R. Langworthy in 1874, and Balfour Stewart (1828–1887) became the first Langworthy Professor of Physics in 1879.[5]

In 1873, Owens College moved to the present site,[6] where it saw a steady development of the physical sciences and engineering, under Arthur Schuster (1851–1934) in physics, and Horace Lamb (1849–1934) in mathematics: Schuster had succeeded Stewart in the Langworthy Chair in 1877. Early students included John Joseph (later Sir J. J.) Thomson (1856–1940) and Arthur (later Sir Arthur) Stanley Eddington (1882–1944). Thomson went on to become Director of the Cavendish Laboratory at Cambridge and Eddington became Director of the Cambridge Observatory. Both were elected as Fellows of the Royal Society. In 1880, the College joined the federal Victoria University, which gained a charter in 1904 as the Victoria University of Manchester.

A campus map from the 1893 Ordnance Survey identifies a medical school; two engineering laboratories; and laboratories for chemistry and for zoology and geology (with an associated museum). Physics was located in the ground floor and basement of the main building. The 1893 map also shows that the campus was surrounded by terraced houses and was adjacent to Oxford Road, which was a main artery for Manchester's busy tramway.

The New Physical Laboratory of the Victoria University of Manchester

By the 1890s, the College was so successful that physics had outgrown the space available. Arthur Schuster, who was familiar with the world of European physics, persuaded the University to establish a new Physical Institute, located on Coupland Street, with facilities that would mirror leading laboratories on the Continent. The new building was designed in collaboration with the Manchester architect J. W. Beaumont[7] and was opened in June 1900 by Lord Rayleigh during a *conversazione* to which the public were invited, and during which the University awarded honorary degrees to a number of senior Fellows of the Royal Society, including James Dewar and J. J. Thomson.

On four levels, the new Institute was shared between Physics and Electro-Technics (later to become Electrical Engineering).[8] The ground floor included a large private laboratory (now called the Rutherford Room), and a workshop and rooms devoted to experimentation in electricity and magnetism. Attached to the main part of the building was the John Hopkinson Electro-Technical Laboratory (now part of the Coupland Building occupied by Psychology), which featured a dynamo-house (now a student common room), a gas turbine engine-room (occupied by the Museum), an electro-chemical laboratory (now demolished) and a switchboard room to control the flow of electricity throughout the building.

The first floor housed the Professor's private room, but was primarily set aside for undergraduate teaching in sound and light and accommodated the 'elementary laboratory' (Fig. 1), where Moseley would have spent time as demonstrator when he arrived in 1910.

The second floor featured a magnificent lecture theatre (Fig. 2) and associated apparatus and preparation rooms. Two rooms were set aside for spectroscopic and astronomic transit research. From outside the building, ledges for mounting a heliostat and the bay window which housed a transit instrument (now on display in the new Schuster Building) can still be seen

Fig. 1 The elementary laboratory in the 1900 building of the New Physical Laboratories of the Victoria University of Manchester, where Moseley would have performed his demonstrating duties. (Source: Archive of School of Physics and Astronomy, University of Manchester, undated, by courtesy of School of Physics and Astronomy, University of Manchester).

today. The lower ground floor or basement (now occupied by the Manchester Museum) was primarily devoted to research in spectroscopy, photometry, photography and low-temperatures.

Discovery of X-rays: A Revolution in Science Begins

Although the new building was designed for research on topics of nineteenth century classical physics, it was built at a time of rapid discovery and revolution in physics and chemistry. Wilhelm Röntgen's experiments on X-rays in 1895[9] were followed by the discovery of radioactive emission from uranium by Henri Becquerel in 1896[10] and the electron by J. J. Thomson in 1897.[11] These developments triggered a decade of frenetic research activity in which many other major discoveries were made. In Cambridge, Thompson was assisted by a young Ernest Rutherford, then a student at the Cavendish Laboratory, who monitored the ionising effects of X-rays on gases by changes in their conductivity.[12] Rutherford subsequently employed such methods to investigate the rays given off by uranium. In his first paper on radioactivity, submitted in

Fig. 2 The lecture theatre in the 1900 building. (Source: Archive of School of Physics and Astronomy, University of Manchester, undated, by courtesy of School of Physics and Astronomy, University of Manchester).

September 1898, he showed that the rays were made up of at least two types with different penetrating powers, which he named α- and β-rays.[13]

That same year, across the Channel, Marie and Pierre Curie extracted two new radioactive elements from pitchblende (an ore of uranium). These were polonium, which separated out with bismuth;[14] and radium, which separated with barium.[15] In 1898, thorium was shown independently by Gerhard Carl Schmidt and the Curies and to be a fourth member of the family of radioactive elements.[16] The following year André-Louis Debierne, working with the Curies, found a third active element in pitchblende residues. He called this actinium.[17] By the close of 1899, Becquerel and others had further shown that radium gave off β-rays. These could be deflected by a magnetic fields like cathode rays and had a charge to mass ratio similar to that of the electron, but with a speed close that of light.

By the time Rutherford published 'Uranium Radiation and the Electrical Conduction Produced by It' in January 1899,[13] he had already moved to Montreal, as MacDonald Professor of Physics at McGill University. There, working either alone or with colleagues, he discovered that some elements including thorium and radium give off a radioactive gas that he christened as the 'emanation', adding radon to the rapidly growing list of radioactive elements. In 1901 he teamed up with Frederick Soddy, then a Demonstrator in Chemistry at McGill, in a collaboration that led to a series of papers published in 1902 and 1903.[18] Their work confirmed that the thorium emanation was a member of the noble gas family and that there must exist another decay product mediating the formation of the emanation from thorium. They called this thorium-X by analogy to William Crookes' uranium-X.[19] It became clear that radioactivity must involve a form of sub-atomic chemical change; in other words transmutation. The outcome was a radical theory of radioactive change that became known as the theory of successive transformations. Every radioactive element could be characterised by its half-life, and at each stage in the series of transformations a chemical change took place whereby a parent atom would give rise to a daughter atom having different chemical properties. This proposal fundamentally challenged the established wisdom that Dalton's chemical atom was immutable.[20]

In 1903 Soddy visited Manchester along with several European scientists, including Ludwig Boltzmann, who were attending the1903 British Association Meeting in Stockport. In the physics section, Rutherford and Kelvin discussed the heating effect of radium and the origin of atomic energy. Soddy had just completed a set of experiments with William (later Sir William) Ramsay at University College London that had demonstrated the evolution of helium from radium, thus providing powerful support for the theory of successive transformations.[21]

Radium had become commercially available at around that time from the factory of Friedrich Oscar Giesel (1852–1927).[22] Rutherford acquired Giesel radium from Soddy during a visit to London in 1903, and much of his later research at McGill involved this highly active new element.

The Revolution Moves to Manchester

In 1906, Schuster persuaded Rutherford to succeed him as Langworthy Professor and Director of the Physical Laboratories. When Rutherford arrived in Manchester in 1907, he was already famous for his work in radioactivity, especially the theory of successive transformations, for which he was to receive the Nobel Prize in Chemistry in 1908.

Rutherford refashioned Schuster's laboratories, and acquired a large stock of radium chloride from Austria, which became central to his experiments. In Montreal, where radium had been stored in the basement, the entire laboratory had been contaminated by radium emanations. So in Manchester, a special room for the radium was set aside at the top of the building, so that any escaping emanation (radon) could be vented outside. Schuster's transit room with its large bay window was ideal for this purpose: the top floor room was therefore dedicated to radium storage and to the preparation of radioactive sources for experiments elsewhere in the building. Because radium chloride as it arrived (in a barium chloride carrier) was useless for experiments, the standard practice was to dissolve the salt in dilute hydrochloric acid in a glass bulb attached to a mercury pump so that the radium emanation could be milked off, purified in a separate glass apparatus and compressed into small glass tubes to make radioactive sources for experimental work. The 'radium room' was the inner sanctum of the laboratory, making possible the work which Rutherford and his Manchester school carried out, including Moseley's research before he turned to X-rays. At the other end of the top floor in the preparation room behind the large lecture theatre resided the key figure of William Kay, the laboratory steward.

One of Rutherford's most important experiments in Manchester, performed in 1908 in collaboration with the young graduate Thomas Royds, showed that α-particles were in fact ionised helium ions. The departmental glass blower Otto Baumbach had perfected techniques for blowing glass tubes with a thickness of only 0.01 mm: this was thin enough for α-particles to penetrate, but strong enough to contain radon at atmospheric pressure. After accumulation in an outer glass tube and compression into another tube, the helium could be detected from characteristic yellow and green lines in its optical emission spectrum.[23] This confirmed the earlier conclusion of Ramsay and Soddy that decay of radium led to a build-up of helium.[21]

With the radon sources from the radium room, Schuster's basement research rooms were soon colonised by Rutherford's team. His assistants included Hans Geiger, who had been hired by Schuster in 1906. Geiger's room was originally intended for low-temperature work, and here he invented an electrical method for detecting radioactive decays, which led to the famous 'Geiger Counter'. Schuster had designated the basement rooms next to Geiger's for photography, but they were well suited for another method Rutherford developed, 'scintillation counting'. This involved counting α-particles by peering through a microscope at a small patch of fluorescent material (sodium iodide) and recording each flash. As it was necessary for stray light to be excluded, a photographic dark room was ideal. Under Geiger's guidance, Ernest Marsden carried out scintillation experiments in 1909 and

1910 which showed that a small number of α-particles were scattered through large angles when directed at ultra-thin gold foils. This led Rutherford to propose his model of the atomic nucleus in 1911.[24]

Moseley in Manchester 1910–1912: Experiments in Radioactivity

This was the scene that greeted Moseley on his arrival in Manchester in the autumn of 1910. Moseley recounted his experiences in an extensive series of letters published by John Heilbron in 1974.[25] Moseley was appointed a Laboratory Demonstrator, and settled in lodgings at 'Dunwood House' in Withington, a suburb south of the city centre and the University. To and from college he would either walk or take a tram along Wilmslow Road and would have daily passed by Rutherford's house, which was north of Withington. The suburbs of Manchester were then rural, so the journey could be pleasant, at least when the city was not shrouded in smog. Fig. 3 shows a photograph of

Fig. 3 Ernest Rutherford and his wife in the Wolseley-Siddeley car bought with the money received for the Nobel Prize in Chemistry for 1908. (Source: Archive of School of Physics and Astronomy, University of Manchester, undated but probably 1910, by courtesy of School of Physics and Astronomy, University of Manchester).

Rutherford outside his house with the 1910 Wolseley-Siddeley he purchased with his Nobel Prize money. Moseley had the pleasure of a ride with Rutherford when he first arrived, although in December 1910 he was to complain that 'remaking the apparatus took a long time as the Laboratory Assistant spent his time mending Rutherford's motor car'.[26]

Within a few months of Moseley's arrival, Rutherford announced the famous scattering law heralding the discovery of the atomic nucleus. He first made his conclusions known to members of the Department during one of the regular Sunday dinners at his home in December 1910. This was dubbed 'the night that changed physics' during the Rutherford Jubilee, held in 1961 to celebrate fifty years of the atomic nucleus.[27] Moseley makes no reference to the discovery in his letters, perhaps because by late December he had already returned to Oxford for Christmas. Indeed, even after the discovery had been presented to the Manchester Literary and Philosophical Society in March 1911, and published in full detail in the *Philosophical Magazine* in May 1911,[24] it did not make much impact. It was only after Bohr had developed his quantum interpretation that its significance was fully appreciated.

It becomes clear from Moseley's letters that the young demonstrator viewed teaching as 'a chore that must be got through'. He disliked students whom he considered to be stupid and, in the words of John Heilbron, who offended his 'patrician sensibilities'. Three letters to his mother and sister illustrate this point:

On Saturday I was disgusted to find a large proportion of coloured students with the thickest heads among those doing Labwork. These seemed to include Hindoos, Burmese, Jap, Egyptian and other vile forms of Indian. Their scented dirtiness is not pleasant at close quarters.[28]

The students are docile but mostly stupid, and repeating the same thing to 100 different students becomes at times monotonous.[29]

Then there is a miserable lecture to Gas Engineers, which I have been bullied into giving early next Term, on a highly technical subject connected with gas lighting. I object to wasting my time looking up a subject of such a kind.[30]

When his teaching duties allowed, Moseley worked on radioactivity in a basement room close to Geiger. His letters relay the precarious nature of these experiments. On 13 November 1910, he described to his sister the difficulties in the work that would lead to his 1912 publication on 'The Number of β-Particles Emitted in the Transformation of Radium'.[31]

> My work is going through much tribulation. On Thursday I at last induced my apparatus to stay at a pressure of 1/400 mm, that is a three-hundred-thousandth of an atmosphere, after plastering with a red sticky stuff which resembles butter. Then I started my experiment, but a glass tube of thickness much less than tissue paper filled with Radium Emanation chose to break off its stalk inside, and everything had to come to pieces to get it out. Fishing for a thing which breaks at a touch was too risky to be tried, for to let Emanation loose upon the Laboratory is a capital offence.[32]

About a year later, on 7 December 1911, he described to his sister the hazards involved in transporting radioactive sources to the basement laboratory.

> I am writing in a Laboratory so steam heated that even with coat off it is hardly bearable, and at the same time I am waiting for an experiment to prepare itself.... At present beta-ray experiments which must be got through while my supply of emanation lasts.... The experiment is now clamoring for attention, and I must go to it. I have to do [so] many things in a minimum of time that it seems more like a conjuring trick than anything else. The experiment begins in the attic and continues in my room on the ground floor, so that I have to race down three flights of stairs in the middle, to the great astonishment of the occasional student.[33]

The risk of accidents and contamination was high, despite the strictest rules laid down by Rutherford. It was inevitable with an ever growing army of researchers that the problem of contamination would become critical and by March 1912 a new extension had been built. As in 1900, the opening was marked by a public *conversazione* and an exhibition of experiments, among them a demonstration of 'counting atoms' by Hans Geiger in the basement. While the extension was needed to accommodate the massive expansion in the number of research students, another reason was that the original laboratory had become so contaminated by radioactivity that it was no longer possible to conduct experiments such as β- and γ-ray spectroscopy using photographic methods – both required long exposures and low back-ground activity. Both Schuster and Rutherford commented on the contamination problem:

> The [new] Physics Research Rooms, are situated on the first floor of the north wing facing Bridge St. In this position they are well outside the range of penetrating radiations from active material in the main building, which is some 30 yards further south. Primarily intended for experiments in connection with radioactivity, they are nevertheless equally well adapted for other branches of Physical work. If necessary, several of the rooms can be darkened for photographic or special radioactive work. ... In addition to the difficulty already mentioned of avoiding the disturbances due to penetrating radiations, a Laboratory in which large quantities of radioactive substances are in continual use gradually becomes contaminated by the distribution of active matter. For example, an invisible trace of radium on a finger suffices to make permanently radioactive every object which is touched. Although precautions have been taken to reduce this infection to a minimum, it has proved sufficiently serious to render difficult, if not impossible, some of the more delicate measurements required in researches on radioactivity and on the ionisation of gases. The new research rooms will prove invaluable for such measurements, and every precaution will be taken against the possibility of radioactive contamination.[34]

> A part of the new extension was set aside for the use of the Physics Department. The rooms so provided have already proved of great service in research work. The new laboratory has the great advantage of being free from all radioactive contamination, and it has thus been possible to carry out refined experiments, which would have been very difficult in the main laboratory.[35]

In March 1912, shortly after the extension opened, Niels Bohr arrived in Manchester, and within just four months had introduced Planck's constant into the Rutherford nuclear atomic model. This is confirmed by the Bohr-Rutherford memorandum of July 1912, which included some early sketches which Bohr sent to Rutherford before returning to Denmark.[36] Moseley may have been unaware of this development. Bohr spent most of his time working in his rooms in Hulme Hall. Although Moseley moved in 1912 to lodgings close to the Hall, there is no evidence that Bohr and Moseley ever met during this period.

In 1912, a departmental photograph (Fig. 4) was taken outside the new extension and shows Moseley in the front row, still a junior member of the group. By this time, however, Moseley had completed his training as a research student, and had published papers jointly with Fajans[37] and Makower[38], as

9. The Manchester University Physical and Electro-Technical Laboratories staff and research students, 1912. L to R: front row: R. Rossi, H. G. J. Moseley, J. N. Pring, H. Gerrard, E. Marsden; row 2: H. Geiger, W. Makower, A. Schuster, E. Rutherford, R. Beattie, H. Stansfield, E. J. Evans; row 3: C. G. Darwin, J. A. Gray, D. C. H. Florance, Margaret White, May Leslie, H. R. Robinson, A. S. Russell, H. Schrader, Y. Tuomikoski; row 4: J. M. Nuttall, W. Kay, H. P. Walmsley, J. Chadwick. Courtesy of Kasimir Fajans.

Fig. 4 Departmental photograph taken in 1912. Moseley is second from left in the front row. The photograph is taken in front of the entrance to the 1912 extension to the New Physical Laboratories of the Victoria University of Manchester. (Source: Archive of School of Physics and Astronomy, University of Manchester, 1912, by courtesy of School of Physics and Astronomy, University of Manchester).

well as his own paper on the β-ray experiment. He would also have been completing work on the so-called 'radium battery', which he published in 1913.[39] This and a further joint paper not published until 1914[40] ended his practical work in radioactivity.

Moseley in Manchester 1912–13: X-Ray Diffraction and X-Ray Spectroscopy

Max von Laue's explanation of the spot pattern observed when X-rays passed through a crystal and his quantitative wave theory X-ray diffraction published in June 1912 created enormous interest worldwide,[41] and won him the Nobel Prize in Physics for 1914. We know from Moseley's letters that he had read von Laue's papers after returning to Manchester in September from his summer vacation.[42] In the meantime, Moseley had been awarded a half-share of the

John Harling Fellowship for the academic year 1912–1913, which relieved him of his demonstrating duties and left more time for research. With this new freedom, Moseley sought to reproduce Laue's experiment with γ-rays – then thought to be similar to X-rays – by passing them through crystals. The attempts to photograph γ-ray spots were not successful, at least in part because of the low-intensity of the diffracted radiation, and perhaps also because of the background contamination within Moseley's room in the 1900 building. However, these failures prompted him to look more closely at von Laue's theory, whereupon flaws in von Laue's explanation became apparent. Moseley teamed up with Charles Galton Darwin (grandson of Charles Darwin, the pioneer of evolutionary biology), then Lecturer in Mathematical Physics.

On 1 November 1912, Moseley gave the first public airing of what came to be known as the Bragg equation. William Henry Bragg was in the Manchester audience, and Moseley recalled:

> It was rather anxious work, as Bragg the chief authority on the subject (Physics Professor at Leeds) was present … I was talking chiefly about the new German experiments of passing the rays through crystals. The men who did the work entirely failed to understand what it meant, and gave an explanation that was obviously wrong. After much hard work Darwin and I found out the real meaning of the experiments, and of this I gave the first public explanation on Friday. I knew privately that Bragg and his son [William Lawrence Bragg] had worked out their explanation a few days before us … We are therefore leaving the subject to them.[43]

The younger Bragg presented the same equation in Cambridge ten days later on 11 November 1912 and was first to publish the work.[44] Moseley and Darwin never quibbled about the precedence given to the Braggs in deriving the equation that led to their award of the Nobel Prize in Physics for 1915. But Moseley abandoned his plan to study γ-ray diffraction and, with Rutherfords's permission, set about the task of investigating the X-rays themselves with Darwin. Their collaboration led to two publications – the first a letter to *Nature* submitted on 21 January 1913,[45] confirming and extending W. L. Bragg's letter to the same journal submitted on 8 December 1912.[46] The Braggs showed that diffraction was observed more easily in a reflection mode. Experiments in Manchester were based on the observation of reflected X-rays by an ionization detector. The rays appeared to possess the contradictory properties of acting both as particle and a wave.

The second Moseley-Darwin paper mapped out the distribution of the reflected X-rays as a function of angle of reflection (and hence of the X-ray wavelength).[47] Their principal conclusion was that the reflected X-rays consisted of a broad and continuous background, superimposed upon which were five sharp lines which they identified as the characteristic X-rays (a term coined by Barkla) of the platinum anti-cathode. The resolution in their experiment was so good that they initially missed the characteristic X-ray lines and required advice from the Braggs as to where the lines were to be found.[48] This second paper was submitted to the *Philosophical Magazine* on 25 May 1913, just over a month after the Braggs had submitted a paper reporting similar results to the *Proceedings of the Royal Society*:[49] both papers were published in July 1913. Having identified the mono-chromatic X-rays as characteristic rays it occurred to Moseley that here was a new form of spectroscopy. This he anticipated in a famous letter to his mother:

Fig. 5 The Manchester physics department photographed in 1913 outside the old 1900 building. (Source: Archive of School of Physics and Astronomy, University of Manchester, 1913, by courtesy of School of Physics and Astronomy, University of Manchester).

We find that an X-ray bulb with a platinum target gives out a sharp line spectrum with five wavelengths which the crystal separates out as if it were a diffraction grating. In this way one can get pure monochromatic X-rays. Tomorrow we search for the spectra of the other elements. There is here a whole new branch of spectroscopy which is sure to tell one much about the nature of an atom.[50]

Fig. 5 shows the departmental photograph of 1913, in front of the 1900 building. Moseley is standing in the second row on the far right, just behind Charles Darwin. This photograph was probably taken at about the time that Moseley and Darwin had published their major paper. Darwin thereafter left the experimental work to Moseley, and a 'division of labour' emerged between Moseley, who concentrated on X-ray spectroscopy, and the Braggs, who concentrated on X-ray diffraction as a tool for structural determination.[51]

The following year 1913 was remarkable for development of a number of ideas that impinged on Moseley's work. First, Antonius van den Broek put forward the hypothesis that the basis of the order of the elements in the periodic table was the intra-atomic charge, and not the atomic weight.[52] Also at the beginning of 1913 a number of chemists, including Georg von Hevesy and A. S. Russell at Manchester, along with Soddy and others, made further important contributions to the displacement law of radioactive decay. Soddy had already argued that chemically inseparable radioactive elements were in fact chemically identical, despite their differing radiological activity. In early 1913 he introduced the term 'isotope' to characterise radioactive elements that were chemically inseparable but which had different radioactive properties and atomic weights.[53] Also in 1913 J. J. Thompson, citing experiments conducted by his assistant Frederick Aston using their positive ray method (a precursor to mass-spectrometry), showed that the light elements could also have different isotopes: in particular, neon yielded two different ions of mass 20 and 22.[54]

In the first week of July 1913, Niels Bohr again visited Manchester, around the time that Moseley and Darwin's paper on X-rays appeared in the *Philosophical Magazine*. Bohr had himself just published in the same volume the first of his famous trilogy of papers, wherein he expressed the Rydberg constant in terms of the fundamental quantities e, m and h, and also explained the spectrum of hydrogen in a simple but elegant fashion in terms of quantised stationary states.[55] Bohr came to Manchester to discuss the second and third of his trilogy with Rutherford. We know there was a meeting between Moseley and Bohr, possibly in Moseley's basement laboratory. Darwin and von Hevesy were also present, and the frequencies of the K rays, isotopy and atomic number were all discussed. Bohr himself later recalled the meeting:

> Yes, but all this came later, you see. Because actually it was me who brought it [i.e. ordering in the periodic table] up to Moseley and Darwin. They wanted to (speak) to me, you see. And then I explained also to them what my view was on the periodic table. I had, of course, thought about it before and had talked to Rutherford. And then, as far as I remember, Moseley said, 'All, right, we will see'. And he did the experiments in a very, very short time.[56]

Following the meeting with Bohr, Moseley was determined to establish whether it was atomic number or atomic weight that controlled the periodic system. He proceeded to measure X-ray spectra of the elements between 20 and 30, the first period of transition metals. Apart from scandium he was able to study a continuous series, including nickel and cobalt. Based on atomic weights, nickel appeared before cobalt but their chemical properties suggested that this order should be reversed. Beginning to assemble his apparatus in August, Moseley soon abandoned the detection of X-rays by the ionization method and returned to photography for recording spectra. By November Moseley was able to write to Bohr of his success. His results were submitted in November and published in the *Philosophical Magazine* in December 1913.[57]

Moseley showed that the Kα rays conformed to a simple pattern with the square root of the frequency of the ray proportional to the atomic number minus 1, thus confirming van den Broek's hypothesis. The constant of proportionality came out as ¾ times the Rydberg frequency appearing in Bohr's model of the hydrogen atom, thus linking at once the quantum explanation of the optical and X-ray spectra of the elements. The Kα and Kβ lines were approximately in constant proportion across the elements and so if the spectra were presented graphically they formed the image now known as Moseley's ladder. After submitting this paper, Moseley moved from Manchester to Oxford at the end of November 1913, where until April 1914 he conducted the work which led to his second paper on the high-frequency spectra of the elements.[58]

Manchester after Moseley: The Revolution Completed
With the outbreak of war in August 1914, the Universities of Europe were drained of half their students and staff. Rutherford's Manchester laboratory did not escape. Volunteering in 1914, Darwin was commissioned in the Royal

Engineers to work on sound ranging and Hevesy was drafted into the Austro-Hungarian army. Hans Geiger and Ernest Marsden found themselves on opposite sides in the same sector of the western front. James Chadwick was interned in Berlin where he had gone to work with Geiger in 1913. Rutherford, with W. H. Bragg, became drawn into war work on anti-submarine research for the Admiralty's Board of Invention and Research.[59] Niels Bohr returned to Manchester from neutral Denmark to succeed Darwin as Reader in Mathematical Physics in autumn 1914, by which time the Department was already reduced to just a very few of his old colleagues, who included Makower and Evans. The laboratory's German glassblower Otto Baumbach was interned. Research became very difficult and publications plummeted. Rutherford managed to keep going, with the help of William Kay the laboratory steward. This eventually led to the discovery of artificial transmutation, but it would not be until after the war in 1919 that this work was published.[60]

Moseley's death on 10 August 1915 came as a great shock around the world and Rutherford himself was deeply affected. The war memorial in the main quadrangle of the 1873 Owens College lists over 500 Manchester men who lost their lives, including 2nd Lt H. G. J. Moseley of the Royal Engineers. In October 1919, a Moseley Memorial committee was set up, chaired by the Vice-Chancellor and including both Rutherford and W. L. Bragg. Rutherford had by then moved from Manchester to the Cavendish Chair at Cambridge, and was succeeded by Bragg, who was to develop a Manchester school of X-ray crystallography. A memorial fund was established – with Darwin and Harold Robinson as secretaries in Cambridge and Manchester respectively – to raise money for a bronze tablet and a Moseley Prize for Physical Research. The

Fig. 6 Left: the plaque currently mounted at the back of the Moseley Lecture Theatre at the University of Manchester. Right: the plaque originally mounted outside the room where Moseley conducted his pioneering X-ray experiments. (Source: Photographs taken by the author, 2015, by courtesy of School of Physics and Astronomy, University of Manchester).

Braggs, J. J. Thomson, and many of Moseley's former Manchester colleagues contributed, including Rutherford, Darwin and Chadwick. The plaque was unveiled on 26 November 1921, with speeches by Rutherford and Darwin, and remained in place until 1968, when the Department of Physics moved to a new building. Then, the plaque was hung at the back of the Moseley Lecture Theatre in the new physics building, where it remains to this day (Fig. 6).

Sometime in the 1950s or 1960s, before physics moved to the new building and during the time when Samuel Devons, the last of Rutherford's Cavendish students succeeded as Langworthy Professor, a number of wooden plaques were placed around the old laboratory marking the locations of key experiments that had taken place during Rutherford's era. Among these was one devoted to Moseley, in the basement room where he carried out his X-ray experiments (Fig. 6).

Looking back, it is remarkable that in Moseley's short lifetime – from the discovery of X-rays in 1895 through to the genesis in Manchester of the nucleus, the quantum atom and the ordering of the periodic table by number – a second scientific revolution was almost complete. It is a cause for celebration that the birthplace of the Dalton atom also spawned the quantum atom. Speaking in the 1930s, on the opening of another new physics building devoted to X-ray crystallography, Rutherford recalled that:

> I owe a great debt to Manchester for the opportunities it gave me for carrying out my studies. I do not know whether the University is really aware that during the few years from 1911 onwards the whole foundation of the modern physical movement came from the physical department of Manchester University.[61]

Today the basement area of the old Schuster Laboratory is occupied by the Manchester Museum and Moseley's wooden plaque is kept in a cardboard box. That part of the building has not been well preserved. Many of the windows have been bricked up and the internal organization almost completely restructured. Apart from the windows, Moseley's room is still intact, but only just. A historic building which was the cradle of modern physics should have been better preserved as a permanent memorial to a unique period in the history of science.

Notes and References

1. Moseley to Rutherford, 17 July 1910, Letter 42 in J. L. Heilbron, *H. G. J. Moseley: The Life and Letters of an English Physicist, 1887–1915* (Berkeley: University of California Press, 1974), 173 (hereafter Heilbron, *Moseley*).

2. S. G. Brush, *The History of Modern Science: A Guide to the Second Scientific Revolution, 1800–1950* (Ames: Iowa State University Press, 1988).

3. B. Jaffe, *Crucibles: The Story of Chemistry* (New York: Dover Publications, Inc., 4th edn, 1976), 84–99.

4. J. P. Joule, 'On the Existence of an Equivalent Relation between Heat and the Ordinary Forms of Mechanical Power'. *Philosophical Magazine Series 3*, 27 (179), (1845), 205–207.

5. J. Thomson, *The Owens College, Its Foundation and Growth* (London: Waterlow and Sons, 1886).

6. P. J. Hartog, *The Owens College, Manchester: A Brief History of the College and Description of Its Various Departments* (London: Waterlow and Sons, 1900).

7. A. Schuster and R. S. Hutton, *The Physical Laboratories of the University of Manchester: A Record of 25 Years' Work* (Manchester: University of Manchester Press, 1906).

8. T. E. Broadbent, *Electrical Engineering at Manchester University* (Manchester: Manchester School of Engineering, 1998).

9. J. A. Crowther, 'Röntgen Centenary and Fifty Years of X-Rays', *Nature*, 155 (3934), (1945), 351.

10. L. Badash, 'Radioactivity before the Curies', *American Journal of Physics*, 33 (2), (1965), 128-135.

11. J. J. Thomson, 'Cathode Rays', *Philosophical Magazine Series 5*, 44 (269), (1897), 293–316.

12. A. S. Eve, *Rutherford, Being the Life and Letters of the Rt. Hon. Lord Rutherford, O.M.* (Cambridge: Cambridge University Press, 1939), 37–47.

13. E. Rutherford, 'Uranium Radiation and the Electrical Conduction Produced by It', *Philosophical Magazine Series 5*, 47 (284), (1899), 109–163.

14. P. Curie and S. Curie, 'Sur un Substance Nouvelle Radioactive, Contenue dans la Pitchblende', *Comptes rendus de l'Académie des sciences*, 127 (1898), 175–178.

15. P. Curie, P. Curie and M. G. Bemont, 'Sur une Nouvelle Substance Fortement Radioactive, Contenue dans la Pitchblende', *Comptes rendus de l'Académie des sciences*, 127 (1898), 1215-1217.

16 Eric R. Scerri, *The Periodic Table. Its History and Significance* (Oxford: Oxford University Press, 2007), 162.

17 A. L. Debierne, 'Sur un Nouvelle Matière Radio-Active', *Comptes rendus de l'Académie des sciences*, 130, (1899), 906-908.

18 T. J. Trenn, *The Self-Splitting of the Atom: The History of the Rutherford-Soddy Collaboration* (London: Taylor and Francis, 1977).

19 W. Crookes, 'Radio-Activity of Uranium', *Proceedings of the Royal Society of London*, 66 (424–433), (1899), 409–423. Uranium-X proved to be the $^{234m}_{91}Pa$ isotope of protactinium, while thorium-X was the $^{224}_{88}Ra$ isotope of radium.

20 J. L. Heilbron, *Ernest Rutherford and the Explosion of Atoms* (Oxford: Oxford University Press, 2003), 32–54.

21 W. Ramsay and F. Soddy, 'Experiments in Radioactivity and the Production of Helium', *Nature*, 68 (1763), (1903), 354–355.

22 H. W. Kirby, 'The Discovery of Actinium', *Isis*, 62 (3), (1971), 290–308.

23 E. Rutherford and T. Royds, 'The Nature of the α Particles from Radioactive Substances', *Philosophical Magazine Series 6*, 17 (98), (1909), 281–286.

24 E. Rutherford, 'The Scattering of α and β Particles by Matter and the Structure of the Atom', *Philosophical Magazine Series 6*, 21 (125), (1911), 669–688.

25 Heilbron, *Moseley, op. cit.* 141–279.

26 Moseley to his mother, [December 1910], Letter 49 in Heilbron, *Moseley, op. cit.* 181.

27 J. B. Birks, *Rutherford at Manchester* (Heywood: London, 1962).

28 Moseley to his mother, [October 1910], Letter 44 in Heilbron, *Moseley, op. cit.* 176.

29 Moseley to his sister Margery, 13 November [1910], Letter 45 in Heilbron, *Moseley, op. cit.* 177–178.

30 Moseley to Margery, [7 December 1911], Letter 53 in Heilbron, *Moseley, op. cit.* 184.

31 H. G. J. Moseley, 'The Number of β-Particles Emitted in the Transformation of Radium', *Proceedings of the Royal Society A*, 87 (595), (1912), 230–255.

32 Moseley to Margery, 13 November [1910], Letter 45 in Heilbron, *Moseley, op. cit.* 177–178.

33 Moseley to Margery, [7 December 1911], Letter 53 in Heilbron, *Moseley, op. cit.* 183–184.

34 A. Schuster, *The Physical and Electrotechnical Laboratories of the University of Manchester* (Manchester: University of Manchester Press, 1912), 6–7

35   E. Rutherford, in *Reports of the Council of the University of Manchester. Appendix XI. Reports on Departments. I Reports on the Physical Department Session 1911–1912* (Manchester: University of Manchester Press, 1912), 170.

36   W. H. E. Schwarz, '100th Anniversary of Bohr's Model of the Atom', *Angewandte Chemie International Edition*, 52 (47), (2013), 12228–12238.

37   H. G. J. Moseley and K. Fajans, 'Radioactive Products of Short Half Life', *Philosophical Magazine Series 6*, 22 (130), (1911), 629–638.

38   H. G. J. Moseley and W. Makower, 'γ Radiation from Radium B', *Philosophical Magazine Series 6*, 23 (134), (1912), 302–310.

39   H. G. J. Moseley, 'The Attainment of High Potentials by the Use of Radium', *Proceedings of the Royal Society A*, 88 (605), (1913), 471–476.

40   H. G. J. Moseley and H. Robinson, 'The Number of Ions Produced by the β and γ Radiations from Radium', *Philosophical Magazine Series 6*, 28 (165), (1914), 327–337.

41   M. Eckert, 'Max Von Laue and the Discovery of X-Ray Diffraction in 1912', *Annalen der Physik* (Berlin), 524 (5), (2012), A83-A85.

42   Moseley to his mother, 10 October [1912], Letter 61 in Heilbron, *Moseley, op. cit.* 193.

43   Moseley to his mother, [4 November 1912], Letter 63 in Heilbron, *Moseley, op. cit.* 194–195.

44   W. L. Bragg, 'The Diffraction of Short Electromagnetic Waves by a Crystal'. *Proceedings of the Cambridge Philosophical Society*, 17 (Read 11 November 1912), (1913), 43–57.

45   H. Moseley and C. G. Darwin, 'The Reflection of X-Rays', *Nature*, 90 (2257), (1913), 594.

46   W. L. Bragg, 'The Specular Reflection of X-Rays', *Nature*, 90 (2250), (1912), 410.

47   H. G. J. Moseley and C. G. Darwin, 'The Reflexion of the X-Rays', *Philosophical Magazine Series 6*, 26 (151), (1913), 210–232.

48   J. Jenkin, *William and Lawrence Bragg, Father and Son* (Oxford: Oxford University Press, 2009). Hereafter Jenkin, *Bragg*. Moseley's interactions with William Henry Bragg are discussed in Chapter 16, 345–349.

49   W. H. Bragg and W. L. Bragg, 'The Reflection of X-Rays by Crystals', *Proceedings of the Royal Society A*, 88 (605), (1913), 428–438.

50   Moseley to his mother, [18 May 1913], Letter 74 in *Moseley*, Heilbron, *op. cit.* 205.

51   The term 'division of labour' was coined by J. S. Wark. John Jenkin argues that withdrawal of the Braggs from the emerging field of X-ray spectroscopy arose in part because of Rutherford's 'bullying' of William Henry Bragg. See Jenkin, *Bragg, op. cit.* 347.

[52] Eric R. Scerri, Chapter 4 this volume.

[53] F. Soddy, 'Intra-Atomic Charge', *Nature*, 92 (2301), (1913), 399–400.

[54] H. Kragh, 'The Isotope Effect: Prediction, Discussion and Discovery', *Studies in History and Philosophy of Science Part B*, 43 (3), (2012), 176–183.

[55] N. Bohr, 'On the Constitution of Atoms and Molecules', *Philosophical Magazine Series 6*, 26 (151), (1913), 1–25.

[56] Interview of Niels Bohr by Thomas S. Kuhn, Leon Rosenfeld, Aage Petersen, and Erik Rudinger on 1962 November 1, Niels Bohr Library & Archives, American Institute of Physics, College Park, MD USA, http://www.aip.org/history-programs/niels-bohr-library/oral-histories/4517-4 (accessed on 17 January 2018).

[57] H. G. J. Moseley, 'The High-Frequency Spectra of the Elements', *Philosophical Magazine Series 6*, 26 (156), (1913), 1024–1034.

[58] H. G. J. Moseley, 'The High-Frequency Spectra of the Elements. Part II', *Philosophical Magazine Series 6*, 27 (160), (1914), 703–713.

[59] John Richardson, Russell Egdell, Nick Hankins and Harold Hankins, 'Rutherford, Geiger, Chadwick, Moseley, Cockroft and Their Role in the Great War – Part I', *The Western Front Association. Stand To!*, 86 (2009), 28–34.

[60] E. Rutherford, 'Collision of Alpha Particles with Light Atoms. IV. An Anomalous Effect in Nitrogen', *Philosophical Magazine Series 6*, 37 (222), (1919), 581–587.

[61] A. S. Eve, *Rutherford, Being the Life and Letters of the Rt. Hon. Lord Rutherford, O.M.* (Cambridge: Cambridge University Press, 1939), 350.

CHAPTER THREE
READING BETWEEN THE LINES: DECONSTRUCTING MOSELEY'S DIAGRAM

KRISTEN M FREDERICK-FROST

Introduction

Order. Linearity. Comprehensiveness. These are all terms that we can use to describe Henry Moseley's famous diagram (Fig. 1), wherein the arrow-straight lines pierce well-behaved data points, giving a convincing display of the relationship between an element's atomic number and the frequency of X-rays it emits.[1] This is Moseley's law in graphical form. It is easy to allow the eye to skip over the gaps in the plot and to extend the lines in order to deduce that all matter will obey this relationship. But this diagram can be used to impart more than a reductive equation. Reconsider the individual points, gaps, and the bounds of the lines. What do they tell us about Moseley's experimental process?

Constraints. Variability. Uncertainty. Each data point (or lack thereof) has a story. This chapter looks at four areas of the diagram that can be used to explore some of the material and professional influences that shaped Moseley's work. We begin by considering ten elements that were analyzed in his first paper on the 'High-Frequency Spectra of the Elements' (Fig. 1, middle shaded section containing nineteen data points). This series may have been chosen to answer a theoretical question, but experimental constraints kept it from being extended in either direction.

The other points that make-up the famous diagram came later, after Moseley left his fellowship at the University of Manchester to become an independent researcher at the University of Oxford. This work was undertaken in three separate phases. Moseley's work with high energy X-rays (Fig. 1, rightmost shaded section) provides an opportunity to look closely at extant artefacts at the Museum of the History of Science to see how they demonstrate variability in Moseley's process. We can also use the work with soft, easily absorbed X-rays (Fig. 1, leftmost shaded section) to discuss the makers who fabricated Moseley's equipment as well as the research environment in which he worked. Lastly, we will look at Moseley's study of the rare earths, specifically the elements for which there is a gap in the plot (Fig. 1, topmost shaded section), to explore how scientists embraced or rejected his process and results.

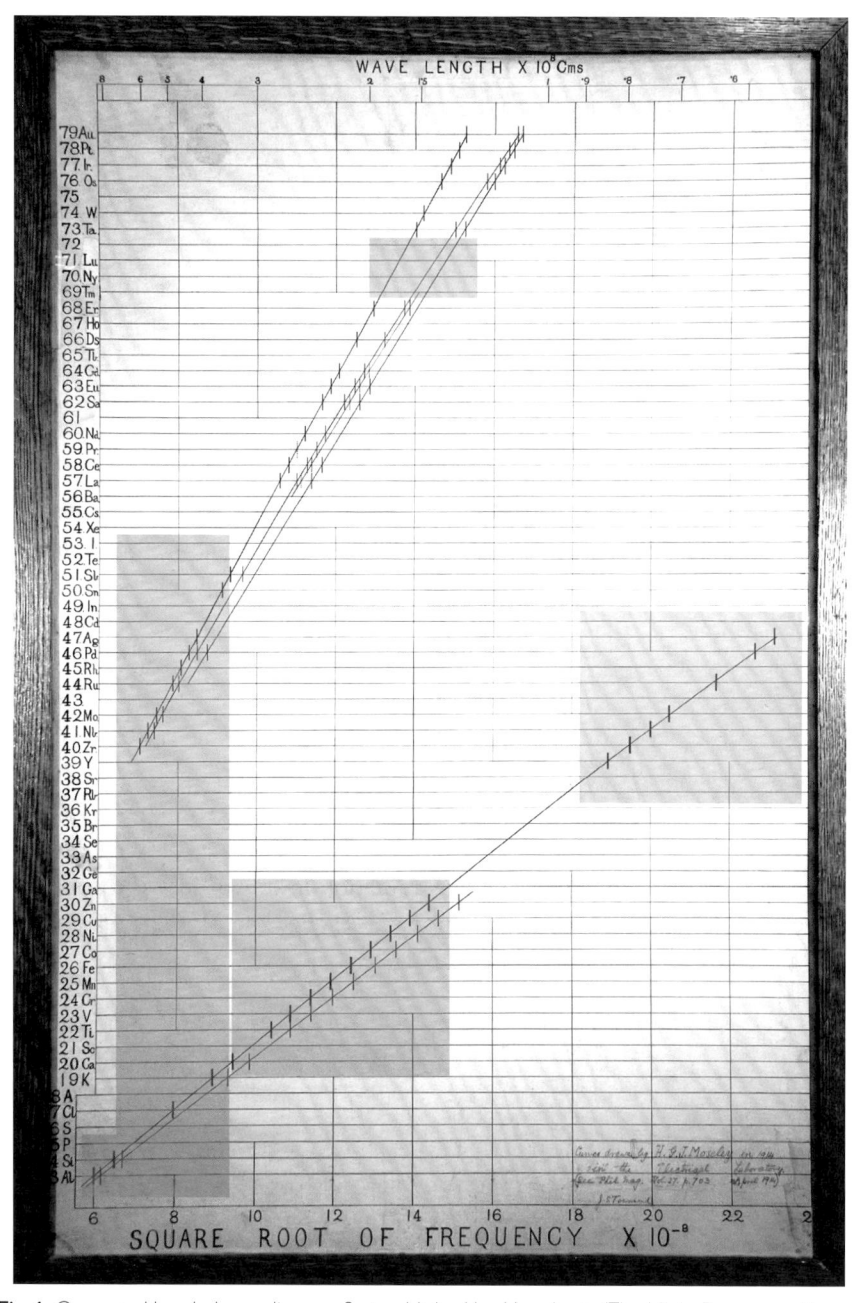

Fig. 1 Corrected hand-drawn diagram first published by Moseley in 'The High-Frequency Spectra of the Elements. Part II'. Four shaded regions have been superimposed on the diagram by the author. (Source: Museum of the History of Science, Oxford, 1914, by courtesy of the Clarendon Laboratory, Oxford).

It is helpful to begin with a sketch of X-ray practice and spectroscopy as it relates to Moseley's research. By 1913, X-ray tubes were widely available in many shapes, sizes, and levels of 'hardness', or capacity to produce penetrating radiation.[2] Moseley used both commercial and custom made X-ray tubes. A tube nominally consisted of two electrodes that were placed inside a partially evacuated glass envelope. In operation, one electrode, called the cathode, was biased at a negative voltage with respect to the other, the target (or anti-cathode). Electrons from the cathode were accelerated across the gap between the electrodes by the bias voltage and hit the target. The collision of the electron with the target created X-radiation.

Scientific understanding of the radiation produced by these tubes underwent several shifts during Moseley's lifetime, from a ray that was thought impossible to reflect or refract, to a major tool of spectroscopy.[3] While Moseley was an undergraduate student at Trinity College, Oxford, Charles Barkla (1877–1944), a lecturer at the University of Liverpool, investigated secondary X-ray emissions from various elements. Barkla discovered that the secondary emission – the radiation produced after a substance is bombarded with X-rays – contained not only the same kind of radiation as the X-ray tube but also included a 'homogeneous' component: homogeneity, defined by Barkla as emission with uniform penetrating power. This was curious, as the X-ray beam used to bombard a substance was known to be 'heterogeneous', or composed of rays of various penetrating power. Barkla also recognised that this homogeneous radiation appeared to be dependent on atomic weight.[4] By 1911, he grouped the homogeneous radiation into two 'series', K and L, defined by similarities in penetration power.[5] At the time Moseley produced the 1914 plot in Fig. 1, the K/L classification was widely used, as shown by his identification of the lower collection of lines as the K rays, and the top lines with the L rays.[6]

While at the University of Manchester, Moseley and C. G. Darwin (1887–1962) started X-ray research on a path parallel to that of two other noted pioneers of X-ray spectroscopy, the father and son W. H. and W. L. Bragg. The use of an X-ray spectrometer made a clear connection between them. The principles of a spectrometer are similar for both visible light and X-rays. Radiation from a source is first collimated and then directed at a substance that interacts with the incoming beam in such a way that the contingent wavelengths are separated. The most familiar example is a prism fanning out incoming white light into its spectrum of colours. X-rays, however, require material different than a prism to produce a similar outcome. W. L. Bragg found the regular crystal lattice of mica allowed for X-rays to be diffracted.[7] He explored the characteristics of reflected X-ray beams with his father, using various crystals. Together, they famously related the wavelength and glancing angle of the radiation with the crystal that reflected the X-rays (Bragg's Law).

In 1913, Moseley and Darwin adapted a prism spectrometer for use with X-rays, in much the same way as had the Braggs (Fig. 2).[8] An ionization detector was positioned on the spectrometer arm that had formerly held a telescope. This detector registered the intensity of the reflected X-rays, giving quantitative measurement of both the 'white' radiation composed of a continuous range of wavelengths as well as the 'monochromatic' radiation, found at specific angles of reflection. When Moseley and Darwin published their work in the *Philosophical Magazine*, they were able to improve upon the Braggs' previous measurement of these characteristic peaks, extending the number identified from three to five (labeled α, β, γ, δ, and ε).[9] Importantly, these characteristic peaks were associated with the platinum target of the X-ray tube.

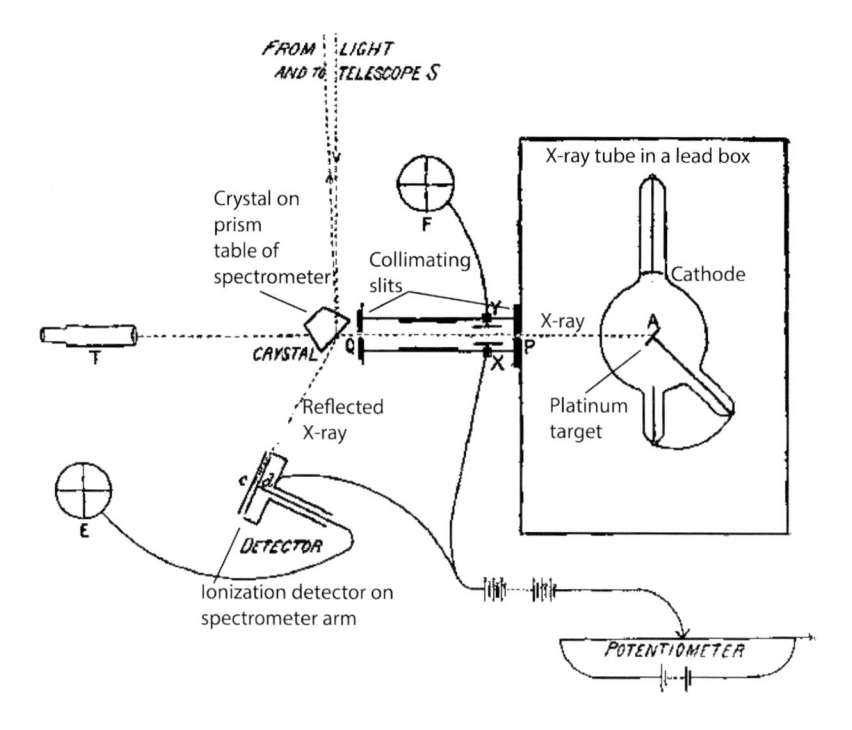

Fig. 2 Annotated diagram of Moseley and Darwin's 1913 experimental setup. (Source: *Philosophical Magazine Series 6*, 26 (1913), 212).

First Data Points

The study of radiation from other elemental targets would bring Moseley fame. His first investigation of the characteristic K and L radiation of the elements came a few months after his collaboration with Darwin on the reflection of X-rays. He was still at Manchester, and very much aware of another burgeoning area of research – that of atomic structure – led by his supervisor, Ernest Rutherford (1871–1937). Rutherford was deeply influential, as was Niels Bohr (1885–1962), who visited Manchester in the summer of 1913. Questions about the make-up of the atoms within elements influenced questions about the organization of elements. Mendeleev's 1869 periodic table of the elements grouped substances according to chemical properties and atomic weight. There were, however, a few sets of neighboring elements in Mendeleev's scheme where the alignment of chemically similar groups did not follow the trend of increasing atomic weight (for example: Ar-K, Co-Ni, Te-I). On the other hand, Antonius van den Broek (1870–1926) developed the concept of the atomic number as a measure of the intra-atomic charge and explored its usefulness as an organisational principle in the years between 1907–1912.[10]

Following his research with Darwin, Moseley was well situated to explore whether the homogeneous X-ray emissions characteristic of the elements followed the sequence of atomic weight or atomic number. However, this research was shaped as much by experimental constraints as it was by theoretical questions. His first K series data points (Fig. 1, middle shaded section containing nineteen data points) included only one pair of the elements that did not follow the trend of increasing atomic weight in Medeleev's scheme. To include the others, Moseley would have had to extend the range of atomic numbers investigated. However, his experimental set-up was not suited for doing so.

For his new line of investigation Moseley constructed a unique X-ray tube that allowed targets to be interchangeable.[11] This permitted the examination of the characteristics of homogenous X-ray emissions from different targets using only one tube. Interchanging elements was done by placing specimens on a cart wheeled under the cathode. The cart positioned the targets at an angle to produce a directed beam and supplied the electrical connection to these elements, permitting the application of a bias voltage between the cathode and the targets.

Such a design lends itself well to metallic elements, but not to non-metals and gases, a fact made manifest in several gaps in Moseley's diagram. The

X-rays produced by Moseley's customised tube were collimated and directed at a crystal of potassium ferrocyanide, which was positioned on the prism plate of a spectrometer.[12] The diffracted beam was then detected on a photographic plate that was placed on the spectrometer arm.[13]

In his first paper on the spectra of the elements, Moseley details difficulty with only one of the elements studied: 'Calcium alone gave any trouble'.[14] Here, he ran into the soft X-ray limitations of his apparatus. The emission from calcium was easily absorbed by both the thin (0.02 mm) aluminum exit window of his X-ray tube and the ambient air between the tube, crystal, and photo plate. So he replaced the aluminum with goldbeater's skin, a thin membrane made from cow intestine. Moseley also used less absorbent hydrogen gas to fill the volume around the crystal and photo plate in order to image the spectrum. He had trouble producing rays with calcium, as his sample reacted poorly to electron bombardment. To achieve a five-minute exposure of a photographic plate, he had to excite his uncooperative target repeatedly for only a few seconds at a time. Moseley's value for calcium's strongest emission in the K series (α) represents a soft X-ray boundary for his first set-up: exploring wavelengths longer than $\sim 3.4 \times 10^{-8}$ cm required him to make material concessions to deal with the easily absorbed X-rays.

The hard end of the K series raises another question. The elements after zinc, being either easily melted or non-metallic, presented other problems in addition to the hardness of their X-rays. Since the ten elements for which Moseley explored K-type emission 'were chosen as forming a continuous series with only one gap' so that he could 'bring out clearly any systematic results', we can only guess at the high energy limitations of his first apparatus. He mentions only two elements for which he investigated L rays (Ta, Pt), and these results were among the harder frequencies he initially explored.[15] This facet of his work was done later at Oxford, with apparatus designed for the necessarily higher voltages.

What we can conclude from the first quarter of Moseley's spectra data is that these elements were the easiest to characterise. All were commercially available, and the experiment was run at ambient pressures and at voltages that did not require unusual equipment. For the next phase of his work, he would have to modify his apparatus and his expectations of support. His transition to life as an independent researcher at the University of Oxford meant leaving behind the research infrastructure he had enjoyed at Manchester. He altered his experimental plan to account for this loss, as is seen in a 17 January 1914 letter to Darwin:

> I am trying to divide the spectrum work into three parts, and do all three at once. This dangerous arrangement is necessitated by the impossibility of getting apparatus mended quickly if it slows down.[16]

These three lines of research were (1) the exploration of hard K radiation, (2) the soft K and L radiation from elements easily absorbed by ambient air, and (3) the characterization of the L radiation produced by the rare earth elements. Moseley's division of work provides a natural framework around which to organise the discussion of his results.

Hard K Lines

Moseley acknowledged some variability of the experimental setup for the hardest K lines in his second paper on the high-frequency spectra of the elements, but he played down its importance by stating 'the general experimental method has remained unaltered'.[17] Instead of trying to diminish the effect of this experimental variability, this section explores the reasons for it. First, there was a limitation associated with voltage. Moseley wrote to Darwin on this point:

> The reason why these hard rays give trouble is firstly [1] that the distance [from] cathode to target in my old apparatus is short. This makes it difficult to get high voltage. – A new apparatus is being made –[18]

The harder the radiation desired, the higher the voltage required. The frequency of the radiation (ν) was linked to the cathode/target bias voltage (V) by $h\nu = eV$, where h is Planck's constant and e is the charge of the electron.[19] As this was the minimum voltage necessary to create a specific frequency X-ray, Moseley operated his tube at even higher voltages. Using this equation and Moseley's published values, the minimum voltage range that he used in his first examination of the elements with atomic numbers 20–30 was roughly 4–10 kV.[20] The voltage range he needed to cover for his hard K ray work was approximately 15–22 kV.[21] But this range ultimately proved too high for the X-ray tube he had used, as the electrode spacing was too short to safely support these higher voltages.

To deal with this problem, Moseley decided to buy a tube from a commercial company. He asked A. C. Cossor to produce a tube to his specifications (Fig. 3, top). The design still used a trolley to pull various elements into the path of the electrons from the cathode. This was done by winding a thread connected

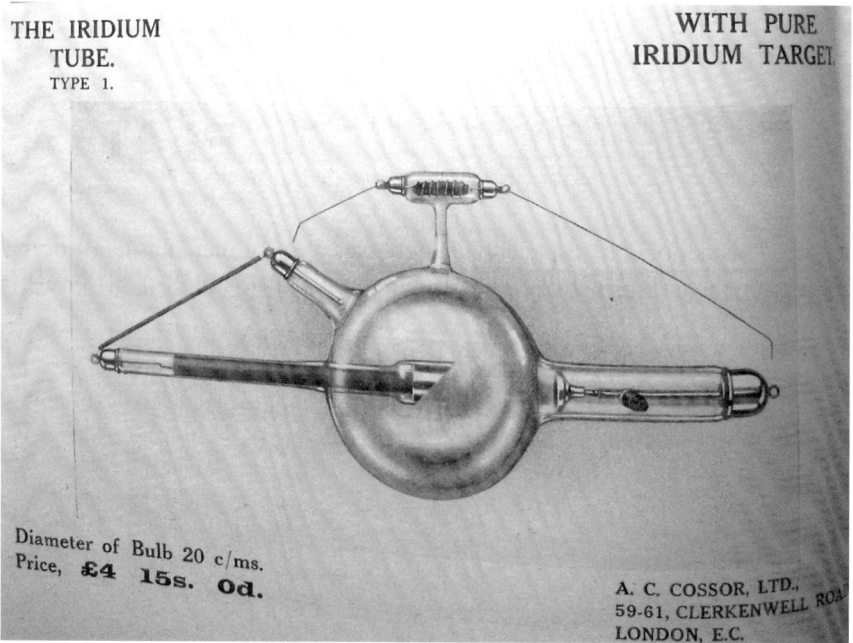

Fig. 3 Top: Photograph of X-ray tube used by Moseley Bottom: X-ray tube from A. C. Cossor Ltd. trade literature. (Source: Top: K. M. Frederick-Frost, 2012, courtesy of the Museum of the History of Science, Oxford. Bottom: *The Journal of the Röntgen Society*, July 1913, vi).

to the trolley around bobbins on either side of the track. The main body of the tube was similar to other tubes that Cossor was making at the time, as can be seen by comparing features (such as the cathode connection) of the tube with pictures in Cossor's trade literature (Fig. 3, bottom). Some of the connecting glassware and wax seals were modifications made by Moseley himself.

Moseley was conscious of the limitations of the cathode-target spacing, especially since he had just witnessed the destructive outcome of a high voltage arc. Twelve days before his letter to Darwin, Moseley wrote to Rutherford that:

> The old apparatus after holding up gallantly all the time at Manchester was smashed up by a discharge which went astray, just as I was beginning on the Tungsten spectrum.[22]

A high voltage discharge (arc) was extremely undesirable. Unlike the low current density of electrons in the cathode stream that produced the X-rays, such discharges contained far more power. In effect, the arc was a mini-lightning strike that leapt between two points (like the cathode and the target electrodes) separated by a high voltage. The result of the discharge was the loss of an X-ray tube, which was not easily mended with the limited Oxford resources.

The voltage Moseley would have used to examine tungsten's L rays was approximately 9 kV. But Moseley had previously pushed his tube to higher voltages.[23] Why did it fail for tungsten? Perhaps the tube fatigued in some way, as he hints to his mother in a letter on 17 November 1913: 'my poor tube is on its last legs'.[24] This is possible, especially if there was a lot of metallic deposition on the surfaces in the well-used tube. The Museum of the History of Science at Oxford has a broken tube in its collection that could be linked with the one that Moseley lost to the discharge (Fig. 4). The break in the glass, the missing cathode, the surface condition, and its similarity to another tube that is linked with a Manchester glass blower, come together to support this possibility.

Another factor that could have caused a discharge was the pressure in the tube. Moseley's tubes were designed to be opened, filled with new elements, and resealed – so they were not hermetic envelopes. Moseley reportedly used a vacuum pump to evacuate the tube to approximately a few millimeters of mercury.[25] To make a vacuum seal, he used a white cement and red wax to join together custom pieces of glasswork. Fig. 5 shows what is left of one of the bobbin winders that was attached to the customised Cossor X-ray tube. Note the white seal conjoining two blown glass pieces and the passage of a wire at the end of one of those pieces. This wire connected the trolley and its elements to the positive terminal of the high voltage

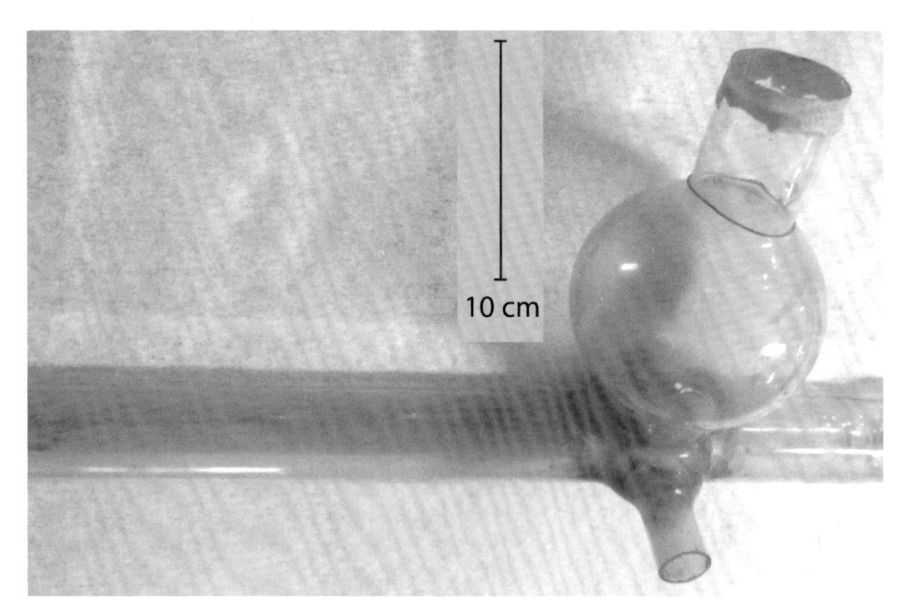

Fig. 4 Photograph of broken X-ray tube associated with Moseley's research. (Source: K. M. Frederick-Frost, 2012, by courtesy of the Museum of the History of Science, Oxford).

supply. Movement would easily disturb these temporary seals, admitting gas. Each time the trolley was taken out and loaded, the tube's vacuum seals would change and potentially create vastly different operating conditions. Pressure affected both the 'hardness' of the tube and the voltage at which the two electrodes broke down and produced an arc.[26] When pushing his first apparatus to the limitations of its voltage capacity, a change in pressure could have made the difference in squeaking out a few more data points and destroying the tube itself. Pressure stability encouraged the use of tubes with larger volume and could have been why Moseley made the transition to the larger Cossor tube.[27]

Even after obtaining a tube that could produce hard K rays, Moseley reported a new problem: trouble in picking the characteristic hard K radiation from the 'most exasperating Barkla fringes'.[28] He was confident that these 'fringes' were caused by the reflected heterogeneous 'white' radiation, as they persisted even when he swapped the target. In the case of the hard K rays, the intensity of the characteristic peaks, which Moseley wanted to measure, was nearly the same as the 'white' background, which would prove problematic on photographic plates.

In print, Moseley changed his tone, 'It is easy to devise methods for getting rid of the fringes,' he wrote. Note that it is easy to 'devise' but 'exasperating'

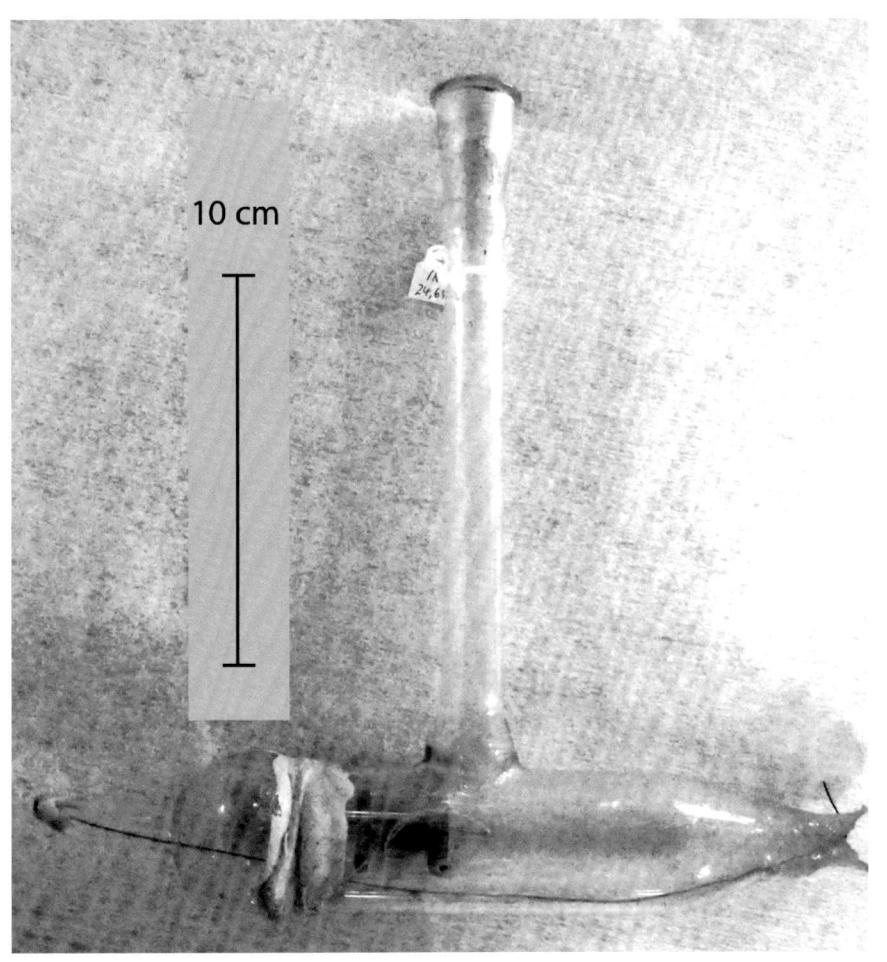

Fig. 5 Photograph of a winding bobbin in one of Moseley's X-ray tubes. (Source: K. M. Frederick-Frost, 2012, by courtesy of the Museum of the History of Science, Oxford).

to enact. In his paper, he detailed many of the methods for dealing with these fringes that he had shared in letters to Darwin. First was the need to increase the distance between the slit, the crystal, and the photographic plate, as compared to the original arrangement. In addition, he suggested narrowing the collimating slit. This method proved successful in his researches with Darwin, where they used a fine 'very narrow pencil' beam to locate characteristic lines.

The parts of the experimental setup might have remained (roughly) the same, but how they were arranged was highly variable. There was no 'definitive' hard X-ray apparatus.[29] There were at least two tubes used to produce the hard K ray data; as well as different arrangements of the spectrometer, tube, and photo plates; various slit widths to pick out the characteristic lines; and several procedures for identifying 'white' radiation (rotating crystal, swapping elements). When Moseley said the 'general experimental method has remained unaltered', he was speaking in the abstract, because he had to 'alter' quite a bit to wrestle the K ray numbers from the background of the heterogeneous radiation. So it is little wonder that he plotted only the Kα rays, and that the curve for the fainter Kβ rays drops away after atomic number 30.

Soft K and L Rays

Moseley's first study of the spectra of elements, conducted when he was at Manchester, ran into trouble when he investigated calcium. The X-rays produced were too soft to penetrate the air surrounding the spectrometer. Moseley solved the problem for calcium by replacing the air around the apparatus with hydrogen. This enabled him to obtain the measurement he sought. But when it came to exploring wavelengths even longer than ~ 3.4×10^{-8} cm, he needed to evacuate the spectrometer itself. This required custom-fabricated equipment, rather than just a re-purposed a prism spectrometer.

To make this apparatus, Moseley turned to the instrument makers he knew from Manchester. As several authors have observed, he had a low opinion of the infrastructure at Oxford. An oft-quoted letter to Rutherford in December 1913 colourfully brings home this point:

> It is too early to yet have an opinion on the prospects of working efficiently here, but I find the professor most obliging and ready to make the way smooth. The mechanic however is likely to prove a thorn in the flesh. Things seem to move slowly here compared to Manchester.[30]

The vacuum spectrometer Moseley used to study soft K and L rays is shown in a photograph taken of the apparatus in the 1940s (Fig. 6).[31] By moving the crystal and photo plate into this iron vessel, Moseley was able to remove most of the ambient air with a vacuum pump attached to a port in the cylinder wall. He reports using gold beaters skin to form a window between the X-ray tube and spectrometer. The X-rays could pass through the skin, but the spectrometer and the tube could be maintained at different levels of vacuum.

This X-ray tube is clearly different from the Cossor tube: the volume is small and the glass walls surrounding the cathode/target region are shaped, not spherical. The electrical connection to the cathode is made by thin bare wire passed through the glass envelope, not by a large metal cap. Since it

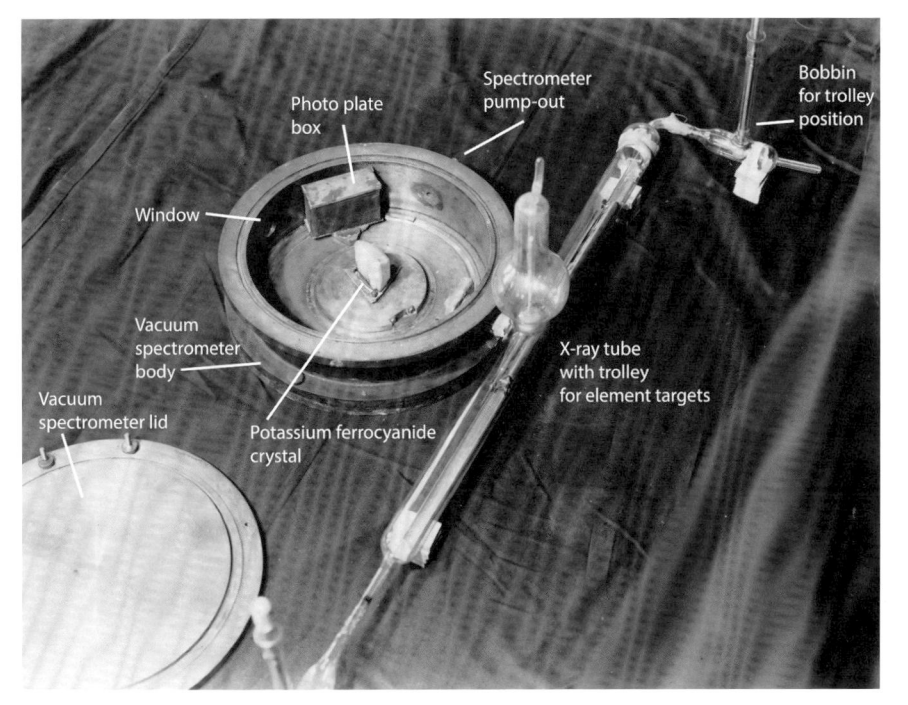

Fig. 6 Annotated photograph showing Moseley's vacuum spectrometer setup. (Source: Museum of the History of Science, Oxford, circa 1940, by courtesy of the Museum of the History of Science, Oxford).

is known that Manchester's instrument maker made the spectrometer, it is likely that this tube was also made there, as the two fit together. Supporting evidence is found in Moseley's report of 23 April 1914 (Fig. 7) to the Institut International de Physique Solvay, which contributed £39.5.4 to his research on 3 December 1913. Among the items he bought was an 'X-ray tube from Baumbach £1.16.0'.[32] Otto Baumbach was the Manchester glass blower.[33] Aside from the costs of the Cossor tubes (which were nearly ten times greater), Baumbach was the only other tube supplier listed in the Solvay report. Since the shape of the vacuum spectrometer X-ray tube is the same as the one Moseley broke in a discharge (Fig. 4), it is possible to ascribe them both to Baumbach.

Baumbach was just one of the instrument makers that provided the University of Manchester with the infrastructure to support a strong experimental programme in physics. When in the autumn of 1906, Sir Arthur Schuster persuaded Rutherford to succeed him, he made a point of commending the Manchester makers:

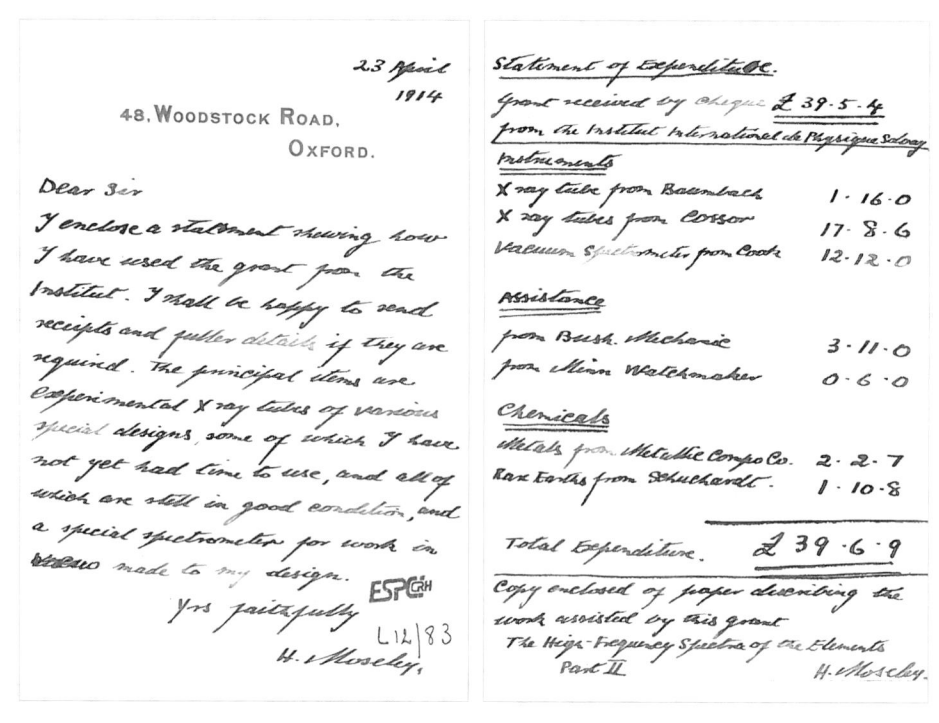

Fig. 7 Moseley's statement of expenditure to the Solvay Institute. (Source: Ecole supérieure de physique et de chimie industrielles de Paris, Centre de ressources historiques, 1914, by courtesy of ESPCI Paris-CRH).

A great acquisition has been the establishment last year of a first class instrument maker within 150 yards of the laboratory. He works by agreement on cheap terms for us and is an excellent man. He was assistant at the Royal Institution and constructed a good deal of Dewar's high pressure apparatus. We also have a tinsmith next door and a very excellent glass blower within easy reach.[34]

This 'excellent man' was Charles Cook (1868–1945), who fabricated Moseley's vacuum spectrometer. Cook was under contract with the University at a guaranteed rate of one shilling and nine pence per hour (£0.1.9/hour) for two thousand hours a year.[35] In addition to making apparatus for Manchester, Cook was free to contract out his services as long as it did not interfere with University orders. His business interests varied widely, as seen in his 1919 *Nature* advert (Fig. 8).[36] The spectrometer set Moseley back twelve guineas (£12.12.0), making it one of the most expensive pieces of his equipment.

While the capabilities and cost of the Manchester shop drew Moseley's custom, the account submitted to the Solvay institute also lists £3.11.0 paid to the University of Oxford Instrument Maker, Sam W. Bush.[37] The work with Bush is important because it contradicts the lone researcher trope that is often applied to Moseley. A letter of recommendation written for Moseley by Oxford's Wykeham Professor of Physics, J. S. E. Townsend, helps perpetuate this myth in commenting Moseley's 'researches are of great interest and importance and he has carried them out most successfully by himself without any assistance'.[38] Although it was not the type of 'assistance' to which Townsend referred, Moseley did identify the help of two tradesmen as such in his dealings with Solvay. Bush's aid was worth more than several Baumbach's X-ray tubes, and for that reason should not be discounted, even if he was the mechanic bemoaned as a 'thorn in the flesh'.[39]

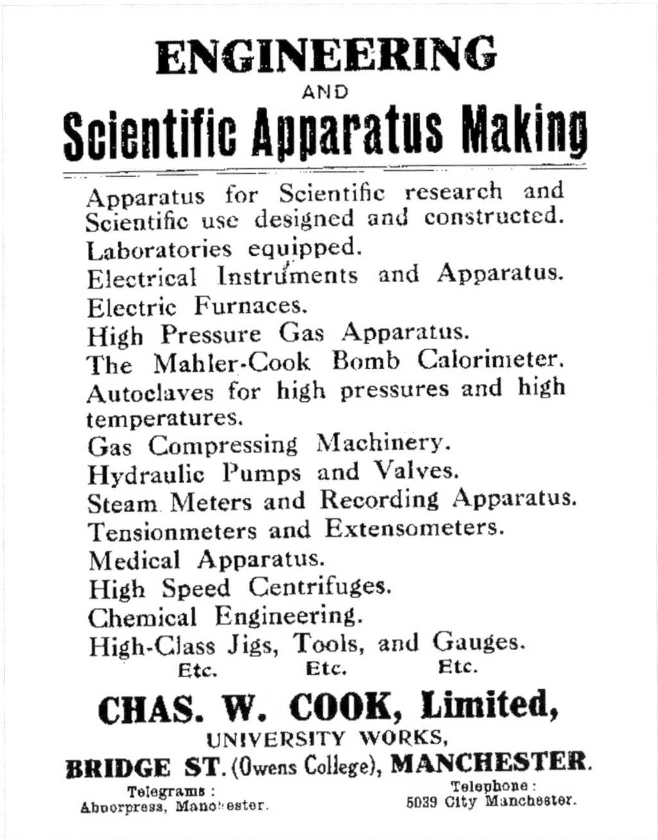

ENGINEERING
AND
Scientific Apparatus Making

Apparatus for Scientific research and Scientific use designed and constructed.
Laboratories equipped.
Electrical Instruments and Apparatus.
Electric Furnaces.
High Pressure Gas Apparatus.
The Mahler-Cook Bomb Calorimeter.
Autoclaves for high pressures and high temperatures.
Gas Compressing Machinery.
Hydraulic Pumps and Valves.
Steam Meters and Recording Apparatus.
Tensionmeters and Extensometers.
Medical Apparatus.
High Speed Centrifuges.
Chemical Engineering.
High-Class Jigs, Tools, and Gauges.
Etc.   Etc.   Etc.

CHAS. W. COOK, Limited,
UNIVERSITY WORKS,
BRIDGE ST. (Owens College), **MANCHESTER.**
Telegrams: Telephone:
Abvorpress, Manchester.   5039 City Manchester.

Fig. 8 Charles Cook trade literature. (Source: *Nature*, 103 (2583), (1919), lxxi.).

Rare Earths

When Moseley published his second paper on the 'High-Frequency Spectra of the Elements', he was quite confused about the existence and appropriate place of various rare earth elements in the periodic table.[40] But by the end of April 1914, he felt that he had it sorted out – thanks perhaps to the rare earth chemists on whom he depended for hard-to-get samples.[41] The hardest elements to obtain were, not surprisingly, those for which there is a gap in Moseley's diagram. Some of these data points were to provide the basis for a third paper in the series, entitled 'High-Frequency Spectra of Elements of the Rare Earth Group', which remained unpublished after his death.[42]

The rare earth elements that Moseley was able to characterise in Fig. 1 were obtained principally from Sir William Crookes or the chemical supply house of Dr. Schuchardt of Görlitz.[43] Setting his sights on filling the data gap between atomic numbers 69–71, Moseley had to turn to the Austrian Carl Auer von Welsbach or the Frenchman Georges Urbain. Both claimed to be the first to separate ytterbium into two different elements (aldebaranium & cassiopeium, or neoytterbium & lutecium, respectively). By the time Moseley's friend and former Manchester colleague Georg von Hevesy offered to obtain samples from Auer, Moseley had already obtained several from Urbain, along with the promise of another for which Urbain was considered the only source: the elusive celtium.[44]

Both Moseley and Urbain hoped celtium would fill in the missing gap in the diagram for the element with atomic number 72. In June 1914, Urbain travelled to Oxford with his sample of celtium, intending to collect X-ray data in support of his discovery of a new element. The story of this research is told uniquely by a small manuscript collection at the Museum of the History of Science – *MS Museum 118* – which contains several pages of Moseley's calculations, notes, letters, and photographs associated with his rare earth research. Recent analysis of these materials by the author has shown that several of the photographs of X-ray spectra in this collection correspond to the samples Urbain brought with him (Fig. 9).[45]

However, Moseley's results for celtium, which he reported to Rutherford, were not what he was expecting:

Celtium has proved most disappointing. I can find no X-ray spectrum in it other than those for Lutecium and Neoytterbium, and so Number 72 is still vacant. My own impression is that the very definite spectrum (visible) given by 'Celtium' is a secondary Lutecium spectrum, which is masked in what

Urbain calls 'Lutecium' by 50 per cent of Neoytterbium, while 'Celtium' is Lutecium with only a small percentage of Ny. These proportions are shewn by the X-ray spectra, but it is naturally impossible to say that there is not a third element present which is the true Celtium.[46]

This statement has been used to illustrate Moseley's knowledge of the true character of Urbain's celtium, which would be confirmed by spectroscopic analysis in 1923 – ultimately degrading Urbain's claim to the discovery of element 72 as well as his claim to 71.[47] But there is more to learn from this. Note that Moseley compared both visible and X-ray spectra. This cross-comparison is not mentioned elsewhere in his work. It would, however, be an important check on the results of two researchers who not only spoke different languages but used different apparatus.

It is possible that Moseley returned the prism spectrometer he previously modified for his X-ray work to its original form, so that he could take visible spectra.[48] The importance of having two spectrometers to crosscheck the results explains why Moseley used the vacuum spectrometer for his analysis of Urbain's samples – a surprising move since the imaging of the X-ray emission in question did not require a vacuum, and the apparatus was far more cumbersome to use.[49] The prism spectrometer Moseley used is not amongst the items in the Museum of the History of Science collection, so one cannot refer to the artefact record to test this theory.

Although they are seldom discussed, the Museum of the History of Science does hold a collection of unidentified visible spectrum plates that are associated with Moseley (Fig. 10), but it is not known whether these correspond to Moseley's work with Urbain's samples. They are, however, illustrative of the relative complexity of visible spectra compared to that of X-ray spectra, wherein a handful of lines can detect both constituent elements as well as relative percentages (see Moseley's breakdown of one of Urbain's samples in Fig. 9).

Moseley was secure enough in his analysis of celtium to be willing to report on it at the British Association meeting in Sydney in 1914. However, his confidence was not universally shared. After Moseley's death, most of the rare earth spectra data and notes went to Rutherford, so it could be assessed for publication.[50] A posthumous paper never came about because Rutherford did not find any 'definite conclusions'.[51] Urbain continued to advocate for celtium, yet waited until 1922 to announce it as the element with atomic number 72, when he could back it up with X-ray data supplied by Alexandre Dauvillier

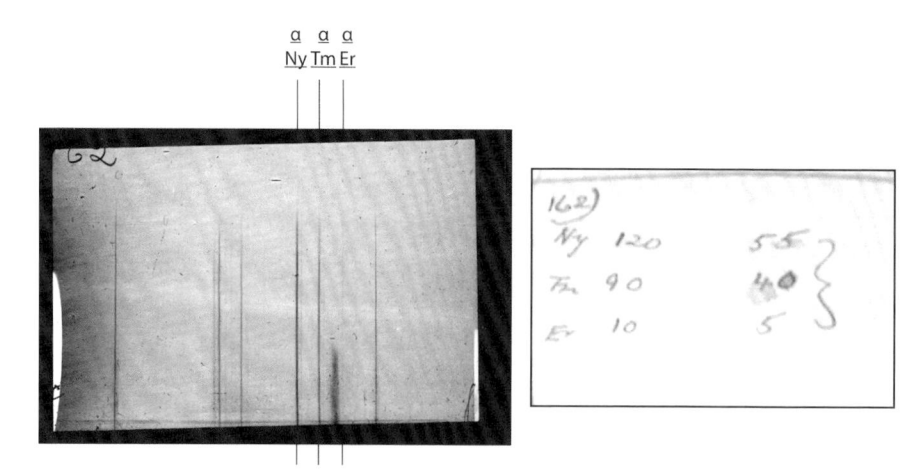

Fig. 9 Left: annotated and inverted negative scan of a spectrum associated with Moseley and Urbain's research. Right: excerpt of page of Moseley's notes. (Source: both panels from *MS Museum 118*, 1914, by courtesy of the Museum of History of Science, Oxford).

(1882–1979). Urbain claimed 'the negative result given by Moseley's method in the case of celtium was due only to the insensitiveness of the method, since the preparation examined by M. Dauvillier is the same as that used in Moseley's X-ray tube'.[52]

While Urbain had a vested interest in explaining away a negative result, Georg von Hevesy – who was ultimately credited with the discovery of element 72, hafnium, in 1923 – wanted to present the discovery and supporting X-ray data as a logical continuation of a clear-cut process. His announcement in *Nature* began on this note: 'Since Moseley's discovery of the fundamental laws of X-ray emission, it has become quite clear that the most simple and conclusive characteristic of a chemical element is given by its X-ray spectrum'.[53]

Fig. 10 Unidentified spectrum slide associated with Moseley. (Source: Museum of the History of Science, date unknown, by courtesy of the Museum of the History of Science, Oxford).

The priority battle that raged over the discovery of celtium has been extensively discussed.[54] The point to be made here is that Moseley's method was presented as being both indefinite and conclusive, but this description depended upon how researchers wanted to present their narrative of success. Instead of reading Moseley's own statement – 'my own impression is that the very definite spectrum (visible) given by "Celtium" is a secondary Lutecium spectrum' – as foreseeing the overthrow of Urbain's claim to 72, it is likely that he acknowledged the uncertainty. While it is tempting to agree with Hevesy that Moseley's X-ray methods ultimately cleared these muddy waters, one cannot ignore the fact that Dauvillier's X-ray analysis supported Urbain. X-ray spectroscopy would certainly become a valuable tool in the discovery of elements, but the debut of data for atomic number 72 did not ultimately agree with the outcome of the priority debate.

Concluding Remarks

This chapter has sought to deconstruct Henry Moseley's diagram from three different perspectives. Analyzing the push and pull of experimental constraints has shown how the higher frequency data points required Moseley to modify his apparatus and alter his process. At the same time, his lower frequency data show his connection with instrument makers and reveals the kind of infrastructure needed to do his research. The gap in the rare earth data, which would persist in the literature for some time despite the results obtained in Moseley's last X-ray investigations, suggests that the conclusiveness of Moseley's rare earth research was a matter of perspective. The diagram with which we began does not have to be interpreted as purely reductive; rather, its points, lines, and gaps provide an illuminating study of how experimental constraints, instrument makers, and conflicting approaches shaped Moseley's work.

Notes and References

[1] The original hand-drawn diagram was published on page 709 in H. G J. Moseley, 'The High-Frequency Spectra of the Elements. Part II', *Philosophical Magazine*, 27 (160), (1914), 703–713 (hereafter Moseley, 'High-Frequency II'). The diagram was modified at a later date to correct errors in naming and numbering the rare earth elements. As published it appeared: 66 Ho, 67 Ds, 69 Tm I, 70 Tm II, 71 Yb, 72 Lu. The modified original, housed in the Clarendon Laboratory in Oxford, now shows: 66 Ds, 67 Ho, 69 Tm, 70 Ny, 71 Lu, and 72 is left blank.

[2] J. H. Gardiner, 'The Origin, History & Development of the X-Ray Tube', *Journal of the Röntgen Society*, 5 (20), (1909), 66–80.

[3] R. Kennedy, *Electrical Installations*, 5 vols., vol. V (London: Caxton Publishing Co., 1902), 192; M. Siegbahn, *The Spectroscopy of X-Rays* (London: Oxford University Press, 1925), 30–87. Hereafter the book is abbreviated as Siegbahn, *Spectroscopy of X-Rays*.

[4] C. G. Barkla and C.A. Sadler, 'Homogeneous Secondary Röntgen Radiations', *Proceedings of the Physical Society of London*, 21 (1907), 336–373.

[5] C. G. Barkla and J. Nicol, 'Homogeneous Fluorescent X-Radiations of a Second Series', *Proceedings of the Physical Society of London*, 24 (1911), 9–17.

[6] Moseley, 'High-Frequency II', *op. cit.* 711.

[7] W. H. Bragg and W. L. Bragg, 'The Reflection of X-Rays by Crystals', *Proceedings of the Royal Society A*, 88 (605), (1913), 428–438.

[8] W. H. Bragg and W. L. Bragg, *X-rays and Crystal Structure* (London: G. Bell, 1915), 22–37; H.G.J. Moseley and C.G. Darwin, 'The Reflexion of the X-Rays', *Philosophical Magazine Series 6*, 26 (151), (1913), 210–232 (hereafter Moseley and Darwin, 'Reflexion of X-Rays'). The diagram appears on 212.

[9] Moseley and Darwin, 'Reflexion of X-Rays', *op. cit.* 220–222.

[10] T. Hirosige, 'The Van Den Broek Hypothesis', *Japanese Studies in the History of Science*, 10 (1971), 143–162.

[11] H. G. J. Moseley, 'The High-Frequency Spectra of the Elements', *Philosophical Magazine*, 26 (156), (1913), 1024-1034 (hereafter Moseley, 'High-Frequency I'). Moseley states on 1025 that his tube was an adaptation of a design used by G. W. C. Kaye.

[12] He most likely used the same 'magnificent specimen' described in Moseley and Darwin, 'Reflexion of X-Rays', *op. cit.* 213.

[13] Moseley, 'High-Frequency I', *op. cit.* 1025-1027.

[14] *Ibid*. 1029.

15 *Ibid*. 1033.

16 American Philosophical Society, Sir Charles Galton Darwin Papers, H. G. J. Moseley to C. G. Darwin, 17 January 1914; Letter 87 in J. L. Heilbron, *H. G. J. Moseley: The Life and Letters of an English Physicist, 1887–1915* (Berkeley: University of California Press, 1974), 221 (hereafter Heilbron, Moseley).

17 Moseley, High-Frequency II, *op. cit.* 706–707.

18 Moseley to Darwin, 17 January [1914], Letter 87 in Heilbron, *Moseley, op. cit.* 221–224.

19 Siegbahn, *Spectroscopy of X-Rays, op. cit.* 30–31.

20 Moseley, 'High-Frequency I', *op. cit.* 1028.

21 Moseley, 'High-Frequency II', *op. cit.* 708.

22 Cambridge University Library, Rutherford Papers, H. G. J. Moseley to E. Rutherford, 5 January [1914]; Letter 85 in Heilbron, *Moseley, op. cit.* 218–219.

23 Moseley had already examined the L rays from Pt, see Moseley, 'High-Frequency I', *op. cit.* He wrote to Darwin before his Cossor tubes were finished that he already had hard K line information for Zr and Mo. See Moseley to Darwin, 17 January [1914], Letter 87 in Heilbron, *Moseley, op. cit.* 221–224.

24 Moseley to his mother 17 November 1913, Letter 82 in Heilbron, *Moseley, op. cit.* 214.

25 Siegbahn, *Spectroscopy of X-Rays, op. cit.* 61.

26 J. Townsend, *The Theory of Ionization of Gases by Collision* (London: Constable & Co., 1910), 51-60; J. Townsend, *Electricity in Gases* (Oxford: Clarendon Press, 1915), 354–356.

27 Siegbahn, *Spectroscopy of X-Rays, op. cit.* 34; Moseley, 'High-Frequency II', *op. cit.* 704.

28 Moseley to Darwin, 17 January [1914], Letter 87 in Heilbron, Moseley, *op. cit.* 222. Compare with description of fringes in Moseley, High-Frequency II', *op. cit.* 707.

29 Heilbron describes the Cossor tube as the 'definitive apparatus for obtaining hard X-rays'. Heilbron, *Moseley, op. cit.* 87.

30 Cambridge University Library, Rutherford Papers, H.G.J. Moseley to E. Rutherford, 7 December [1913]; Letter 84 in Heilbron, *Moseley, op. cit.* 216–217. Quotation discussed in the following: Heilbron, *Moseley, op. cit.* 87; P. M. Heimann, 'Moseley and Celtium: The Search for a Missing Element', *Annals of Science*, 23 (4), (1967), 249–260 (hereafter Heimann, 'Moseley and Celtium'). The professor Moseley refers to is J. S. E. Townsend, who was Wykenham Professor of Physics at Oxford.

31 The various components are in the Museum of the History of Science in Oxford, and were reassembled as part of the 'Dear Harry' Exhibition.

[32] Centre de Ressources Historiques de l'ESPCI ParisTech, Cote L12/83, Lettre: H. Moseley, 23 Avril 1914.

[33] A. Gall, 'Otto Baumbach – Rutherford's Glassblower', *IOP History of Physics Newsletter*, 23 (2008), 44-55; A. Gall, 'Rutherford's Glassblowers – Otto Baumbach & Felix Niedergrass', *Science Technology: The Journal of the Institute of Science Technology* (2004), 3–8

[34] University of Manchester Archives, Sir Arthur Schuster Papers, A. Schuster to E. Rutherford, 7 October 1906.

[35] University of Manchester Archives, Sir Arthur Schuster Papers, Charles Cook Contract, undated.

[36] C. Cook, 'Chas. Cook Trade Literature', *Nature*, 103 (2583), (1919), lxxi.

[37] B. LeLong, 'Translating Ion Physics from Cambridge to Oxford', in R. Fox and G. Gooday (ed.), *Physics in Oxford, 1839–1939 : Laboratories, Learning, and College Life* (Oxford: Oxford University Press 2005), 213.

[38] Museum of the History of Science, MS Ludlow-Hewitt, Moseley's application for Birmingham, Recommendation Letter by J.S. Townsend, undated.

[39] Heilbron suggests G.A. Bennett is the mechanic in question. Heilbron, *Moseley, op. cit.* 218n.

[40] Moseley to von Hevesy, [20 March 1914], Letter 93 in Heilbron, *Moseley, op. cit.* 232–233; Niels Bohr Institut, Hevesy Papers, Moseley to Hevesy, 23 April [1914]; Letter 94 in Heilbron, *Moseley, op. cit.* 234.

[41] *Ibid*. Before Moseley obtained Ny, Lu, Tb, Dy from Urbain, he asked Hevesy to write to Auer for 'Pr.Ad.Cp.Tm.Ho.Ds.Tb'. He also appealed to a chemist at New Hampshire College in Durham New Hampshire for holmium and thulium.

[42] Museum of the History of Science, MS Ludlow-Hewitt, M. Ludlow-Hewitt Note, 12 June 1933 (hereafter Ludlow-Hewitt, 'M. Ludlow-Hewitt Note').

[43] Moseley, 'High-Frequency II', *op. cit.* 706.

[44] Moseley to Hevesy, 23 April [1914], Letter 94 in Heilbron, *Moseley, op. cit.* 234; G. Urbain, 'Sur un Nouvel Element qui Accompagne le Lutécium et le Scandium dans les Terres de la Gadolinite: le Celtium', *Comptes rendus de l'Académie des Sciences*, 152 (1911), 141–143 (hereafter Urbain, 'Celtium').

[45] K.M. Frederick-Frost, 'For the Love of a Mother—Henry Moseley's Rare Earth Research', *Historical Studies in the Natural Sciences*, 47 (2017), 529–567 (hereafter Frederick-Frost, Love of a Mother).

⁴⁶ University of Cambridge Library, Rutherford Papers, Moseley to Sir Ernest [Rutherford], 5 June [1914]; Letter 96 in Heilbron, *Moseley, op. cit.* 236–237.

⁴⁷ Heilbron, *Moseley, op. cit.* 237n; H. M Hansen and S. Werner, 'On Urbain's Celtium Lines', *Nature*, 111 (2788), (1923), 461; H. Kragh, 'Elements No. 70, 71, and 72: Discoveries and Controversies', in C. H. Evans (ed.), *Episodes from the History of the Rare Earth Elements* (Dordrecht: Kluwer Academic Publishers, 1996), 81.

⁴⁸ 'Visible' here is used more to denote the use of a prism spectrometer. Urbain claimed to have found celtium lines between 2459.4 and 3665.6 Ångstroms, which is in the ultraviolet; Urbain, Celtium, *op. cit.*

⁴⁹ Moseley, 'High-Frequency I', *op. cit.* 706; Frederick-Frost, Love of a Mother *op. cit.*

⁵⁰ Ludlow-Hewitt, 'M. Ludlow-Hewitt Note' *op. cit.*

⁵¹ E. Rutherford, 'H. G. J. Moseley, 1887–1915', *Proceedings of the Royal Society A*, 93 (655), (1917), xxvi.

⁵² G. Urbain, 'Les Numéros Atomiques du Néo-Ytterbium, du Lutécium et du Celtium', *Comptes rendus de l'Académie des sciences*, 174 (1922), 1349–1351. Translated in E. Rutherford, 'Identification of a Missing Element', *Nature*, 109 (2746), (1922), 781.

⁵³ D. Coster and G. Hevesy, 'On the Missing Element of Atomic Number 72', *Nature*, 111 (2777), (1923), 79.

⁵⁴ Heimann, 'Moseley and Celtium', *op. cit.* 249–260; H. Kragh and P. Robertson, 'On the Discovery of Element 72', *Journal of Chemical Education*, 56 (7), (1979), 456–459.

CHAPTER FOUR
ANTONIUS VAN DEN BROEK, MOSELEY, AND THE CONCEPT OF ATOMIC NUMBER

ERIC R SCERRI

Introduction

All scientific discovery depends on the work of prior contributors. In his pioneering experiments in X-ray spectroscopy, Henry Moseley set out to verify the hypothesis of Antonius van den Broek (1870–1926), an obscure Dutch lawyer and amateur scientist. Moseley had read van den Broek's papers, but never met the man who was the first to publish a series of precise statements that the chemical elements should be ordered on a whole number scale, later termed the atomic number. This chapter will examine the background and the contribution of Moseley's remarkable contemporary.

In so doing, this story will take issue with the widely-known but deeply flawed idea that science develops as a result of singular and dramatic discoveries, performed by gifted individuals who produce scientific revolutions.[1] For the most part, science actually develops as an organic whole, comprising large numbers of small contributions.[2] There is also a strong case for seeing major advances in science as not so much revolutionary, but as part of a gradual evolutionary development in which important steps are taken in an almost piecemeal fashion.[3] Indeed, science may be regarded as a giant living organism, in a sense not unlike Lovelock's Gaia.[4] The history of the concept of atomic number affords a striking illustration of this idea, one in which the approaches of chemistry were in key respects sidelined by the approaches of physics.

In 1914, Moseley states that he undertook his experiments 'with the express purpose of verifying van den Broek's hypothesis'.[5] Chemistry may well have been Moseley's first passion. It was obviously the science that enthused him most at Eton and when he arrived at Trinity College in Oxford, he was probably intending to read chemistry after completing his Mathematical Moderations. His Millard Scholarship in Natural Science was usually earmarked for chemists and his College tutor at Trinity, David Nagel, was a physical chemist. It is unclear why Moseley took up physics after completing Mathematical Moderations in his first year.[6] Whatever the reason, Moseley's knowledge of chemistry would have helped him appreciate van den Broek's chemical approach to atomic structure, which focused on the periodic table as a whole, rather than on the specific properties of particular elements, which the leading physicists favoured.[7] Moseley was to give the concept of atomic number firm experimental grounding. But the idea that the order number in the periodic table may be equated with the nuclear charge originates with van den Broek.

A Brief Biography

Antonius van den Broek (Fig. 1) was born in Zoetermeer, in the province of South Holland in 1870 and graduated from the University of Leyden with a law degree in 1895. The following year, he married Elisabeth Mauve, the daughter of a painter and a cousin of Vincent van Gogh.

Van den Broek worked in The Hague as a solicitor until 1900, after which his interests turned towards the sciences. He never held a university appointment, but became skillful at handling numerical information. Between the years 1901 and 1902 he studied mathematical economics in Vienna, with Carl Menger, while also spending time in Leipzig and Berlin. He turned to the structure of the atom in 1903, and in 1907 published his first article on the periodic table, 'Das α-Teilchen und das periodische System der Elemente', in *Annalen der Physik*.[8]

Fig. 1 Antonius van den Broek. (Source: Photograph provided to the author by the late Jan Van Spronsen, undated, courtesy of the author).

Van den Broek and the Periodic Table

Van den Broek's article of 1907 takes up a question posed by Ernest Rutherford. After discovering the α-particle, and while attempting to elucidate its nature, he set out the options that the alpha particle was either:

(1) a *molecule* of hydrogen carrying the same charge as an ionized hydrogen atom,
(2) a helium atom carrying twice the ionic charge of hydrogen, or
(3) *one half* of the helium atom carrying a single ionic charge.[9]

Van den Broek favoured the third possibility, proposing a periodic table in the spirit of the early nineteenth century Prout hypothesis.[10] However, instead of all atoms being composites of hydrogen as in the Prout model, van den Broek envisaged the basic building block to be the 'alphon', which weighed twice as much as a hydrogen atom. Van den Broek adds:

> Where the experiment fails, only pure speculation remains, and so it would seem reasonable to try to see whether this helium atom, or the 'half helium atom' (let's say free Alphon as a 'half' atom is an absurdity) would be better suited to act as a primary element, than the Prout H-atom ever could.[11]

TABLE 1

	VII	0	I	II	III	IV	V	VI
1	2* (α)	4 He	6 Li	8 Be	10 B	12 C	14 N	16 O
2	18 F	20 Ne	22 Na	24 Mg	26 Al	28 Si	30 P	32 S
3	34 Cl	36 Ar	38 K	40 Ca	42 Sc	44 Ti	46 V	48 Cr
4	50 Mn	52	54	56 Fe	58 Co	60 Ni	62	64
5	66	68	70 Cu	72 Zn	74 Ga	76 Ge	78 As	80 Se
6	82 Br	84 Kr	86 Rb	88 Sr	90 Y	92 Zr	94 Nb	96 Mo
7	98	100	102	104 Ru	106 Rh	108 Pd	110	112
8	114	116	118 Ag	120 Cd	122 Jn	124 Sn	126 Sb	128 Te
9	130 J	132 Xe	134 Cs	136 Ba	138 La	140 Ce	142 Nd	144 Pr
10	146	148	150 Sa	152	154 Gd	156	158 Tb	160
11	162	164	166 Er	168 Tu	170 Yb	172	174 Ta	176 W
12	178	180	182	184 Os	186 Ir	188 Pt	190	192
13	194	196	198 Au	200 Hg	202 Tl	204 Pb	206 Bi	208
14	210	212	214	216	218	220	222	224
15	226	228	230	232 Ra	234	236 Th	238	240 U

* Theoretical atomic weight.

Fig. 2 Van den Broek's periodic table of 1907. (Source: *Annalen der Physik*, 4 (23), (1907), 199–203).

Curiously, the hydrogen atom, the staple of the Prout hypothesis, is absent from this scheme, and is omitted completely from van den Broek's periodic table (Fig. 2). But this would be consistent with van den Broek's dislike of the idea of 'half-atoms'. For his periodic table, the hydrogen atom would have to be half an alphon, which appears to be nonsensical. However, the complete omission of an atom as fundamental as hydrogen would seem to fly in the face of all subsequent developments – such as the Bohr model of the atom, which was soon to appear in the literature in the famous trio of papers published in 1913.[12]

In van den Broek's table of 1907, there are many empty spaces, a feature that did not seem to disturb him. Many new isotopes were being discovered, and it was plausible that some of these might turn out to be new elements.

After a four-year silence, in 1911 van den Broek published another article, in which the alphon particle was no longer mentioned.[13] What was retained was the notion of elements that differed from each other by two units of atomic weight. Van den Broek proposed another periodic table (Fig. 3), which he claimed had first been suggested by Dmitri Mendeleev. This is said to be a three-dimensional periodic table, although it was clearly not literally so, given that the third dimension was only implied by sets of three elements that were assumed to recede into a plane perpendicular to the plane of the printed page.

TABLE 2

		0 1	0 2	0 3	I 1	I 2	I 3	II 1	II 2	II 3	III 1	III 2	III 3	IV 1	IV 2	IV 3	V 1	V 2	V 3	VI 1	VI 2	VI 3	VII 1	VII 2	VII 3
A	1	He			Li			Be			B			C			N			O			F		
A	2		Ne			Na			Mg			Al			Si			P			S			Cl	
A	3			Ar			K			Ca			Sc			Ti			V			Cr			Mn
B	1	Fe			Co			Ni			Cu			—			—			—			Zn		
B	2		—			—			—			Ga			Ge			As			Se			Br	
B	3			Kr			Rb			Sr			Y			Zr			Nb			Mo			Ru
C	1	Rh			Pb			—			—			Ag			—			—			Cd		
C	2		—			—			—			In			Sn			Sb			Te			J	
C	3			Xe			Cs			Ba			La			Ce			Nd			Pr			(Sm)
D	1	(Eu)			(Gd₁)			(Gd₂)			(Gd₃)			(Tb₁)			(Tb₂)			(Dy₁)			(Dy₂)		
D	2		(Dy₃)			(Ho)			(Er)			(Tu₁)			(Tu₂)			(Tu₃)			(Yb)			(Lu)	
D	3			—			—			—			—			—			Ta			W			Os
E	1	Ir			Pt			Au			Hg			Tl			Bi			Pb			—		
E	2		—			—			—			—			—			—			—			—	
E	3			—			—			Ra			—			Th			—			U			—

Fig. 3 Van den Broek's periodic table of 1911. (Source: *Physikalische Zeitschrift*, 12 (1911), 490–497).

In the same year, van den Broek also published a letter in *Nature*, which is remarkable for several reasons, one of them being its sheer brevity (just twenty lines).[14] Here he turned to the researches of Ernest Rutherford and Charles Barkla, both of whom had concluded that the atomic charge of any atom was approximately half of its atomic weight – or, to use modern notation,

$$Z \approx A/2$$

where Z denotes atomic charge and A denotes atomic weight. Connecting this fact with his earlier work of 1907, van den Broek reasoned as follows. If successive elements differ from each other by two units of atomic weight, and if $Z = A/2$, then it follows that successive elements differ from each other by one unit of charge. Herein lies the birth of the notion of atomic number, which Moseley was to establish experimentally and report in his articles of 1913[15] and 1914.[16]

Many years after the event, the physicist and historian Abraham Pais made the following comment on van den Broek's papers of 1911:

> Thus based on an incorrect periodic table and on an incorrect relation ($Z \approx A/2$), did the primacy of Z as an ordering number of the periodic table enter physics for the first time.[17]

But there is more to van den Broek's contribution than this vague anticipation of the atomic number. Why had the physicists of the day not arrived at this 'vague anticipation', as I have described it, whereby successive elements differ from each other by one unit of charge? One response might be that the 'amateur' van den Broek took a more holistic approach, or perhaps even a chemical approach, to considering all the elements together within a periodic table.

What van den Broek did not Achieve in his 1911 Articles

What van den Broek failed to achieve in his studies of 1911 was to draw a clear distinction between atomic number and atomic weight. As would later emerge, the relationship of $Z \approx A/2$ holds approximately only for the first twenty or so elements in the periodic table. As atomic number increases beyond calcium ($Z = 20$), the relationship becomes increasingly inaccurate. This is because, as the atom contains a larger number of protons, an increasingly larger number of neutrons are required for stability; and as a result, successive atoms in the periodic table have a correspondingly larger value of the ratio between atomic weight and atomic number. Had van den Broek left the stage in 1911, his research would have amounted to only a 'vague anticipation' of the importance of atomic number.

However, van den Broek went a good deal further, and argued explicitly for the need to distinguish between the atomic number and the atomic weight of any particular atom, and for the need to focus on atomic number in order to characterise the elements. How did he take this crucial step?

How and Why did van den Broek Separate Z and A

The year 1913 saw a landmark paper from van den Broek published in *Physikalische Zeitschrift*.[18] The earlier cubic table was replaced by a rather elaborate two dimensional system (Fig. 4) and for the first time a serial number for each element was proposed, that could be equated with the nuclear charge. In translation we have: 'The serial number for every element in the sequence ordered by increasing atomic weight equals half the atomic weight and therefore the intra-atomic charge'.

TABLE 3

0	I	II	III	IV	V	VI	VII	VIII		
2* He	3 Li	4 Be	5 B	6 C	7 N	8 O	9 F			
10 Ne		11 Na	12 Mg	13 Al	14 Si	15 P	16 S	17 Cl			
18 —	19 Ar	20 K	21 Ca	22 23 Sc —	24 Ti	25 V	26 Cr	27 Mn	28 Fe	29 Co	30 Ni
			31 Cu	32 Zn	33 Ga	34 Ge	35 As	36 Se	37 Br	38 —	39 —	40 —
41 —	42 Kr	43 Rb	44 Sr	45 46 Y —	47 Zr	48 Nb	49 Mo	50 —	51 Ru	52 Rh	53 Pd
			54 Ag	55 Cd	56 In	57 Sn	58 Sb	59 Te	60 J	61 —	62 —	63 —
64 —	65 Xe	66 Cs	67 Ba	68 69 La —	70 Ce	71 Nd	72 Pr	73 —	74 Sa	75 Eu	76 Gd
			77 Tb	78 (Tb₂)	79 Dy	80 Ho	81 Er	82 Ad	83 AcC	84 TuI	85 TuII	86 AcA
87 —	88 AcEm	89 AcX	90 TuIII	91 92 RAc Cp	93 Ct	94 Ta	95 Wo	96 —	97 Os	98 Ir	99 Pt
			100 Au	101 Hg	102 Tl	103 Pb	104 Bi	105 RaF	106 ThC	107 RaC	108 ThA	109 RaA
110 ThEm	111 RaEm	112 ThX	113 Ra	114 115 116 RTh Io Th		117 UII	118 U	119 —	120 —	121 —	122 —

* Atomic number.

Fig. 4 Van den Broek's periodic table of 1913. (Source: *Physikalische Zeitschrift*, 14, (1913), 32–41).

This paper was followed by two brief letters to *Nature* published on 27 November[19] and 25 December 1913.[20] The first letter was radical in jettisoning the connection with atomic weight altogether. Van den Broek's point of departure was a set of experiments that had been conducted by Hans Geiger and Ernest Marsden, two of Ernest Rutherford's co-workers. They had examined the scattering of α-particles by samples of various elements.[21]

As seen in Fig. 5, the degree of scattering given by the quantity N increases with increasing atomic weight A. The ratio $N/A^{3/2}$ shows less variation, although there is a small increase as the atomic weight of the element decreases.[22]

Variation of Scattering with Atomic Weight.
(Example of a set of measurements.)

I.	II.	III.	IV.	V.	VI	VII.
Substance.	Atomic weight. A.	Air equivalent in cm.	Number of scintillations per minute corrected for decay.	Number N of scintillations per cm. air equivalent.	$A^{3/2}$.	$N \times A^{2/3}$.
Gold	197	·229	133	581	2770	0·21
Tin	119	·441	119	270	1300	0·21
Silver	107·9	·262	51·7	198	1120	0·18
Copper	63·6	·616	71	115	507	0·23
Aluminium..	27·1	2·05	71	34·6	141	0·24

Fig. 5 Geiger and Marsden's values of alpha particle scattering in several elements, using a radium C alpha source. The header for column VII is given as $N \times A^{2/3}$. This is incorrect and should be $N \times A^{-3/2}$. The numbers in the column correspond to the latter. (Source: *Philosophical Magazine Series 6*, 25 (148), (1913), 604–628).

In discussing these results, Geiger and Marsden observed that 'This ratio [$N/A^{3/2}$] should be constant according to the theory [of Rutherford]. The experimental values show a slight decrease with atomic weight'. It appears that the disagreement between theory and experiment did not bother them unduly. Having performed the experiment using both 'Radium C' (the bismuth isotope $^{214}_{83}Bi$) and 'Radium emanation' (the radon isotope $^{222}_{86}Rn$) as the alpha sources on similar groups of elements, they offer the following comment:

> These results are similar, and indicate the essential correctness of the assumption that the scattering per atom is proportional to the square of the

atomic weight. The deviations from constancy of the ratio are nearly within experimental error.[23]

The discrepancies which Geiger and Marsden regarded as being due to experimental error, were the starting point for van den Broek's first *Nature* letter of 1913. Here van den Broek was starting to think that electric charge was a better criterion for explaining the periodic table and the structure of the atom.

He was now in a position to demonstrate this by drawing on Geiger and Marsden's data. In performing his analysis, van den Broek took values of the scattering per atom divided by A^2: these values had been derived by Geiger and Marsden from empirical measurements of the atom density rather than assuming the atom density scales as $A^{-1/2}$ as in Fig. 5. Van den Broek multiplied mean values for the scattering per atom divided by A^2 by a numerical factor of 5.4 to allow for the fact that A is typically just over twice M. He then multiplied these rescaled values by A^2/M^2 to give what was basically the scattering per atom divided by the square of the order number M. The values emerging from this final step showed much less variation than in the previous analysis using A^2.[24] Only then was van den Broek ready to take the crucial step of severing the connection between atomic weight and charge, or atomic number. In van den Broek's own words:

> If now in these values the number M of the place each element occupies in Mendeléeff's [sic] series is taken instead of A, the atomic weight, we get a real constant (18.7 ± 0.3); hence the hypothesis proposed holds good for Mendeléeff's [sic] series, but the nuclear charge is not equal to half the atomic weight.[25]

In his analysis of the scattering data, van den Broek discretely opted (without explanation) to exclude the experimental results for aluminium, which clearly represented a major outlier that did not conform to his arguments. Nonetheless, his whole analysis was repeated in a more transparent fashion in a paper published in *Philosophical Magazine* in 1914.[26] Here van den Broek first divided the scattering per atom by the factor $A^2/5.4$. Next he gave the atomic number M for each element and then divided the scattering by M^2, the square of the charge of an atom instead of the square of its atomic weight (See Table 1). It was clear that the constancy predicted by Rutherford was better reproduced by using M instead of A – but only if scattering data for aluminium were excluded.

	A	M	Scattering per atom/A^2 (Radium C)	Scattering per Atom/A^2 (Radium emanation)	Average scattering per atom/A^2	5.4× Scattering/A^2	Scattering/M^2
(Aluminium	27.1	13	3.6	3.4	3.5	18.9	15.2)
Copper	63.6	29	3.7	3.95	3.825	20.6	18.5
Silver	107.9	47	3.6	3.4	3.5	18.9	18.4
Tin	119	50	3.3	3.4	3.35	18.1	19.0
Platinum	195.1	82	3.2	3.4	3.3	17.8	18.6
Gold	197	83	3.4	3.1	3.25	17.5	18.4
Mean						18.6	18.6

Table 1 Table adapted from van den Broeks's papers (*Nature*, 92 (2300), (1913), 372–373 and *Philosophical Magazine Series* 6, 27 (159), (1914), 455–457) and the data presented by Geiger and Marsden (*Philosophical Magazine Series* 6, 25 (148), (1913), 604–628). Note that the atomic weights for aluminium and tin differ from currently accepted values (27.0 and 118.7 respectively), as do the atomic numbers for platinum and gold (current values are 78 and 79 respectively). Van den Broek did not include the data for aluminium in his analysis but they are reproduced in the adapted table.

Preempting Moseley

Writing in *Nature*, on 25 December 1913, van den Broek appeared to preempt Moseley in using the concept of atomic number to predict the existence of new or missing elements.

He began with the following explanation:

> I am grateful to Mr. Soddy (*Nature*, December 4, p. 399) that in accepting in principle the hypothesis that the intra-atomic charge of an element is determined by its place in the periodic table, he directed attention to the possible uncertainty of the absolute values of intra-atomic charge and of the number of intra-atomic electrons. Surely the absolute values depend on the number of rare earth elements; but if to the twelve elements of the rare earth series, the international table contains between cerium and tantalum the new elements (at least four) discovered by Auer von Welsbach in thulium (*Monatshefte für Chemie*, 32, Mai, S. 373), further keltium discovered by Urbain (*Comptes rendus d. l'Acad. des sciences*, 152, 141–3), and an unknown

one for the open place between praeseodymium and samarium be added, this long period too becomes regular. Moreover if only twelve instead of eighteen elements existed here, the ratio of the large-angle scattering per atom divided by M^2 is no longer constant, the values of copper, silver, tin platinum and gold then being 1.16, 1.15, 1.19, 1.26, and 1.24 respectively instead of 1.16, 1.15, 1.19, 1.17, and 1.15 and the same holds for the following relation concerning the number of intra-atomic electrons.[27]

Van den Broek proceeded to do away with the irregularities in Mendeleev's system by removing hydrogen and helium from the body of the table, and by 'condensing each triad of group VIII and ... likewise all the elements from Ce to Ta ... into one place'.[28] It was then possible to assign a 'periodic number P' to each element, the discrepancy between M and P starting at two and increasing after successive traverses through the three transition series and the lanthanides. He then derived an empirical relationship linking the atomic weight A, the periodic number P and the order number M in 'Mendelejeff's [sic] series'. The relationship was $A_{\text{calculated}} = 2(M + kP^2)$, where k is an empirically derived constant equal to 0.00468. Van den Broek then compared his calculated values of atomic weight with the experimental values, finding only a few minor discrepancies, as shown in Table 2.

	C	Mg	Ar	Cr	Zn	Kr	Mo	Cd	Xe	W	Hg	U
M	6	12	18	24	30	36	42	48	54	78	84	96
P	4	10	16	22	26	32	38	42	48	54	58	70
kP^2	0	0	1	2	3	5	6	8	11	14	16	23
A(calc)	12	24	38	52	66	82	98	112	130	184	200	238
A(exp)	12	24	40	52	65	82	96	112	130	184	200	238

Table 2 Comparison of A calculated by van den Broek with experimental values. Discrepancies occur only for Ar, Zn and Mo. (Adapted from *Nature*, 92 (2304) (1913), 476–478 and *Philosophical Magazine* Series 6, 27 (159), (1914), 455-457).

Somewhat confusingly, in his 1914 paper van den Broek re-introduced the idea that the *total* number of electrons in an atom was half the atomic weight and that 'kP^2 must be the number of electrons *in* the nucleus'. We know Moseley had read some of these papers. To what extent did they shape his ideas about the importance of atomic number?

What Bohr Knew about Atomic Number

Given the publications of 1913 and 1914 and the interest they generated, it is worth asking how much Niels Bohr knew about the nature of atomic number, in developing his model of the atom. The history of modern physics seems to downgrade van den Broek's contribution. For example, Heilbron has claimed that Bohr understood atomic number independent of the work of van den Broek.[29] But although by 1912 Bohr may have understood the importance of charge as a means of identifying the number of electrons in hydrogen and helium, especially, he did not fully appreciate the need to separate clearly the concepts of atomic weight and atomic number.

This view is confirmed in the *Collected Papers of Niels Bohr*, where he discusses his views on atomic number and the researches of Richard Whiddington.[30] Bohr mentions Whiddington for the first time in the January 1913 issue of the *Philosophical Magazine*, where he discusses J. J. Thomson's theory on the passage of alpha and beta articles through matter, and how their velocity decreases in the process. He states that Whiddington's experiments have confirmed Thomson's theoretical work.[31] In the same article, Bohr mentions an earlier work by Whiddington, in which the minimum velocity of a particle required to excite the characteristic X-radiation from any target element is proportional to the atomic weight of the element.[32] More precisely,

$$v_{minimum} = 10^8 \times A \text{ m/sec}$$

where A represents the atomic weight of an element. Bohr then applies Whiddington's law in the following way. He proposes that the energy of such a particle (kinetic energy = $1/2mv^2$) should be:

$$\text{Kinetic Energy} = (m/2) \times 10^{16} \times A^2$$

and invokes Planck's radiation law, which he gives in the form of

$$E = vk$$

where v is the number of vibrations per second in a Planckian atomic vibrator and k is a constant equal to 6.55×10^{-27}. Equating these two expressions for energy, Bohr obtains:

$$vk = (m/2) \times 10^{16} \times A^2$$

and solving for v after inserting the values of k and m gives,

$$v = 6.7 \times 10^{14}$$

He then considers the case of an oxygen atom, substituting the atomic weight for this element or $A = 16$ to obtain,

$$v = 1.7 \times 10^{17}$$

and finally for the frequency of vibration n, using the simple expression,

$$n = 2\pi v$$

he obtains: $\quad\quad\quad n = 1.1 \times 10^{18}$

Bohr states that this value bears a 'remarkable' agreement with the value for the vibrational frequency calculated by a quite separate approach from the absorption of α-rays, namely 0.6×10^{18} – despite the fact that there is a factor of two disagreement in the two values.

But for elements heavier than oxygen, even this sort of agreement is less 'remarkable'. Using the measured velocity of particles passing through a number of other elements, Bohr calculates a quantity that he calls r, the number of electrons per atom, which should be about half of the atomic weight of each element, according to Rutherford's theory. This is, of course, the approximate relationship $Z \approx A/2$,[33] but using r in place of Z. Bohr finds the results shown in Table 3.

Element	Atomic weight A	$A/2$	r
Aluminium	27	13.5	14
Tin	119	59.5	38
Gold	197	98.5	61
Lead	207	103.5	65

Table 3 Table compiled by author, based on Table on p. 27 of Bohr's paper of 1913 (*Philosophical Magazine Series 6*, 25 (145), (1913), 10–31).

Bohr writes:

> According to Rutherford's theory we shall expect values for r equal to half of the atomic weight; we see that this is the case for aluminium, but that the values of r for the elements of higher atomic weight are considerably lower.[34]

He then discusses corrections that would be necessary, while seemingly continuing to believe that r must eventually turn out to be equal to $A/2$.

At this stage, it appears that Bohr had no inkling that Z and A would have to be sharply distinguished from each other. Like Geiger and Marsden, Bohr showed no great concern for the discrepancy between the experimental findings and the relationship $Z \approx A/2$ as it applies to elements with high atomic weights. But this discrepancy was the crucial feature that motivated van den Broek to unravel the question, by severing the connection between Z and A, since the relationship only applies to elements with low atomic weights.[34]

Nevertheless, Bohr claimed with confidence that hydrogen has one electron while helium has two. About elements with higher atomic weights, all he can say is:

> For elements of higher atomic weight, it is shown that the number and frequencies of the electrons, which we must assume, according to the theory, in order to explain the absorption of α-rays are of the order of magnitude to be expected.[35]

As with Geiger and Marsden, agreement within an order of magnitude seemed to be good enough for Bohr. *Pace* Heilbron, I do not believe this work suggests that Bohr had a good understanding of the concept of atomic number at the start of 1913. That honour I insist belongs to van den Broek, whose contribution was more influential than we have been led to believe. Bohr's focus on individual elements, as in the experiments that he directed Geiger and Marsden to perform, would not have given him a global view of the elements as a whole comparable to that glimpsed by van den Broek.

Conclusion

Van den Broek was an outsider to the scientific establishment of his day. This may be one of the reasons for his neglect, but it cannot fully explain the fact that even today he remains virtually unknown. *En route* to the recognition of the importance of atomic number, van den Broek entertained many ideas which, in retrospect, were quite wrong. But such notions as the existence of the alphon particle, and the need for a three-dimensional periodic table, were steps forward. His failure to include scattering data for aluminium in the key papers which led to his correct identification of atomic number as the charge on the nucleus was a major sleight of hand, which no doubt would have been picked up by referees in a contemporary setting. Nonetheless, the evolution of his thinking highlights the manner in which scientific developments may propagate. Such apparently shaky foundations led to progress, to the extent that van den Broek outpaced Bohr, Rutherford, Barkla, and others. When Moseley started his experiments, it was only van den Broek who had clearly articulated in print the need to jettison atomic weight as an ordering principle, and to replace it with the concept of atomic number, as identified with the charge on the nucleus of any particular atom. The 'logic of discovery' seemed to follow a path that was not necessarily logical. Instead, science proceeded in a unified and organic manner, in this case, producing an evolutionary outcome.

Notes and References

[1] T. S. Kuhn, *The Structure of Scientific Revolutions* (Chicago: University of Chicago Press, 1962).

[2] This chapter is adapted from Eric Scerri, *A Tale of Seven Scientists and A New Philosophy of Science* (New York: Oxford University Press, 2016), 41–62.

[3] S. Toulmin, 'Does the Distinction between Normal and Revolutionary Science Hold Water?' in Imre Lakatos and Alan Musgrave (eds.), *Criticism and the Growth of Knowledge* (Cambridge: Cambridge University Press, 1970), 39–47, 1; D. Campbell, in P.A. Schlipp (ed.), *The Philosophy of Karl Popper* (Chicago: Open Court, 1974), 412–463; K. R. Popper, *Objective Knowledge, An Evolutionary Approach* (Oxford: Clarendon Press, Oxford, 1972).

[4] J. Lovelock, *Gaia: A New Look at Life on Earth, 3rd Edition* (Oxford: Oxford University Press, 2000). I disagree with any hint of teleology associated with Lovelock's Gaia, which has been one of the main sources of criticism by Dawkins. Dawkins critique of Lovelock's Gaia appears in R. Dawkins, *The Extended Phenotype: The Gene as the Unit of Selection* (Oxford: W.H. Freeman, 1982).

[5] H. Moseley, 'Atomic Models and X-Ray Spectra', *Nature*, 92 (2307), (1914), 554. This short letter responds to criticisms of his paper 'The High-Frequency Spectra of the Elements', *Philosophical Magazine Series 6*, 26 (156), (1913) 1024–1034 voiced by Dr. F.A. Lindemann. In his 1913 paper Moseley acknowledges that the idea that 'N [the nuclear

charge] is the same as the number of the place occupied by the element in the periodic system … [was] … originated by Broek'.

⁶ In his biography of Moseley, Bernard Jaffe states that 'In his second year at the college he received a first in mathematics moderations and shifted from chemistry to physics. One his classmates recalled the day Moseley turned up in his room and asked if he would think it wise if he made such a change. Others, too, had apparently given him the same advice for it was not long after the visit that he made the switch'. [There seems to be a sentence missing in this narrative]. The 'classmate' (not a term that would have been used at the time) is identified in Jaffe's Preface as R. T. Lattey who was '[one] of the people who were kind enough to talk or write to me about Moseley'. See Bernard Jaffe, *Moseley and the Numbering of the Elements* (London: Heinemann, 1972), 13 and vi. The Archives of Trinity College, Oxford, have no formal record of Moseley's initial intentions and he embarked on the study of physics in Michaelmas Term 1907, immediately after completion of Mathematical Moderations. However, Robert Tabor Lattey (1881–1967) was a Millard Scholar, who took a first class degree in chemistry in 1903. He evidently changed discipline thereafter and from Michaelmas Term 1906 onward worked as a demonstrator in Townsend's Electrical Laboratory. Thus Jaffe's account of Moseley's change of subject is credible, although Moseley must have decided to study physics during his first year in Oxford, or very early on in his second year.

⁷ Eric R. Scerri, *A Tale of Seven Scientists and A New Philosophy of Science* (New York: Oxford University Press, 2016).

⁸ A. van den Broek, 'Das α-Teilchen und das periodische System der Elemente', *Annalen der Physik*, 4 (23), (1907), 199–203 (hereafter AvdB I).

⁹ E. Rutherford, 'The Mass and Velocity of the α particles expelled from Radium and Actinium', *Philosophical Magazine Series 6*, 12 (70), (1906), 348–371

¹⁰ Further information on the Prout hypothesis can be found in Eric R. Scerri, *The Periodic Table: Its Story and Its Significance* (New York: Oxford University Press, 2007), 38–42.

¹¹ AvdB I, *op. cit.* 19

¹² N. Bohr, 'On the Constitution of Atoms and Molecules', *Philosophical Magazine Series 6*, 26 (151), (1913), 1–25; N. Bohr, 'On the Constitution of Atoms and Molecules. Part II – Systems Containing a Single Nucleus', *Philosophical Magazine Series 6*, 26 (153), (1913), 476–502; N. Bohr, 'On the Constitution of Atoms and Molecules. Part III – Systems Containing Several Nuclei', *Philosophical Magazine Series 6*, 26 (155), (1913), 857–875 (hereafter Bohr I, Bohr II and Bohr III respectively).

¹³ A. van den Broek, 'Das Mendelejeffsche "kubische" periodische System der Elemente und die Einordnung der Radioelemente in dieses System', *Physikalische Zeitschrift*, 12 (1911), 490–497. Quotation on 491.

[14] A. van den Broek, 'The Number of Possible Elements and Mendeleeff's "Cubic" Periodic System', *Nature*, 87 (2177), (1911), 78.

[15] H. G. J. Moseley 'The High-Frequency Spectra of the Elements', *Philosophical Magazine Series 6*, 26 (156), (1913), 1024–1034.

[16] H. G. J. Moseley, 'The High-Frequency Spectra of the Elements. Part II', *Philosophical Magazine Series 6*, 27 (160), (1914), 703–713

[17] A. Pais, *Inward Bound* (Oxford: Clarendon Press, 1986), 227.

[18] A. van den Broek, 'Die Radioelemente, das periodische System und die Konstitution der Atome', *Physikalische Zeitschrift*, 14 (1913), 32–41. Van den Broek's own translation was 'if all elements be arranged in order of increasing atomic weight, the number of each element in that series must equal its intra-atomic charge'. See note. 19.

[19] A. van den Broek, 'Intra-Atomic Charge', *Nature*, 92 (2300), (1913), 372–373 (hereafter AvdB II).

[20] A. van den Broek, 'Intra-Atomic Charge and the Structure of the Atom', *Nature*, 92 (2304), (1913), 476–478 (hereafter AvdB III).

[21] H. Geiger and E. Marsden, 'The Laws of Deflexion of α-Particles through Large Angles', *Philosophical Magazine Series 6*, 25 (148), (1913), 604–628.

[22] *Ibid*. Table IV on 617. The factor of 3/2 arises as follows: The scattering of α-particles was taken to be proportional to the square of the atomic weight of any element. Moreover, for different foils of the same 'air equivalent' the number of atoms per unit area is inversely proportional to the square root of atomic weight. Combining these two relations, $A^2 \times A^{-1/2} = A^{3/2}$. Both table IV on 617 (reproduced here as Fig. 5) and table VI on 619 incorrectly head the final column with $N \times A^{2/3}$, where it should be $N \times A^{-3/2}$.

[23] *Ibid*. 619. The experiments with radium emanation (which was a stronger α emitter) also included results for carbon, a weak scatterer.

[24] AvdB II. *op. cit*. 373. The numerical factor of 5.4 is introduced to give the ratios derived from the A^2 division the same mean value as those derived from the M^2 division.

[25] *Ibid*. 373. The standard deviation in the value of scattering divided by $(A^2/5.4)$ is 1.1, whereas the standard deviation in the analysis using M^2 is only 0.2. However if the scattering data for Al is included alongside that for Cu, Ag, Sn, Pt and Au, the standard deviation in the A^2 analysis is less than that in the M^2 analysis, with respective values of 1.0 and 1.3. Van den Broek does not explain why he ignored the scattering data for Al.

[26] A. van den Broek, 'On Nuclear Electrons', *Philosophical Magazine Series 6*, 27 (159), (1914), 455–457.

[27] AvdB III, *op. cit*. 477.

[28] The triads are those appearing in his 1913 periodic table. They are Fe, Co, Ni; Ru, Rh, Pd; and Os, Ir, Pt.

[29] Public lecture by John Heilbron in Manchester, England, October, 2013. https://www.youtube.com/watch?v=EYIhx5YFNg0 (accessed on 29 November 2017).

[30] Richard Whiddington, (1885–1970), was a physicist who was at Cambridge when he published the work discussed here and who moved to the University of Leeds after the First World War.

[31] N. Bohr, 'On the Theory of the Decrease of Velocity of Moving Electrified Particles on Moving through Matter', *Philosophical Magazine Series 6*, 25 (145), (1913), 10–31 (hereafter Bohr IV).

[32] R. Whiddington, 'The Production of Characteristic Röntgen Radiations', *Proceedings of the Royal Society A*, 85 (579), (1911), 323–332.

[33] The relationship that $A/Z = 2.00$ applies only to the first 20 or so elements, from helium to calcium, after which the ratio increases as Z increases. For gold (the heaviest element for which scattering data were available to van den Broek), A/Z is equal to 2.49.

[34] Bohr IV, *op. cit.* 27.

[35] Bohr IV, *op. cit.* 30.

CHAPTER FIVE
SACRIFICE OF A GENIUS: MOSELEY THE SIGNALS OFFICER AND SIGNALLING IN GALLIPOLI

ELIZABETH BRUTON

Introduction

By the start of the summer in 1914, Henry 'Harry' Moseley, aged twenty-six, was already an internationally recognised physicist. His work on X-ray spectra had given a new way of viewing the Periodic Table of the elements and was contributing to the development of the nuclear model of the atom. For this ground-breaking research, conducted over a mere twenty months, Harry was and is best known and remembered. However, his lesser-known roles as Harry the student, soldier and signaller must also be recognised. His brief military career was in many ways deeply significant, and his death, a defining moment, which cast a long shadow over his and later generations. This aspect of Harry's story is the focus of this chapter.[1]

At the outbreak of war in August 1914, Harry was in Australia, attending a congress of the British Association for the Advancement of Science (BAAS). He presented a paper to the combined Physics and Engineering Sections, but returned to Britain as quickly as he could to enlist after war had been declared. Harry's war lasted a mere ten months, from October 1914 to August 1915, when his life was cut short, killed in action in Gallipoli, aged twenty-seven. He was one among many thousands of British, French, Indian and Commonwealth troops, as well as Ottoman men, lost during the ill-fated Dardanelles campaign.

Harry's death was widely lamented, with contemporary headlines citing the 'Sacrifice of a Genius' and a man 'Too Valuable to Die'.[2] The international scientific community was fleetingly re-united in its condemnation of the loss of such outstanding talent. It was small satisfaction that Moseley's death gave fresh impetus to movements already underway to recognise and make use of the special skills that scientists could bring to the conduct of Britain's war.

Eton and Oxford

We know today that the beginnings of Moseley's military career predated the Great War, but were shaped by the wars and threats of war that coursed through his generation. On Thursday 11 October 1906, Harry began his studies for a four-year degree in Natural Sciences at Trinity College, Oxford, on a Millard Scholarship,[3] and went on to win a First Class in Mathematical Moderations, followed three years later with a BA (second class) in Natural Science.[4] However, science was not his only pursuit at Oxford. As Clare Hopkins has shown in Chapter 1, his non-scientific occupations included rowing for Trinity and the Oxford University Officer Training Corps (OUOTC). The latter is often neglected, but is of special relevance here.

In 1908, following the disasters of the Anglo-Boer War, and in light of fresh tensions with France and Germany, the British Government established the Officer Training Corps (OTC) to ensure the supply of trained officers for the regular and reserve forces. The OTC was divided into a Senior Division for universities and a Junior Division for schools, with Oxford's OTC being part of the initial cohort of Senior Division OTCs.[5]

In May 1908, Lord Haldane, the Liberal Government's Secretary of War, whose initiative led to this and other military reforms, visited Oxford as part of a recruitment drive, and spoke to University students at Oxford Town Hall on the subject of 'The Relation of the University to the Army'.[6] Haldane was accompanied by General Ian Hamilton, who would seven years later command the Allied Mediterranean Expeditionary Force at Gallipoli. While this event is not recorded in Harry's correspondence, or in his mother's diary, it is quite possible that Harry attended this talk and was inspired to enlist in the OUOTC.

In November 1908, Oxford – along with the Universities of Cambridge, Edinburgh, Durham, Birmingham and Manchester, Aberystwyth College, and Clifton and Winchester Colleges – saw the transfer of units from the old Volunteer Corps to the new OTCs.[7] Oxford's OTC was based at 9 Alfred Street, alongside the University Delegacy for Military Instruction.[8] In a letter to his sister in December 1909, twenty-one-year-old Harry said he enlisted because 'he could find no sound argument with which to confute the advocate of universal service'. In his own words, he was 'antimilitarist [sic] ... by conviction, a soldier by necessity'.[9] These lines proved to be prophetic.

Harry had not joined the popular Rifle Volunteers at Eton, where he kept busy with scientific activities and sport, including rowing and the traditional game of Fives, interests he brought from Summer Fields School in North

Oxford.[10] Harry described his first experience of military service to his sister:

> We march about the fields – two men to a skipping rope make a section – imagining ourselves a large army; then the section commanders find themselves behind their 'men', and advance hastily over the skipping rope to the front – a manoeuvre which would be worth watching if practiced on parade. The skipping rope wheels, and trys [sic] to form fours, in which it necessarily fails. Then we skirmish open into extended order, and fire rapidly wooden cartridges into trees.[11]

In September 1910, after graduating from Oxford, Harry began three years with Ernest Rutherford at the University of Manchester. As he was not formally a student of the University, he did not become involved with Manchester's OTC, nor could he rejoin the OTC when he returned to Oxford in November 1913. The War would be Harry's next experience of military life.

Fig. 1 Harry in a Royal Engineers officer's uniform. As a Royal Engineer signaller, Harry wore the distinctive signallers' armbands, as shown in this photograph (taken between October 1914 and Harry's departure from England in June 1915). Harry had duplicates of this photograph distributed to friends and family including his mother Amabel, as well as Henry Jervis, his close friend from his Trinity college days. This image, along with others taken between 1908 and 1914, was donated to the Royal Society around 1958/1959 by A.M. Ward, who received them from Henry Jervis in 1938. (Source: The Royal Society, undated copy, by courtesy of the Royal Society of London).

Australia and Enlistment

In the summer of 1914, Harry was sufficiently confident about his research that he began gathering material to apply for a chair at the University of Birmingham.[12] He was also prepared to put his research on hold for three months to present a paper to the 84th BAAS meeting, to be held in Adelaide, Sydney, Melbourne and Brisbane from 28 July to 31 August 1914.[13] On 8 August, four days after Britain entered the war and following a lengthy holiday trip via Canada and Hawaii, Harry and his mother arrived in Australia.[14]

In Melbourne on 18 August, Harry played an active part in a 'Discussion on the Structure of Atoms and Molecules', where his research was frequently referred to and discussed.[15] In a letter to his sister Margery, he described the session as '[doing his] duty to the B. Ass [a common, tongue-in-cheek sobriquet accorded the BAAS] by talking in an oily general mixed Physics and Chemistry discussion'.[16] With a week before his paper, and accompanied by his mother, Harry went on some of the excursions for delegates.[17] On 25 August, he spoke in Sydney on 'High-Frequency Spectra of the Elements' to Section A (Mathematics and Physical Sciences) and Section G (Engineering).[18] With his paper presented, Harry quickly left Australia, sailing with fellow scientist, Oxford Alembic Club member, and friend Henry (later Sir Henry) Tizard. They departed Sydney by ship on 29 August and sailed to San Francisco before crossing the United States by train to New York, and onward home to England aboard the *Lusitania*.[19]

Crossing the Pacific, Harry practised signalling, describing in letters to his mother and sister, 'becoming more or less an expert in signalling Morse and semaphore' and 'reading up a smattering from War Office manuals, and practising flag wagging Morse and semaphore while crossing the Pacific'.[20]

On returning to England, Harry abandoned research, eager to enlist in Kitchener's New Army. He did not, however, abandon the idea of a scientific career. In early October 1914 and a mere two days before departing for Royal Engineer office training, Harry submitted an application for the Poynting Professorship of Physics at the University of Birmingham. In his application, Harry noted that he was hoping to obtain a commission in the Royal Engineers, and his services would not be available for the duration of the war.[21] Less than a month later, while Harry was in the midst of officer training, Professor Oliver Lodge, the distinguished physicist, responded from Birmingham that the professorship would not be filled 'until conditions are more normal'.[22]

In early October 1914, Harry took a two-fold approach, drilling with Territorials and those unable to serve in Oxford, whilst also attempting to enlist. Faced with a War Office that recruited only civil engineers into the Royal Engineers, and saw no use for a physicist, Harry persisted, and 'bullied' the War Office until he was given a commission. Less than a week after he was accepted, Harry left the family home at St Giles, Oxford for the Royal Engineers depot at Aldershot, where he began signal officer training.[23] Harry's mother returned in Oxford from Australia on Saturday 17 October 1914, missing her son's departure by a few hours.[24]

Harry signed up for the duration of the War and, in a letter to his mother, gave an outline of military life: five months training through the winter, followed by attachment to an RE company, and departure to the Front in the Spring of 1915.[25] Harry became one of the best-paid officers in the British Army: Royal Engineer officers received a bonus 'Engineers' Pay' which put the daily rate for Second Lieutenant H.G.J. Moseley at 9s 6d a day.[26]

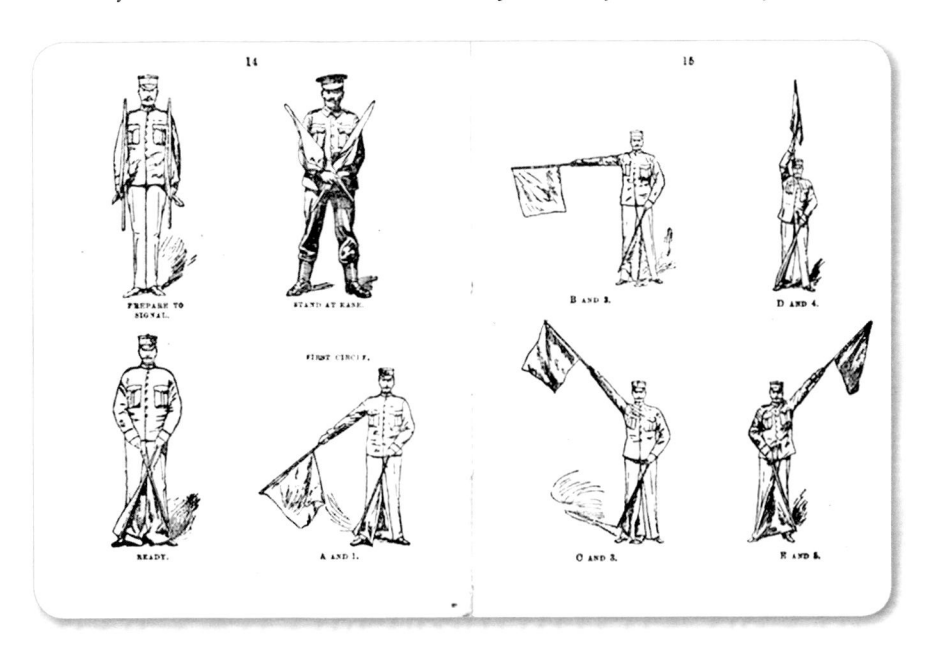

Fig. 2 Image from British Army signal training manual (1914), showing semaphore positions. (Source: Royal Signals Museum Archive, 1914, by courtesy of the Royal Signals Museum).

Harry's training lasted one month, beginning with a week of drilling and physical exercises while waiting for other trainees to arrive, during which time Harry was attached to the reserve signalling company.[27] Thereafter, real training began, with Harry kept 'extremely busy' riding and grooming horses, and being shown a wide range of signalling techniques, including traditional methods such as heliographs and telescopes, and more modern methods such as telegraphs and telephones. Telephones had been widely adopted by the army, but older methods of communication remained vital.

The signals training was 'strenuous', with Harry and his fellow Second Lieutenants spending five hours a day reading Morse messages on a telephone buzzer at six words per minute; reading Morse sent from a signalling lamp at night; and reading and receiving semaphore signals, which Harry described as 'the sending by no means easy to do in good style'.[28] The 'flag-wagging' that Harry had practised on the journey home from Australia was now carried out using pairs of white flags with a distinctive blue stripe.[29] In a letter to his mother, Harry noted the practical difficulties:

> Up every day at 6.45 I am kept fairly busy until 7.15 P.M. signalling: reading Morse by ear and eye, and sending and reading semaphore. The signalling is easy except the reading of messages sent by lamp. In this case all that is seen is a twinkling light, and I have the greatest difficulty in distinguishing one group of flashes from another.[30]

Fig. 3 Royal Corps of Signals training in the use of telescope and Lucas Lamps taken about 1920. The signalling lamp remained in service until well after World War II. (Source: Royal Signals Museum Archive, undated but probably around 1920, by courtesy of the Royal Signals Museum).

On the last weekend of November 1914, Harry was given a leave pass and visited his mother before departing for Bulford Camp, Salisbury Plain, built on 750 acres purchased by the British Army in 1897.[31] Here Harry was assigned to a company as a junior subaltern and allocated a section of about twenty-six men.[32] Harry described the men of the company as 'so keen and quick witted in most cases that it is a pleasure to have to do with them'.[33] Harry learned more advanced skills, including methods of setting up telegraphs and telephones, and the practice of communication by despatch riders. In early January 1915, Harry had his first motorbike ride, a circuit around Stonehenge, an experience he did not enjoy.[34]

In mid-December 1914, Harry was given two days leave to attend his mother's wedding to Professor William Sollas, which took place at St Giles in Oxford.[35] A week later, he was given Christmas Day to visit his mother and her new husband, who spent the first two weeks of their marriage at the Moseley family home at Pick's Hill. Here Harry raised the possibility of transferring from the Royal Engineers to the Royal Flying Corps (RFC).[36] Harry noted that it was safer than being a Signals Officer and his mother raised no objection; indeed, in early January 1915, she sent him a pair of luxurious rabbit-skin gloves as a sign of her support.[37]

On 27 December 1914, three days after discussing the topic with his mother, Harry wrote to his sister about his application to become an RFC observer.[38] The experience that led to this change of heart had taken place on Christmas Eve, nearly a month after Harry arrived at Bulford Camp. Harry had travelled to RFC Netheravon to be taken on a flight over Salisbury Plain in a Bleriot Experimental (B.E.) aircraft piloted by Edgar Ludlow-Hewitt, Margery's brother-in-law.[39] Throughout the war, Netheravon was operated as a forming-up point for squadrons destined for the frontline and as a training organisation for aircrew, ground crew, specialist signallers, and fitters.[40] Like many of his contemporaries, Ludlow-Hewitt (who became an Air Chief Marshal and Air Officer Commander-in-Chief of Bomber Command during World War II) had only learnt to fly a few weeks before the outbreak of war at Central Flying School, Upavon, a few miles away from Netheravon, but nonetheless became a probationary member of the RFC shortly after war broke out.[41]

After his flight, Harry decided that his most valuable contribution to the War would be as an Observer, the second man in early two-seater aeroplanes who observed and operated weapons while the pilot flew the aircraft. Just as less than six months earlier he had worked hard to join the Royal Engineers, Harry now worked hard to leave. However, there was one obstacle that could not be overcome by 'bullying' the War Office: the weight limit. The RFC had a 10st 7lb weight limit on pilots and observers. A photograph taken in his last year of rowing at Oxford in 1910 during Torpids (a type of rowing race) shows that at 11st 7lb, Harry was over this limit, and remained so in late 1914.[42] In any case, his attempts to reduce his weight to the 'proper figure' were

Fig. 4 Close-ups of Harry in the Trinity College Boat Club photographs of the 1st Torpid Crew 1907 and the 2nd Torpid Crew 1910, where his weight gain (from 10st 9lbs to 11st 7lbs) is evident from his facial profile. (Source: Archive of Trinity College, Oxford, 1907 and 1910, by courtesy of the President and Fellows of Trinity College, Oxford).

overtaken by an edict from Kitchener in January 1915, prohibiting transfers out of the New Army.[43]

In mid-February 1915, Harry and his company were sent to Woking, where they were attached to the 13th (Western) Infantry Division of the Third Army, one of Kitchener's New Army Divisions, formed of volunteers.[44] The same month, Harry turned down an offer of a transfer to the Royal Aircraft Factory at Farnborough. As he wrote to his mother, 'the aircraft factory is run entirely by civilians, and if I was physically unfit I would like nothing better than to work there'.[45] In any case, he was determined to see frontline service and to resist all attempts from well-meaning friends, family members, and former colleagues to do otherwise.

Sometime between mid-March and early April 1915, Harry moved from Woking to Cowshot Common training groups at Brookwood, Surrey, in anticipation of departing for the Front.[46] Harry was assigned as a Signalling Officer to the 38th Brigade (commanded by Brigadier General Anthony Hugh Baldwin) of the 13th Division, and was put in charge of the twenty-six men of

No 2 section.⁴⁷ A divisional signals company consisted of about 150 men, in sections led by a Second Lieutenant.

Harry assumed he would be departing for France later in the month and set his affairs in order. But Ernest Rutherford was determined that Harry's scientific talents should not be wasted. In July 1915, Rutherford tried to have him transferred to wartime research, by asking Sir Richard Glazebrook, Director of the National Physical Laboratory (NPL), whether he had work that that would justify Harry's recall from active services, but by then Harry had been shipped off to Gallipoli.⁴⁸ Family friend Sir Edwin Ray Lankester also tried to get Harry's 'great scientific skill' put to better use in England, but this also came to naught.⁴⁹

In late May 1915, Harry and his division assembled at Blackdown in Hampshire, where they were issued tropical kit and told that they would form part of the Mediterranean Expeditionary Force (MEF) led by Sir Ian Hamilton. They were shortly to depart for the Dardanelles, where they would form a key part of the campaign to take the high ground in the middle of the Gallipoli peninsula.⁵⁰ Supplies and support were not ready. In a brief, hasty letter, presumably to his mother, Harry was critical of the equipment he was given, including 'most inferior and badly made substitute [telephones]'. He further complained that 'anything is good enough for a division which is not going to Flanders'.⁵¹ Harry duly set out to secure better equipment for his men: he bypassed the Quartermaster at Aldershot and commandeered wire at the field stores, and relieved a nearby stationmaster of helmets and supplies intended for the Worcestershire regiment.⁵²

Harry saw his mother for the last time on 6 June 1915, an occasion recorded by Amabel in her social diary: 'Harry went 5.30. I walked with him to the top of Romsey Hill [near the family home at Pick's Hill]'. After her son's death, she added a note: 'The last time I saw him'.⁵³ On 15 June 1915, Harry and his division departed Avonmouth for the Dardanelles, travelling via Gibraltar, Malta, and Alexandria. It was at Alexandria – which Harry described as being 'full of heat, flies, native troops, and Australians' – that Harry wrote his last will, using an active service form, and sent a copy to his mother.⁵⁴

In this will, dated 27 June 1915, Harry left his estate to the Royal Society of London 'to be applied to the furtherance of experimental research in pathology, physics, physiology, chemistry or other branches of science, but not in pure mathematics, astronomy or any branch of science which aims merely at describing, cataloguing, or systematising'.⁵⁵ A week later, in early July 1915, Harry and his division arrived at Mudros, a town with a large harbour on the island of Lemnos. The island had been captured from the Ottomans by the Greeks in the First Balkan War in 1912, and from early 1915, Allied forces were granted permission by technically neutral Greece to use it as a base. The harbour and the island became a staging and evacuation post for British and French forces. In June 1915, the island was connected to the British-controlled Eastern Telegraph Company Mediterranean telegraph line by spur lines to Helles directly and via the island of Tenedos.[56] This enabled the Allied Forces to use secure and reliable wired telecommunications to communicate with Gallipoli and nearby Lemnos, as well as the long-distance 'All Red Line'.[57] Wireless telegraphy was also used in a limited fashion for inter-service signalling, especially between the British Army and the Royal Navy.

At Lemnos, Harry and his division had a few days of training in preparation for landing on the Gallipoli peninsula. Between 6 and 16 July 1915, the 13th Infantry Division took their first steps onto Ottoman soil at Cape Helles, the southernmost tip of the peninsula.[58] Harry landed with his brigade on 9 July 1915, and his Division relieved the 29th Division.

Gallipoli[59]

At Helles, Harry's Division experienced its first three weeks of combat against Ottoman forces before being readied for a fresh invasion further north. Signalling at Gallipoli was significantly different from that on the Western Front in three important respects. First, the terrain was more difficult, and different communications technologies and systems were required. Second, there was a need for inter-service signalling, relying mostly on wireless telegraphy and visual signals such as semaphores. Thirdly, much of the signalling was provided not by Regular Army signal units (as it was, for the most part, on the Western Front) but rather by signallers such as Harry from Kitchener's New Army, as well as by the Territorial Army and ANZAC forces.[60]

As a Brigade Signals Officer, Harry's role was to provide a communications system using a mixture of telephone and telegraph cables, messengers ('runners'), and visual signalling. Harry also provided telephone communications between Divisional and Brigade HQ as well as telephone links between forward observation posts and rear artillery. Messages were sent by field telephones and Morse key telegraphs. Brigade signals were hampered by regular bombardments by Ottoman artillery occupying higher ground, which required regular repair of lines.

Supplies were short and much of the equipment was inferior to that being sent to the Western Front, as Harry himself had found before leaving Britain.

Harry's men had communications equipment designed for use in the thick, heavy mud of Flanders rather than in the sandy gullies of Gallipoli. For example, cable drums, heavy large reels, with a centre spindle grasped either side by two signallers, were not suitable for the conditions in Gallipoli and there were many complaints and pleas for smaller, lighter reels. Australian signallers working alongside the British simply cut their cable into shorter lengths, often throwing the cables down ravines – where they lay visible in the scrub – then scrambled down and joined the ends together.

There was, however, one piece of equipment that stood in good stead for signals officers at Gallipoli, at least when it was available: the Telephone D Mark III (sometimes referred to as the DIII), which became the standard British Army field telephone from its introduction in 1915 through the Second World War. This robust and adaptable telephone set could be used on poor-quality lines damaged by bombardment and stretched by limited supplies of wire. This model proved as successful in the flat terrain and muddy trenches of the Western Front as in the arid hills and gullies of Gallipoli.

Writing to his mother and sister, Harry did not discuss signalling matters, knowing that details would be censored. Instead, his letters reflect the general experience of a soldier, interspersed with less general references to the natural world around him. Harry commented on the animals and insects of the dry and arid landscape of Gallipoli at the height of summer, with occasional references to joyful swims in the sea (omitting the risk of being shot by Ottoman snipers).[61] Harry also described the endless flies, sand, heat, limited water, and illness among his men.[62] Shortages of equipment are evident from his frequent requests for personal, medical, and even signalling supplies. Harry also asked for things he could share with his men and fellow officers, including his close friend Captain G.E. Chadwick.[63] For example, on 4 July, Harry asked to be sent quantities of chlorodyne, the 'miracle drug of the 19th century', to treat dysentery and relieve pain, along with two suits of khaki, two pairs of pyjamas, and a pair of stocking puttees.[64] Even more indicative of shortages, in the same letter of 4 July, he asked to be sent dry cells and batteries – 'v. important', he noted.[65] A month later, on 4 August, and on the day he left Murdos for Anzac Cove, Harry spent the last page of the short letter that would be his last, requesting tools and equipment required for signalling work, including a dozen pliers for wire cutting (Harry included a neat drawing of the type of pliers he wanted) along with a mosquito net for his head.[66] These shortages foreshadowed the wider failures of the Allied campaign.

In late July, after three weeks of fighting, Harry's Division returned to

Murdos. The harbour had become a staging area for the August offensive, in which Harry would take part. The plan was to deploy the fresh units of Kitchener's New Army to break the stalemate that had set in after the first landings in April; and to gain a new position at Suvla Bay, five miles north of Anzac Cove, which was relatively lightly defended. The first objective of the August campaign, planned and spearheaded by Sir Ian Hamilton, was the important high ground in the middle of the Gallipoli peninsula.

On 4 August 1915, Harry's Division departed Murdos, and early on the morning of 5 August, landed at Anzac Cove near the base of Chailak Dere, as part of the signalling reserve in support of the breakout attack. The 13th Division's signals section spent the day at Anzac Cove providing communications in support of the Allied attack.[67] They left for what became their No 3 Outpost later that evening, and the following day, were put to work to support the attack. Cables were laid to 40th Brigade at Damakjelik Bair, and to support General Cox, commanding the Indian Brigade and the Left Assaulting Column. Finally, a set was put in place for General John (later Sir John) Monash, commander of the 4th Australian Infantry Brigade. Harry was kept busy setting up and maintaining telephone lines and telegraph cables – especially the latter – which were laid to the various brigades in advance of and during the attack.

Two days later, on 6 August 1915, Harry's 38th Brigade (known as 'Baldwin's Brigade' after its commander, Brigadier Anthony Hugh Baldwin) alongside their Commonwealth allies – Australians, New Zealanders, the 39th, the 40th, and Indian Brigades – were set the task of taking the strategic high ground held by the Ottomans at either side of Chunuk Bair. This attack came to be known as the Battle of Sari Bair, a murderous four-day battle, resulting in thousands of deaths on both sides.[68] The battle, and the failure of the Allied attack, remains discussed and debated to the present day, with much of the blame often being placed at the feet of Harry's commander, Brigadier Baldwin.

Baldwin's No.3 Column was to attack a feature called Hill Q, *via* an area called 'The Farm', the site of a shepherd's stone hut on the western slopes of Chunuk Bair, 300 yards short of the summit.[69] Baldwin's arrival would signal to Nos. 1 and 2 Columns on either side the start of the attack, timed to follow an Allied artillery barrage at 5.15 am. In terms of signalling, the 13th Division War Diary for 8 August 1915 recorded that cable communication to the 38th Brigade was interrupted by the New Zealand Brigade and there was general confusion owing to the attack.[70] Furthermore, telegraphic communications were compromised by the practice of staff officers speaking on telephones

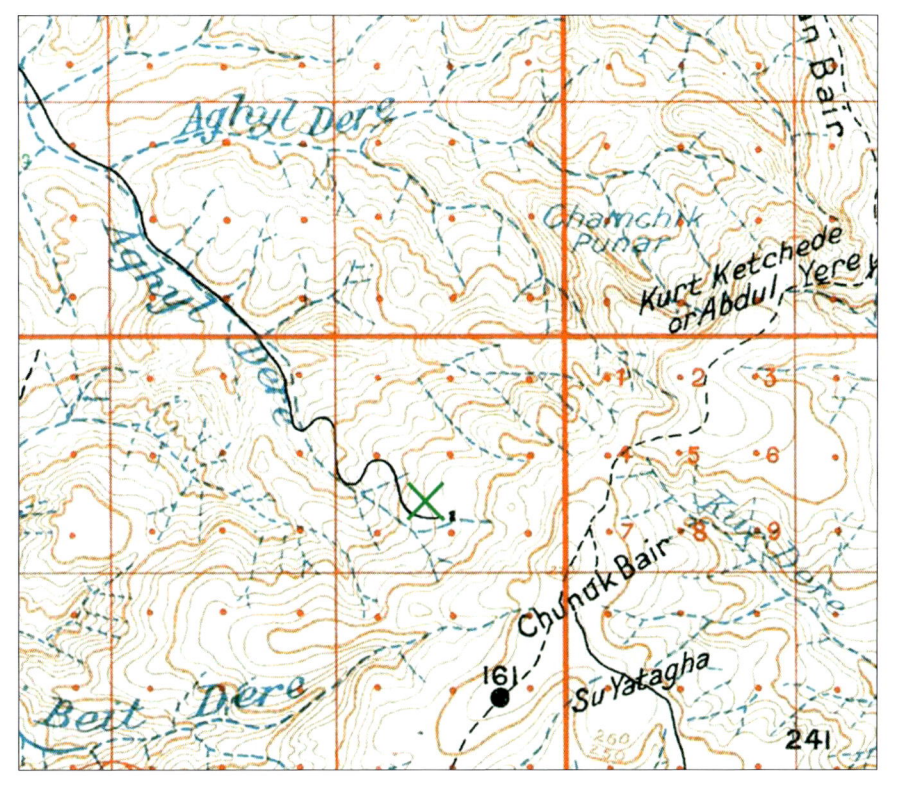

Fig. 5 Detailed map of 'The Farm' from the British 1:20,000 map entitled 'KURIJA DERE', issued to officers for the battle of Sari Bair. The 'X' (not on the original map) marks where Harry was killed. (Source: D.F. Harding, 2016, Crown Copyright).

instead of writing messages. Given Harry's complaints about equipment, the 13th Division War Diary also recorded that the cable and drums were 'most unsatisfactory', and that the Telephone D Mark III should have been used in place of the much less reliable Stephens telephone.[71] Lack of supplies was a problem.[72] But there is no evidence that this had an immediate effect upon the outcome. By all accounts, Harry was an efficient Signals Officer and managed Divisional signalling throughout the battle.[73]

The traditional account of Harry's last battle, based on the British Official History, alleges that on the morning of 9 August, Baldwin's No.3 Column – an *ad hoc* force led by Baldwin's 38th Brigade HQ and including his 6th East Lancashire and a mixture of other brigades – got hopelessly lost on their way to 'The Farm', and attacked in the wrong place several hours too late, precipitating the failure of the third and final assault on Sari Bair.[74] However, careful study of the primary sources, including Australian archives, has recently established that the 'official' account is a travesty.[75] Baldwin's 38th Brigade HQ carried out a thorough reconnaissance of the Chailak Dere valley during the night of 7–8 August, with Harry providing telephone communications between Baldwin and his staff as they made their way up and back down. At the top, a staff officer showed them the position occupied by the New Zealanders, whom Baldwin expected to reinforce. Because of this recce, Baldwin was at no stage lost.

As part of the attack at 05.15 hrs on 9 August, Major General Godley ordered Baldwin to assault a feature called Hill Q *via* a route going 'east of "The Farm"' and recommended the simplest route, heading straight up the Chailak Dere and over the top of Cheshire Ridge to 'The Farm'. At a meeting on the evening of 8 August, Baldwin and Cooper discussed the route with Brigadier Johnston, commander of the New Zealand Infantry Brigade and of No. 1 Column, and between them confirmed Godley's choice of route. Baldwin's column set off from its current position two-thirds of the way up the Chailak Dere, but was blocked when, in disobedience of orders, a New Zealand staff officer allowed a mule train of 100 animals to be sent down the Chailak Dere. The way up was totally blocked in the narrowest part of the Dere, but thanks to his recce, Baldwin was able to turn his column around and reach 'The Farm' via a side valley. His and Cooper's Brigade HQs and their two leading battalions were in place before 04.30 on 9 August, and they attacked on time at 05.15, just east of 'The Farm' as Godley had ordered: this was confirmed by 10th Gurkha Rifles to their left, as well as by other eyewitness accounts. However, Baldwin's assault was shattered by cleverly-sited Ottoman machine guns, riflemen, and artillery. The Allied attack failed because the Ottoman defence rendered the ridge unassailable. Harry was in the thick of these events on 8-9 August. The following day was his last.

Traditional accounts of what happened at 'The Farm' on the morning of 10 August are also inaccurate. During Kemal Atatürk's famous massed counter-attack, which threw the British off Sari Bair ridge, the Ottoman soldiers did not overrun the British and Gurkha troops at 'The Farm'. What happened is that the British battalion holding Chunuk Bair resisted the first two waves of the massed attack, but was overwhelmed by the third and fourth waves. Allied reserves held in trenches at a place called the Pinnacle, due south of 'The Farm', also gave way, and from the Pinnacle, the Ottoman soldiers could enfilade Baldwin's trenches on the edge of 'The Farm' plateau. They all but wiped out 38th and 29th Brigade headquarters.

Baldwin and Harry were two of those killed. Harry was shot through the head while calmly sending a telephone report to Divisional HQ.[76] He died, aged just twenty-seven, and his body, like that of so many on both sides, was never recovered. Nonetheless, Captain Chadwick (mentioned earlier) was able to search his friend's body and remove his Sam Browne belt and other effects, which he could do only because the position was not overrun.[77] Chadwick was one of the few officers from both HQs to survive, and without a letter he wrote to his mother, which she later copied for Amabel, we would not know what really happened in and around Baldwin's HQ on 9 and 10 August.[78]

The Ottoman counter-attack was stopped by Allied naval and land-based artillery, and by Chadwick's and the New Zealanders' machine guns. Yet the British were obliged to abandon 'The Farm', which became no-man's-land until the Ottoman soldiers re-occupied and fortified it two days later. Neither side had time to bury the countless dead, and their bones were still there when the War Graves Commission first visited the peninsula in 1919. Remains from 'The Farm' position – all known to be Allied soldiers, but none individually identified – were gathered into what is now 'The Farm' Cemetery. This Cemetery contains markers for seven officers and men 'believed to be buried in this cemetery', while all the others – the remaining 645 who are unidentified – are commemorated by name on Memorials to the Missing, chiefly the main British and Indian memorial at Cape Helles, close to where Harry first set foot on the peninsula.[79] Harry's name appears there, although his bones probably lie somewhere in 'The Farm' Cemetery.

Aftermath and Legacy

News of Harry's death did not reach his family in England until Monday 30 August, almost three weeks after he was killed. This came to Amabel in a War Office telegram which eventually arrived at the family home at 48 Woodstock Road at about 2.15 pm.[80] The telegram was wrongly addressed and was first delivered to 43 Woodstock Road. When the mistake was corrected, Amabel immediately sent for Margery – Harry's sister and her last remaining child of four – to share her grief. Afterwards, Amabel edited and amended her diary for 1915, to gather and record the last moments she spent with her son before he went to war.[81]

On 5 September 1915, less than a month after her son's death, Amabel wrote to the War Office requesting further details, stating simply and starkly that 'any information that you can give will be gratifying to me'.[82] The War Office reply two days late provided no further details.[83] Meanwhile, two officers who had served with Harry, and witnessed his death, wrote informally to Amabel, giving details and describing how he had died instantly when shot through the head by an Ottoman sniper while telephoning through an order, as will be discussed in more detail below.[84] It seems likely that Amabel received at least one of these letters a few days after the official War Office telegram informing her of her son's death, and possibly around the time of her letter to the War Office, as these details were given in Ernest Rutherford's moving obituary, published in *Nature* on 9 September 1915.[85]

Following the Battle of Sari Bair, the news had been making its way its way through the system. On 13 August, three days after the battle, Harry's death warranted just a single line in the 13th Division War Diary, and that merely in reference to his being replaced: 'Lt Musson reorganised No. 2 Section vice Lt Moseley killed using remnants of No 2 Sec'.[86] The same day, 13 August, Harry's commanding officer, Captain Crocker (who had himself been slightly injured earlier in the battle) wrote an emotional letter informing Harry's family of his death:

> I very much regret to inform you that Lieut. Moseley commanding one of my sections was killed in action during the recent fighting.
>
> He was shot through the head and died instantly. I am glad to know that he was spared any suffering from wounds.
>
> He was a most promising officer – extraordinarily hard working and took an immense amount of interest in his work. I am very sorry indeed to lose him, and I beg to offer you my very sincere sympathy. This sorrow at his loss is felt by all the officers and men of my Company equally with myself.[87]

On 16 August 1915, Captain G.E. Chadwick, 38th Brigade Machine Gun Officer and Harry's close friend, wrote to Amabel promising a detailed account of the battle and of Harry's death.[88] Chadwick was one of the few survivors from the 38th Brigade HQ and had witnessed Harry's death firsthand. He wrote to Amabel promising he would provide a detailed account of her son's death in the near future, being concerned that at present 'it would [weigh] on [her] already heavy burden' to give her all the details. Instead he wrote a letter of consolation, confirming and echoing Captain Crocker's account of Harry's death as well as his skills as a Signals Officer:

> [Harry] died the death of a hero, sticking to his post to the last. He was shot clean through the head & death must have been instantaneous.
>
> In him the Brigade has lost a remarkably capable Signalling Officer & a good friend.

> To him the work always came first, & he never let the smallest detail concerning it go by him unnoticed.
>
> When he was shot, the Turks had got round our right flank & were within 200 yards of us firing into our right […] I was close to your son & he was calmly telephoning a message to the Division….

On 27 September 1915, more than a month after his initial letter, Chadwick sent a second letter to Amabel briefly describing the battle and providing the address of his mother, from whom Amabel could request a transcript of Chadwick's detailed account of the Battle written on 14 August 1915.[89] Chadwick suggested this option as he could not bear to 'pen again a description of those ghastly days' in which he lost 'so many friends … that [he was] doing [his] best to forget'. Chadwick wrote that he found nothing of importance on Harry's body but had carefully packed up his 'Sam Browne belt & personal effects' and sent them back to Amabel. It would appear that Harry's personal effects did not make it back to Amabel: on 13 October 1915, Amabel called the War Office about her son's effects.[90] The War Office representative noted that none had been received but they would be passed on following receipt, and the will had been checked to confirm that Harry's effects were to be forwarded on to his mother as next of kin.[91]

In the remainder of his letter of 27 September, Chadwick focused on Harry's personal attributes and skills as a Signals Officer:

> [Harry] always put his work before everything … No sacrifice was too great for him to make so long as that which he was given to do, was done. Ever cool calm & collected in dangers, & there were many here, he had, besides the brains of a genius, the courage of a lion & a determination such as is seldom found in the young man of today.
>
> As a signalling officer he was perfection itself, I speak technically as I am a Signal Service man myself though not officially as such out here.…
>
> [Harry] was a man, a hero, & a son any Mother would be proud of.[92]

Chadwick's letters described Harry's life and death as a soldier, and the impact of Harry's death on his men. Harry's death had a strong impact, personally and scientifically. One of his most direct legacies was his bequest to the Royal Society: his entire estate, valued at £1799 6s 1d.[93] Amabel sold the family holiday retreat at Pick's Hill and topped up the legacy to £10,000. This resulted in the foundation of the Moseley Research Studentship – which, in

1945, became the Alan Johnston, Lawrence and Moseley Research Fellowship – awarded regularly by the Royal Society.

The changing recognition of the value of scientific research in wartime also featured strongly among the many enduring legacies of Harry's death. His case clearly showed that, at least until 1915, men of science were not always being put to good use. Harry's fate forced public and political attention on the fact that investment in research might win wars more certainly than sacrifice on the battlefield. The general conclusion was compelling. In late 1915, but probably unrelated specifically to Harry's death, the British government established the Department of Scientific Industrial Research (DSIR) to hire scientists for laboratory research and encourage private firms to establish co-operative industrial research associations.[94] The impact of DSIR was felt more strongly in 1916 and it was in the same year that a reserved class was created in which scientists and engineers were given special consideration if they were in designated occupations. This contrasted with Harry's experiences in 1914 and 1915 when he and others had a choice of how to serve: in the frontline on the battlefield; or on the homefront, using their scientific skills for wartime research and military goals. Harry chose the former while his friend Henry Tizard eventually chose the latter.

More generally, while Harry's death on the battlefield was truly a 'Sacrifice of a Genius', it was a sacrifice (and those of other prominent 'scientific men' on the battlefield) that encouraged the British government to establish permanent scientific services in the army and the Royal Navy, and in what was to become the Royal Air Force, and to strengthen the military and naval use of scientific advice.

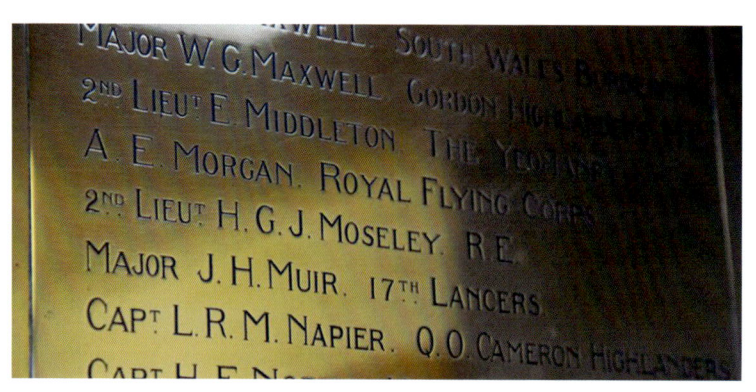

Fig. 6 Photograph of part of Summer Fields school war memorial featuring Harry's name. (Source: Museum of History of Science, Oxford, 2015, by courtesy of Summer Fields School, Oxford).

Notes and References

1. This chapter builds upon my earlier research for a project on 'Innovating in Combat: Telecommunications and intellectual property in the First World War', 2013–2014, supported by the UK Arts and Humanities Research Council (AHRC), and led by Professor Graeme Gooday, University of Leeds. See http://blogs.mhs.ox.ac.uk/innovatingincombat/ (accessed 29 November 2017). This work continued with a Heritage Lottery Fund (HLF) funded centenary project and exhibition, 'Dear Harry: Henry Moseley, a Scientist Lost to War', which was shown at the Museum of the History of Science, Oxford in 2015 and 2016. For further details, see http://www.mhs.ox.ac.uk/moseley (accessed 29 November 2017).

2. Oliver Lodge was quoted in 'Sacrifice of a Genius: Life Story of a Brave Student' in *Daily Express*, 1919 and similarly with the headline of 'Too Valuable to Die' from *Evening News*, 23 October 1919. Both are quoted in John Heilbron, *H. G. J. Moseley: The Life and Letters of an English Physicist, 1887–1915* (hereafter Heilbron, *Moseley*) (Berkeley: University of California Press, 1974), 124–125.

3. Manuscript held by Museum for History of Science, University of Oxford Ludlow-Hewitt Archive (hereafter MHS MS Ludlow-Hewitt. The holdings are in two boxes labelled 1 and 2), Amabel Moseley's social diary, 11 October 1906.

4. E. Rutherford, 'Henry Gwyn Jeffreys Moseley', *Nature* 96 (2393), (1915), 33–34.

5. For further information on the establishment and early history of OUOTC, see the Bodleian Library's 'Officer's Training Corps' at https://www.bodleian.ox.ac.uk/__data/assets/pdf_file/0003/192738/Officer-Training-Corps-OT.pdf (accessed on 29 November 2017). Copy available from author.

6. 'Mr. Haldane and The Army', *The Times*, Friday 15 May 1908 (issue 38647), 12.

7. G.J. Eltringham, *Nottingham University Officers' Training Corps 1909–1964*. Nottingham: Hawthornes, 1964, 11–12 cited in Edward M. Spiers, Council of Military Education Committees of the United Kingdom (COMEC) Occasional Paper No. 4: University Officers' Training Corps and the First World War (2015), 11.

8. Bodleian Library, Oxford, 'Officer's Training Corps', https://www.bodleian.ox.ac.uk/__data/assets/pdf_file/0003/192738/Officer-Training-Corps-OT.pdf (accessed online 29 November 2017)

9. Moseley to his sister Margery, [4/5 December 1909], Letter 37 in Heilbron, *Moseley, op. cit.* 166–168.

10. For details of Harry's sporting interests, see Elizabeth Bruton. Henry 'Harry' Moseley and his sporting achievements (2016) at https://www.mhs.ox.ac.uk/moseley/2016/02/08/henry-moseley-sporting-achievements/ (accessed online 29 November 2017). Copy available from author.

[11] *Ibid.*

[12] MHS MS Ludlow-Hewitt 1, job application, testimonials, and published papers of H. G. J. Moseley, gathered in a paper binding to form his application for the post of Professor of Physics at Birmingham University, 1914.

[13] See *Report of the Eighty-Fourth Meeting of the British Association for the Advancement of Science, Australia: 1914* (London: John Murray, 1915); available at http://www.biodiversitylibrary.org/item/95821 (accessed 29 November 2017). For further details, see Rebekah Higgitt. The H-Word blog at the *Guardian* 'The big Australian science picnic of 1914' at https://www.theguardian.com/science/the-h-word/2014/sep/03/big-australian-science-picnic-1914-history (accessed 29 November 2017)and John Scheckter, '"Modern in Every Respect": The 1914 Conference of the British Association for the Advancement of Science', *The Journal of the European Association for Studies of Australia*, 5 (1), (2014), 4–20.

[14] MHS MS Ludlow-Hewitt 2, Amabel Moseley's social diary, August 1914.

[15] *Report of the Eighty-Fourth Meeting of the British Association for the Advancement of Science, Australia: 1914* (London: John Murray, 1915), 293–301.

[16] Moseley to Margery, 28 September [1914], Letter 106 in Heilbron, *Moseley, op. cit.* 250–251,

[17] MHS MS Ludlow-Hewitt 2, Amabel Moseley's social diary, 22 August 1914.

[18] *Report of the Eighty-Fourth Meeting of the British Association for the Advancement of Science, Australia: 1914*. (London: John Murray, 1915), 305.

[19] Heilbron, *Moseley, op. cit.* 116. Harry's mother was left to travel at a more sedate pace, and was accompanied by an old family friend, William Sollas, a Cambridge graduate and Professor of Geology at Oxford since 1897; by the time the pair reached Britain in October, they were engaged to be married.

[20] Moseley to his mother, 4 September [1914], Letter 105 in Heilbron, *Moseley, op. cit.* 249–250; Moseley to Margery, 28 September [1914], Letter 106 in Heilbron, *Moseley, op. cit.* 250–251.

[21] MHS MS Ludlow-Hewitt 1, Letter 107 in Heilbron, *Moseley, op. cit.* 251.

[22] Oliver Lodge to Moseley, 4 November 1914, Letter 115 in Heilbron, *Moseley, op. cit.* 256–257.

[23] Moseley to his mother, 16 October [1914], Letter 111 in Heilbron, *Moseley, op. cit.* 254–255. For the Royal Engineers Signal Service on the Western Front, see R.E. Priestley, *The Signal Service in the European War of 1914 to 1918 (France)* (Chatham: W. & J. Mackay & Co., 1921). The author's preface (p.v) raised the possibility of a second volume, focussing 'on signal practice radically different from those of the main battle front [including]

the amphibious nature and the cramped conditions of the warfare in Gallipoli'. All that remains of further volume(s) are drafts on Salonica and East Africa at: Royal Signals Museum Archive 938.00/9 R.E. Priestley Priestley, *[Drafts of] The Signal Service in the War of 1914-1918: Part 2. Salonica and Part 3: East Africa.* n.d. [before 1933].

[24] MHS MS Ludlow-Hewitt 2, Amabel Moseley's social diary, 17 October 1914.

[25] Moseley to his mother, 16 October [1914], Letter 111 in Heilbron, *Moseley, op. cit.* 254–255.

[26] British Army rates of pay, as defined by War Office Instruction 166 (1914). See http://www.1914-1918.net/pay_1914.html and http://blogs.mhs.ox.ac.uk/innovatingincombat/british-army-royal-engineers-rates-pay-1914-1915/ (accessed on 29 November 2017).

[27] Moseley to his mother from RE Mess at Aldershot, 18 October [1914], Letter 112 in Heilbron, *Moseley, op. cit.* 255; Moseley to Margery from RE Mess at Aldershot, 1 November [1914], Letter 114 in Heilbron, *Moseley, op. cit.* 256.

[28] Moseley to Margery from RE Mess at Aldershot, 1 November [1914], Letter 114 in Heilbron, *Moseley, op. cit.* 256.

[29] Moseley to Margery from Cunard RMS *Lusitania*, 28 September [1914], Letter 106 in Heilbron, *Moseley, op. cit.* 250–251.

[30] Moseley to his mother from RE Mess at Aldershot, [31 October 1914], Letter 113 in Heilbron, *Moseley, op. cit.* 255–256.

[31] BBC Domesday Reloaded: Bulford Camp Then and Now (1986) http://www.bbc.co.uk/history/domesday/dblock/GB-416000-141000/page/5 (accessed on 29 November 2017)

[32] Moseley to his mother from RE Mess, Bulford Camp, Salisbury, [7 December 1914], Letter 118 in Heilbron, *Moseley, op. cit.* 259.

[33] *Ibid.*

[34] Moseley to his mother from RE Mess, Bulford Camp, Salisbury, 8 January [1915], Letter 120 in Heilbron, *Moseley, op. cit.* 261.

[35] MHS MS Ludlow-Hewitt 2, Amabel Moseley's social diary, 17 December 1914. Amabel referred to Sollas as 'Professor Sollas' in her social diaries, both before and during her second marriage.

[36] Military aircraft before and during the First World War were commonly two-seater, with an observer who assisted with aerial reconnaissance and aerial photography. Later in the war, observers operated aircraft guns and dropped bombs. For details, see C.G. Jefford, *Observers and Navigators: And Other Non-Pilot Aircrew in the RFC, RNAS and RAF* (London: Grub Street Publishing, 2014).

[37] Moseley to Margery from RE Mess, Bulford Camp, Salisbury, 27 December [1914],

Letter 119 in Heilbron, *Moseley, op. cit.* 260–261; Moseley to his mother from RE Mess, Bulford Camp, Salisbury, 8 January [1915], Letter 120 in Heilbron, *Moseley, op. cit.* 261.

[38] Moseley to Margery from RE Mess, Bulford Camp, Salisbury, 27 December [1914], Letter 119 in Heilbron, *Moseley, op. cit.* 260–261.

[39] *Ibid*.

[40] 'LZ', 'Airfield Camp, Netheravon, 1912–2012', *Army Air Corps Journal,* Regimental Headquarters Issue No 52 (Spring 2012), 40–42.

[41] Max Hastings, 'Hewitt, Sir Edgar Rainey Ludlow (1886–1973)', *Oxford Dictionary of National Biography* (Oxford: OUP, 2004); online ed. May 2011, http://www.oxforddnb.com/view/article/31380 (accessed 29 November 2017). In early 1915, Ludlow-Hewitt was posted to No 1 Squadron in France and went on to win a Military Cross in 1916.

[42] Thanks to Professor Russell Egdell for pointing out the significance of Harry's rowing weight, and for providing digitised versions of the rowing photographs, held by the Archives of Trinity College, Oxford.

[43] Moseley to Margery from RE Mess, Bulford Camp, Salisbury, 17 January [1915], Letter 121 in Heilbron, *Moseley, op. cit.* 261–263; Moseley to his mother from RE Mess, Bulford Camp, Salisbury, 25 January [1915], Letter 122 in Heilbron, *Moseley, op. cit.* 262–263.

[44] Moseley to his mother from RE Mess, Bulford Camp, Salisbury, February [1915], Letter 124 in Heilbron, *Moseley, op. cit.* 265.

[45] Moseley to his mother from RE Mess, Bulford Camp, Salisbury, February [1915], Letter 125 in Heilbron, *Moseley, op. cit.* 265–266.

[46] MHS MS Ludlow-Hewitt 2, Amabel Moseley's social diary, 28 March and 25 April 1915. The earlier entry refers to 'Camp Cowshott Common' with the later entry corrected to 'Calshott Camp'. This matches the address in Moseley to Sir Ernest [Rutherford] from 13 Signal Coy RE, 4 April [1915], Letter 127 in Heilbron, *Moseley, op. cit.* 266–267.

[47] Moseley to Sir Ernest [Rutherford] from 13 Signal Coy RE, 4 April [1915], Letter 127 in Heilbron, *Moseley, op. cit.* 266–267. Heilbron gives Harry's address incorrectly as {Cornhill} Common, 'P??kwood', Surrey.

[48] W.S. Eve, *Rutherford; Being the Life and Letters of the Rt. Hon. Lord Rutherford, O.M.* (Cambridge: Cambridge University Press, 1939), 247.

[49] Heilbron, *Moseley, op. cit.* 125, footnote 4 to letter in Ludlow-Hewitt archive, Lankester to Margery Ludlow-Hewitt (née Moseley), 13 September 1915.

[50] There are a multitude of publications about the Dardanelles campaign, some marking its centenary in 2015 and 2016. See L.A. Carlyon, *Gallipoli* (London: Bantam Books, 2003); Peter Hart, *Gallipoli* (Oxford: OUP, 2013); Alan Moorhead, *Gallipoli* (London: Aurum

Press, 2015); and Eugene Rogan, *The Fall of the Ottomans: The Great War in the Middle East, 1914–1920* (London: Allen Lane, 2015). Of special relevance is Arthur Beecroft, *Gallipoli: A Soldier's Story* (London: Robert Hale, 2015), a vivid post-war account by a Royal Engineer Signal Officer who survived the war.

51 Moseley [to his mother] from Alexandria, [c. 26 May 1915], Letter 128 in Heilbron, *Moseley, op. cit.* 267–268.

52 Moseley to his mother from HQRS Walton Bay, 15 June 1915, Letter 131 in Heilbron, *Moseley, op. cit.* 269–270.

53 MHS MS Ludlow-Hewitt 2, Amabel Moseley's social diary, 6 June 1915.

54 Moseley to his mother from Alexandria, 27 June [1915], Letter 134 reproduced in part in Heilbron, *Moseley, op. cit.* 271–272. Heilbron, *Moseley, op. cit.* 271–272; Moseley to his mother from B. Med. Exp. Force, 2 July [1915], Letter 135 in Heilbron, *Moseley, op. cit.* 272.

55 Moseley to his mother from Alexandria, 27 June [1915], Letter 134 reproduced in part in Heilbron, *Moseley, op. cit.* 271–272.

56 A. L. Spalding, *Cable Laying at Gallipoli 1915–1916* (1917). Unpublished manuscript available at http://atlantic-cable.com/CableStories/Spalding/ (accessed 29 November 2017). See also: TNA WO 106/704 Signals, Gallipoli 1915: Account by Major H.C.B. Wemyss DSO MC, Royal Signals (1920). Major Wemyss served with the Royal Engineers during World War One.

57 For further details of the 'All Red Line' and the 'Cable Wars' during World War One, see Elizabeth Bruton, 'The "Cable Wars": Military and state surveillance of the British telegraph cable network during World War One', in Andreas Marklund and Rüdiger Mogens (eds.), *Historicizing Infrastructure* (Aalborg: Aalborg University Press, 2017).

58 The Long, Long Trail: 13th (Western) Division at http://www.longlongtrail.co.uk/army/order-of-battle-of-divisions/13th-western-division/ (accessed 29 November 2017).

59 Much of this section on the Battle of Sari Bair was written with the kind assistance of D. F. Harding, via telephone and email, in November 2016. For details of his work, see D. F. Harding, *The Battle of Sari Bair, Gallipoli August 1915. A Radical Study of the Main Attack, from Primary Sources* (London: Foresight Books, forthcoming).

60 The National Archives (Kew), WO 106/704 Signals, Gallipoli 1915. Account by Major H.C.B. Weymss, DSO MC, Royal Signals (1920), 4.

61 Moseley to his mother, 4 July [1915], 8 July [1915], 8 July [1915], 12 July [1915], 14 July [1915], 26 July [1915], 3 August [1915] and 4 August [1915], Letters 137, 139, 140, 141, 142, 143, 144 and 145 in Heilbron, *Moseley, op. cit.* 272–279. Harry's mother shared these letters with his sister Margery.

⁶² *Ibid*.

⁶³ Chadwick was an officer from the East Lancashire Regiment, serving as Machine Gun Officer in 38th Brigade HQ with Harry, who was promoted temporarily to Captain on 25 May 1916. See 'Notice', The *London Gazette,* 20 June 1916, 668. He was mentioned in despatches by General Ian Hamilton alongside hundreds of other soldiers whose service in the initial Allied evacuation of Gallipoli on 11 December 1915 was highlighted. See The *London Gazette* 28 January 1916, 1197. See also The National Archives, Kew, WO 372/24/10780 Medal card of Chadwick, G E, East Lancashire Regiment [1916].

⁶⁴ Moseley to his mother, 4 July [1915] June 1915, Letter 137 in Heilbron, *Moseley, op. cit.* 272–273.

⁶⁵ *Ibid*.

⁶⁶ Moseley to his mother, 4 August [1915], Letter 145 in Heilbron, *Moseley, op. cit.* 279.

⁶⁷ The National Archives, Kew, TNA WO 95/4301 13 Division [PART III Gallipoli, Dardanelles.]: Divisional Signal Company: vol. 2: Mediterranean Expeditionary Force War Diary. Unit: Signal Co, 13th Division, from 4-8-15 to 31-8-15.

⁶⁸ See Harding, *op.cit.* for a detailed account of the battle.

⁶⁹ See https://www.awm.gov.au/collection/REL/00698.003 (accessed on 29 November 2017) for further details.

⁷⁰ The National Archives (Kew), TNA WO 95/4301 13 Division [PART III GALLIPOLI, DARDANELLES.]: Divisional Signal Company: Volume no. 2: Mediterranean Expeditionary Force War Diary. Unit: Signal Co, 13th Div from 4-8-15 to 31-8-15.

⁷¹ *Ibid*.

⁷² Moseley to his mother, 4 August [1915], Letter 145 in Heilbron, *Moseley, op. cit.* 279.

⁷³ Ludlow–Hewitt Family Archive, handwritten letter from G.E. Chadwick, HQ 38th Brigade, 13th Infantry Division, 27 September 1915.

⁷⁴ Brig-Gen. C.F. Aspinall-Oglander, *Military Operations Gallipoli* (London 1932), vol. 2, 217–221.

⁷⁵ For further details, see Harding, *op. cit*.

⁷⁶ The circumstances of Harry's death were set out by surviving fellow officers after the engagement. See Ludlow-Hewitt Family Archive: handwritten note from Captain Crocker, 13th Signal Company, 13th Infantry Division, to Harry's family, 13 August 1915; Ludlow-Hewitt Family Archive; also handwritten letter from G. E. Chadwick, HQ 38th Brigade, 13th Infantry Division, 27 September 1915; and Ludlow-Hewitt Family Archive: handwritten excerpt of a longer letter from G. E. Chadwick to his mother, Mrs Spencer

Chadwick transcribed by Mrs Spencer Chadwick, [original letter from Chadwick to his mother dated 14 August 1915] sent by request to Amabel, presumably after 27 September 1915, as Chadwick's letter of this date includes his mother's address to which Amabel could write and ask for a copy of his original letter. The third letter was cited in Heilbron, *Moseley, op. cit.*, 122, but the three letters from Chadwick were not included in Heilbron's appendix, possibly because they were not written by Harry himself. These letters are transcribed in Appendix II of this volume.

[77] Ludlow-Hewitt Family Archive, handwritten letter from G.E. Chadwick, HQ 38th Brigade, 13th Infantry Division, 27 September 1915.

[78] Ludlow-Hewitt family archive, handwritten excerpt of a longer letter from G.E. Chadwick to his mother Mrs Spencer Chadwick transcribed by Mrs Spencer Chadwick. The original letter from Chadwick to his mother was dated 14 August 1915. According to family records, Amabel transcribed the excerpted content and passed it onto an 'Aunt Tye' and this copy is also held in the Ludlow-Hewitt family archives. According to D.F. Harding, the transcribed account (which goes up only to Harry's death and hence is incomplete) is nonetheless the best primary account of Baldwin's Column during the Battle of Chunuk Bair.

[79] CWGC Cemetery Details: 'The Farm', ANZAC – http://www.cwgc.org/find-a-cemetery/cemetery/66606/THE%20FARM%20CEMETERY,%20ANZAC (accessed on 29 November 2017).

[80] MHS MS Ludlow-Hewitt 2, Amabel's social diary, 30 August 1915.

[81] MHS MS Ludlow-Hewitt 2, Amabel's social diary, 1915.

[82] The National Archives, Kew, WO339/32272 Amabel Sollas to the War Office, 15 September 1915, taken from the file of Second Lieutenant H. G. J. Moseley RE, Signalling Officer to 38 Brigade, 13th Division.

[83] The National Archives, Kew TNA WO339/32272 War Office to Amabel Sollas, 17 September 1915, taken from the file of Second Lieutenant H. G. J. Moseley RE, Signalling Officer to 38 Brigade, 13th Division.

[84] Ludlow-Hewitt Family Archive, 3 page handwritten note from Captain Crocker, 13th Signal Company, 13th Division, to Harry's family, dated 13 August 1915 and 4 page handwritten letter from G.E. Chadwick, HQ 38th Brigade, 13th Infantry Division, 16 August 1915.

[85] E. Rutherford, 'Henry Gwyn Jeffreys Moseley', *Nature* 96 (2393), (1915), 33–34.

[86] The National Archives, Kew, TNA WO 95/4301 13 Division [Part III Gallipoli, Dardanelles]: Divisional Signal Company: Volume no. 2: Mediterranean Expeditionary Force War Diary. Unit: Signal Co, 13th Div from 4-8-15 to 31-8-15.

[87] Ludlow-Hewitt Family Archive, handwritten note from Captain Crocker, 13th Signal Company, 13th Division, to Harry's family, dated 13 August 1915. For a brief description of Captain Crocker's injury on 11 August 1915, see National Archives, Kew, TNA WO 95/4301 13 Division [Part III Gallipoli, Dardanelles]: Divisional Signal Company: vol. 2: Mediterranean Expeditionary Force War Diary. Unit: Signal Co, 13th Div from 4-8-15 to 31-8-15.

[88] Ludlow-Hewitt Family Archive, handwritten letter from G.E. Chadwick, HQ 38th Brigade, 13th Infantry Division, 16 August 1915.

[89] Ludlow-Hewitt Family Archive, handwritten letter from G.E. Chadwick, HQ 38th Brigade, 13th Infantry Division, 27 September 1915.

[90] The National Archives, Kew, TNA WO 339/32272 War Office Schedule of Correspondence for 2/Lieutenant H.G.J. Moseley, Royal Engineers (Headquarters Staff, 38th Infantry Brigade). Killed in action 10 August 1915.

[91] *Ibid*.

[92] Ludlow-Hewitt Family Archive, 5 page handwritten letter from G.E. Chadwick, HQ 38th Brigade, 13th Infantry Division, 27 September 1915.

[93] National Archives, Kew, WO 339/32272 Lieutenant Henry Gwyn Jefferys Moseley. Royal Engineers. 1914-1915; 1976: Extract from Letters of Administration Register No 60977/2, 27 May 1915.

[94] Roy MacLeod and E.K. Andrews, 'The Origins of the DSIR: Reflections on Ideas and Men, 1915–16', *Public Administration*, 48 (1), (1970), 23–48.

PART TWO
LEGACY

CHAPTER SIX
HENRY MOSELEY AND THE POLITICS OF NOBEL EXCELLENCE

ROBERT MARC FRIEDMAN

Introduction

Whether Henry Moseley was destined to win a Nobel Prize is a question that has stimulated endless speculation. Moseley's tragic death at Gallipoli in August 1915, so soon after publishing ground-breaking studies that transformed both chemistry and physics, prompts a number of questions: would Moseley have been recognized with a Nobel Prize had he lived; could the Royal Swedish Academy of Sciences have awarded him the Prize during the year of his death; and might the award given to C. G. Barkla in 1918 be considered a posthumous tribute to Moseley. The details are few; the evidence is inconclusive. Moseley received one nomination for the 1915 Nobel Prize for Physics and one for Chemistry. The statutes may have opened the possibility of a posthumous Prize. Yet, a prize for Moseley was not to be.

The fate of Moseley's candidacy cannot be divorced from the scientific competence and disciplinary priorities of the Nobel Committees of the Royal Swedish Academy of Sciences. That being so, it is instructive to consider the circumstances surrounding the award in 1918 of the previously reserved Physics Prize for 1917 to C. G. Barkla for his discovery of the characteristic X-ray radiation of elements. This, in turn, leads us to consider the claim that the Physics Prize awarded in 1925 (which had been reserved from 1924) to Manne Siegbahn for his application of the tools of X-ray spectroscopy can be viewed as a recognition (and valorisation) of Moseley's work a decade earlier. More widely, these circumstances give us pause to reflect on the wisdom of overstating the face value of the Prize in the culture of contemporary science.

Moseley Nominated

On 30 January 1915, just before the annual deadline by which prize nominations could be accepted, Svante Arrhenius sent identical letters to the Nobel Committees for Chemistry and for Physics.[1] The letter was oddly formulated. Rather than simply proposing one or more candidates, Arrhenius politely requested that, in their deliberations, the committees should take into consideration H. G. J. Moseley's 'extraordinarily important' papers in the *Philosophical Magazine*, vols. 26 (1913) and 27 (1914), on 'High frequency spectra of the elements'. These papers, he wrote, provided 'a rational explanation of Mendelejew's [*sic*] periodic system'. Noting that Moseley had collaborated with C. G. Darwin, and had acknowledged a debt to W.H. Bragg, Arrhenius suggested that a division among them might also be appropriate.

Arrhenius, a founding father of modern physical chemistry, was the winner of the 1903 Prize in Chemistry. Notwithstanding, physics was his disciplinary identity of choice. He was a member of the Nobel Committee for Physics and held a professorship in physics at the Stockholm Högskola (a non-degree giving, research-oriented institution, which later became Stockholm University).[2] Around 1900, physical chemistry lacked a clear disciplinary profile, and in any case, Arrhenius included in his portfolio the field of so-called cosmical physics, a *fin-de-siècle* programme that sought to unify the macro-physical phenomena of earth, sea, atmosphere, sun, and cosmos. Through his extraordinary rich network of correspondents and frequent travel abroad, Arrhenius possessed greater knowledge of European and American science than his colleagues in either committee. This enabled him to wield considerable influence, especially on the Chemistry Committee where he frequently played a pivotal behind-the-scene role, including as an insightful schemer.[3]

When both committees received nominations for the same candidate, representatives met to decide which committee should take responsibility for an evaluation. The chemists accepted the task, no doubt as a result of Arrhenius's influence. Moreover the five-member Chemistry Committee took the unusual step of formally recruiting Arrhenius as an *ad hoc* extra member that year for the purpose of evaluating 'Moseley's investigations concerning the periodic system' in a 'special report'.

In what had become customary practice in the decade and a half since the first prizes were awarded in 1901, a few candidates were selected for detailed evaluation, while all nominees were briefly discussed in an annual 'general report', which concluded with a recommendation to the Academy. In his typed ten-page text on the recent advances made possible by X-ray spectroscopy, Arrhenius summarized the difficulties with Mendeleev's system and the recent challenges to atomic theory posed by studies in radioactivity. He then traced the development of X-ray spectroscopy and Moseley's establishment of atomic number, including this concept's implications for chemistry and for Ernest Rutherford and Niels Bohr's recent atomic models. Arrhenius concluded by noting that a number of persons contributed to this 'extremely significant' line of investigation on fundamental principles of physics and chemistry, Moseley being foremost. Almost as an afterthought he asked whether the roles of Darwin and W. H. Bragg for their preliminary physical and mathematical studies were of sufficient importance that some form of division among the three might be considered. Given that Bragg had been nominated for a prize in physics for other contributions, Arrhenius wondered, in a parenthetical comment, whether it might make sense to recognize this research with a physics prize. But clearly the report highlighted the extraordinary significance of Moseley's contributions for chemistry.[4]

In the Chemistry Committee's general report, presented at its meeting of 29 September, Moseley's work along with Darwin and Bragg's contributions received a cautious acknowledgment. Drawing upon Arrhenius's detailed evaluation, the Committee concluded that apparently if these results 'prove to be correct, [they] unquestionably promise to be of fundamental significance for theoretical chemistry'. But given the recent date of the findings, the Committee suggested waiting for yet further confirmation and, similarly, a postponement was desirable to seek greater clarity as to which of the various contributors deserved most credit.[5]

The general report of the Physics Committee touched briefly on Moseley's 'important work concerning the spectrum of X-rays' which brought 'a rational basis' to the so-called Mendeleev scheme of chemical elements. Referring to the fact that Moseley's nominator (Arrhenius) considered this work to be more suitable for a prize in chemistry, the committee report provided no further detail: Moseley was not considered a candidate for a physics prize.[6]

These reports were written during the summer of 1915, but by the time the committees came to vote, in late September, news of Moseley's death in August had been widely reported. According to the Nobel statutes at this time, 1915 would be the only year in which Moseley could be awarded, a consideration that, in late October, was put to the ten-member Physics and Chemistry Sections of the Academy, who acted on committee proposals (members of the Nobel committees at this time were often also members of the 'sections'). There was also an opportunity to consider Moseley's candidature at the mid-November meeting of the full Academy, at which the final vote was taken. Occasionally, the sections and the Academy overturned committee proposals. But there is no evidence in the archives of any last minute effort to award Moseley a prize.

To understand why and how Moseley's contributions were passed over, it will be helpful to step back from the specifics of his candidacy, and to seek insight from the leading features of the early Nobel institution.

A Swedish Prerogative

The Nobel Prizes may well be international in scope, but from the start the Royal Swedish Academy of Sciences determined the science awards, based largely on the evaluations and recommendations of its respective five-member committees. The record shows that winning a prize has never been an automatic process of reward for having attained an extraordinary level of achievement. The Swedish committees' own scientific understanding was critical to the outcome.[7] Some committee members were dispassionate, others championed their own agendas promoting a wide range of disciplinary, political, and personal interests.

No juggling of statistics related to nominations – number, frequency, or origin – can adequately explain the awards. How and why committee members responded to particular nominations cannot be generalized. During the first half century of deliberations, committees hardly ever selected those candidates who achieved a rare consensual or even majority status from the nominators. Some winners received only one nomination.

Finally, and of particular relevance to the case of Moseley, some committee members occasionally submitted nominations just prior to the 1 February deadline, when it had become clear that names of particular interest were missing. These were typically researchers whom the committee desired to evaluate, but who had not yet been nominated. Such letters were formalities; they need not necessarily mean that the author intended to endorse the candidate for a prize.

Indeed, on the same day that Arrhenius sent his letter to bring Moseley's accomplishments to the attention of the committees, he also sent a letter to the Nobel Physics Committee proposing that a prize that year be awarded to astrophysicist Henri Deslandres, for his contributions to solar physics.[8] This makes sense only when understood in relation to the fact that in 1914 the Physics Committee had declared the three most worthy candidates to be Max von Laue, Max Planck, and George Ellery Hale. But one committee member insisted that Hale's contributions to solar physics could not be considered without also taking into account the work of Deslandres, who had not been nominated that year.[9] When, for 1915, Hale alone again received a nomination, Arrhenius sent in the proposal for Deslandres, which would enable the committee to examine their closely related contributions together.

But, his parallel proposal for Moseley was not related to prior committee actions; on the contrary it represents an initiative from Arrhenius himself.

To make sense of Arrhenius's nomination, and its reception, we must look to the early history of the Prize, and the disruptive effect of World War I.

Committee Concerns

During the early decades of awarding the Nobel Prize, Swedish physicists with a strong experimentalist orientation associated with Uppsala University dominated the proceedings. For these physicists, precision measurement and experiment were paramount, and mathematically sophisticated, theory-driven work was conspicuously less valued. Intellectually, institutionally, and often politically at odds with the majority, Arrhenius accepted the need for compromise. At times, however, he drew upon his popularity in the Academy to advance candidates of his own against majority wishes.[10]

The award in chemistry, to a greater extent than in physics, did not reflect any idealized consensus on disciplinary priorities and standards of significance. By 1900, chemistry had fragmented into relatively autonomous sub-disciplines, with a multitude of national orientations, hierarchies, priorities, and methods. Few candidates stood head and shoulders above others. Not surprisingly, nominators seldom agreed on candidates or standards. In spite of efforts to avoid disunity and controversy, the committee frequently sank into gridlock.

In this context, Arrhenius was both an asset and an obstacle. In the Academy, he could successfully challenge committees as in 1906 when he blocked a prize for Mendeleev. When the Chemistry Committee was at a loss, Arrhenius found opportunities. In this way, he was responsible for recognition of Ernest Rutherford, Wilhelm Ostwald, Marie Curie, Alfred Werner, and T.W. Richards. But for over a decade he blocked Walther Nernst, the one candidate who for several years commanded a solid mandate.[11]

Arrhenius's influence might have brought the 1915 Chemistry Prize to Moseley. But once war began, considerations outside science came to feature in the calculations of both the committees and the Academy.

Biased Neutrality

The Nobel establishment was poorly prepared to meet the challenges brought about by the war. The subsequent breakdown of international relations in science and culture threatened the entire basis for awarding prizes. Neutrality proved ambiguous. Impartiality revealed itself to be highly political.[12]

The Nobel establishment received ample warning, but failed to prepare.

The Nobel statutes stipulate that prizes can be withheld for only one reason: when no candidate is found worthy. They offer no provision for suspending the awards in case of war. When the Nobel committees came to vote at their September meetings, the newspaper *Aftonbladet* raised a question that nobody had earlier dared to ask: should the Prizes be awarded during wartime?[13] As hundreds of thousands marched to war, the committees were reduced to debating, once again, whether the annual Nobel Day should be moved from dark and dismal 10 December to a date in mid-June, when it might also be designated Sweden's national day. Clearly, the Nobel committees preferred the tradition of business as usual.

There were, however, larger questions at issue. When war came, the first casualty was impartiality.[14] In Stockholm, the moderate-conservative government of Hjalmar Hammarskjöld sought to keep Sweden formally out of the war by declaring official neutrality. This decision angered many upper-class conservatives, including members of the Academy, who had hoped Sweden would give direct support to Germany. For the Academy, the Prize threatened to become a victim of its own success.

The achievement of the early prizes in bringing Sweden international prestige created a dilemma. Would announcing winners create problems for the prize institutions and more generally for Sweden? Impartiality had been touted as Sweden's claim to moral superiority; awarding the Nobel Prizes allegedly embodied this national racial trait. Would the announcement of prize-winners be interpreted at home and abroad as a Swedish endorsement for one or another of the warring alliances? Even the best intentions can easily be misinterpreted.

Olof Hammarsten, chair of the Chemistry Committee and a man of principle, feared that the Academy's decisions could be taken as expressions of sympathy, and 'perhaps even misconstrued as a departure from the impartiality, we are obligated to observe'.[15] He knew that the Chemistry Committee had already agreed that two candidates, the American T. W. Richards and the German Richard Willstätter, had both been declared worthy of recognition. Perhaps the only way out of the dilemma lay in the government permitting a one year suspension, by which time, in any case, the war might be over.

Hammersten saw no problem in waiting until 1915, especially if this would resolve impending conflict among Academy chemists over which of the two should be first awarded. Following Arrhenius's advice, the Chemistry Committee proposed Richards for the 1914 Prize, as he had been waiting longest. At the same time, they agreed that Willstätter was in line for the next

Prize. A postponement of the 1914 Prize, such as Hammarsten and others proposed, could have the happy consequence of crowning both a fellow neutral and a *Kulturträger*.[16]

Whilst such considerations preoccupied the Chemistry Committee, the Physics Committee was trying to surmount some of the challenges facing classical physics. By 1914, the Academy's refusal to acknowledge Max Planck and quantum theory had become embarrassing. The physicists had always found ways to ignore radical new ideas that allegedly were 'in conflict with physical reality'.[17] Fortuitously, the announcement in 1912 that Max von Laue, Paul Knipping, and Walter Friedrich had obtained a diffraction pattern when sending X-rays through a crystal seemed to provide an answer to controversial questions about the nature of Röntgen's phenomenon. Their experiment also yielded a new tool for studying the molecular structure of crystals. Meanwhile, the work of father and son W. H. and W. L. Bragg soon underscored the importance of von Laue's discovery, while opening new methods for investigating the structure of matter.

When the Physics Committee met, Max Planck received twelve nominations, but von Laue received only two – from the ageing organic chemist, Adolf von Baeyer, and Emil Warburg, for a divided prize to be shared with W. H. Bragg. Their collaborators were ignored. Coming from Germany, these two recommendations held appeal for the Committee, and von Laue was immediately elevated to the list. In a detailed evaluation, dated 3 July, Allvar Gullstrand of Uppsala concluded that W. H. Bragg could not be awarded without also recognizing W. L. Bragg; and for this reason, recommended that von Laue alone should receive the Prize. Gullstrand overlooked Friedrich and Knipping, who took von Laue's idea and designed, constructed, and conducted the difficult experiment, but who were omitted from consideration.

At the September meetings, both committees considered the question of postponement. The physicists, like the chemists, decided that this would not be too painful: two prizes could then be awarded to leading researchers in the emerging field of X-ray crystallography, who could be fêted together in a grand display of even-handedness: one to Germany and one to an Entente nation.[19]

The two committees sent their reports to the Academy, along with the proposals for Richards and von Laue, and for the next few weeks, all was quiet: perhaps there was to be no further debate about giving prizes during wartime. Science could remain transnational, and above political conflict. But this prospect was not to last. When, in late October, the Academy's Physics and Chemistry Sections prepared to vote, the situation had deteriorated. Europe seemed intent on division.

Nobody foresaw such a rapid collapse of the international spirit in science that Nobel had presupposed. At first, Allied scientific circles tended to separate German culture from the Prussian-dominated military-state that was held responsible for the atrocities in Belgium, and the bombing of civilians. This view survived until October 1914, when ninety-three leading German academics, intellectuals, and artists issued a 'Proclamation to the Civilized World', in which they denied German wrongdoing. Moreover, they asserted, *Kultur* was an organic element of German society. To attack German militarism was to attack German *Kultur*: the two nourished each other.[20]

The violent reception the Manifesto produced across the world, and especially in the United States, dispelled the illusion that the world of learning was internationalist.[21] For Swedish academicians, the escalating hostile public response raised the stakes. On 20 October, representatives of the three Swedish prize-awarding institutions agreed to petition the Swedish government for a postponement of the Nobel Prizes for one year, and to delay the award ceremony for the 1914 and 1915 laureates to 1 June 1916. Many in the Academy were against postponement.[22] But in early November, the government agreed to postpone the next award until 1915.

The government's decision sparked another round of protest. Virtually all the major Swedish newspapers regretted the decision to postpone, but the political right was most vociferous. Calls to pause for reflection, and to avoid the appearance of being pro-German, were branded as 'politically motivated'.[23] The pro-German Conservative media favoured awarding the prizes: the influential *Svenska Dagbladet* claimed it would be better to risk criticism than accept '*smälek* [disgrace]'. If prizes were awarded according to Nobel's spirit and statutes, Sweden would be morally invincible. But if prizes were not awarded to deserving candidates, Swedish institutions would be criticised for capitulating to politics.

In March 1915, as the war on the Western Front ground to a stalemate, *Svenska Dagbladet* published a survey of foreign scientists, humanists, and artists who were asked their views on the future of cultural cooperation.[24] This contained much of interest to the Nobel establishment, perhaps not least because the newspaper's pro-German editor, Fredrik Böök, slanted the report in favour of Germany, and exaggerated the importance of Swedes in the restoration of international relations.[25] Even so, several British and French respondents warned neutral nations that conflict between Germany and France went deeper than political and national hatred. 'Prussian predatory savageness' had violated civilization. A resumption of international cooperation could come about only by excluding Germans and their allies. If neutrals attempted to re-establish international relations without appreciating these feelings, their efforts would be doomed.

No member of the Swedish establishment could read such views without alarm. Given his close ties to scientific communities on both sides, Arrhenius was especially despondent and took upon himself the task of convincing his pro-German colleagues of the need to respond pragmatically rather than emotionally. As a Germanic people, Swedes must endeavour to defend Germans from being cast as barbarians. By taking a leading role in re-establishing relations in learning, Sweden would inevitably work to include German participation.[26] In the meantime, the machinery of processing and evaluating carried on, and during the spring and summer of 1915, the Committees continued their work. The outcome seemed all but inevitable – for chemistry, prizes would go to Richards and Willstätter. For physics, the winners would be Max von Laue and a division between W. H. and W. L. Bragg. Evaluators limited their comments to how prizes would be distributed, if they were to be awarded at all.

In spite of his nomination of Moseley, Arrhenius became persuaded that making awards would be a mistake while emotions were so frayed; surely it was best for everybody to wait for the end of the war. At an Academy meeting in October 1915, he proposed that the Prizes again be withheld, and argued that by postponing now, there might be a chance for Stockholm to host a multi-national Nobel celebration once the war ended. A truly untarnished Sweden could then begin the work of bringing the hostile factions together again. Those few members in attendance gave Arrhenius a slight majority (21–19), and instructed the Academy to petition the Swedish government for permission to postpone yet again.[27]

Many in the Academy's leadership refused to accept postponement. Hammarsköld, the prime minister, who was also a member of the Academy, sided with them, and at the November meeting, declared there was no political reason not to award. As the committees had proposed, so the Academy voted – for Von Laue and the Braggs; Richards and Willstätter. For some, this was a vote of solidarity with Germany; others proclaimed it a wise balance: two prizes to Germany, one to a neutral, and one to the Entente.[28]

And what of Moseley? In a letter written a few weeks later, Sir William Ramsay wrote to Arrhenius, noting, 'I should like to say that young Moseley should have been the recipient [of the Physics Prize], had he not lost his life as a soldier. That is one of the most severe blows that science has suffered. I knew him fairly well: he was a capital fellow in every way. However, it is too late'.[29] Actually it had not been too late, as the Nobel statutes allowed for awarding a candidate who had been nominated before his death, but only for that same year.

The Physics Committee had met this situation before. In 1910, Committee member Knut Ångström was nominated by Wilhelm Röntgen, and his Uppsala colleagues jumped at the opportunity to see a Prize awarded to 'one of us'. The committee was set to back Ångström when he suddenly died. There was a quandary. Would the full Academy go along with its recommendation? Or would it reject the Committee's view and select another candidate? Informal enquiries revealed an unwillingness on the part of the Academy to make posthumous awards, even when allowed by the statutes. The Committee scrambled to find another candidate, and in the end, reversed its earlier rejection of J. D. van der Waals.[30]

The experience of the Ångström case may well have featured in the reluctance of those within the committees and sections of the Academy in relation to Moseley. So, too, there was the prospect of taking a decision that amounted to an eulogy for a brilliant British scholar who had fallen heroically in combat against an ally of Germany.

Under the circumstances, Moseley's one chance to receive a Nobel Prize was thus derailed. Before his death, the Chemistry Committee had assumed that he could safely be kept in a holding pattern while the reserved 1914 and 1915 Prizes were awarded to Richards and Willstätter. The comment that, when 'fully confirmed', Moseley's contribution would be of fundamental importance to 'theoretical chemistry', could be seen as another reason for delay as the Chemistry Committee only reluctantly gave priority to theory.

A Proxy Prize for Moseley?

With the coming of Spring 1918, following the Russian Revolution, and despite the German offensive and continued fighting on the Western Front, the political climate of science was changing. International cultural and scientific cooperation had to be re-established, and among the Allies and the United States, plans for a post-war future were being aired. The British and Americans met few French or Belgian voices for reconciliation. By the Armistice in November, it seemed inevitable that the Central Powers would be excluded from international science for an indefinite period, and neutral nations would be permitted to join only if they accepted the Allied boycott.

In these circumstances, Sweden's position could easily become a point of contention. Arrhenius saw German defeat at the Marne as evidence that Germany could never win, and foresaw the possibility of a negotiated peace with the Allies by the Spring of 1919.[31] To participate in the post-war reconstruction of European science and to assist their German colleagues, Sweden and the Nobel establishment must act quickly. He called for increased communication with scientists in Britain, France, and Western Europe; only by being seen as truly neutral could Swedes fulfil this mission.

In the physics nominations for 1918, Max Planck received solid backing from the German and Austrian nominators, as did Albert Einstein. But quantum theory and relativity were unacceptable to a majority of the Physics Committee. Arrhenius and Vilhelm Carlheim-Gyllensköld – the one member who was both anti-German and anti-experimentalist – had to steer support away from ultra-nationalist Johannes Stark, who had been declared as first in line for an award. Stark received but one nomination in 1918 – from his fellow nationalist and laureate (1905) Philipp Lenard. Although one nomination might seem a slender mandate, it was sufficient for C. G. Barkla, the candidate who ultimately received a prize in 1918.[32]

The one nomination for Charles Glover Barkla came from Ernest Rutherford, whom Arrhenius knew as Britain's leading physicist. Rutherford had not engaged in anti-German propaganda, as had some of his colleagues. On the contrary, his letters during the war to Arrhenius revealed an open-minded approach towards international relations. Therefore, his nomination of Barkla – 'for his discovery of the characteristic Röntgen radiation of the elements' – attracted Arrhenius's attention. Barkla had not previously been nominated, and was a problematic candidate. He was actually burnt out as a researcher, and had conspicuously failed to participate in the exploitation of his discovery, while also dogmatically rejecting well-received theory, and promulgating unfounded speculations as fact.[33]

Just what was Rutherford thinking when he nominated Barkla? His letter of nomination is succinct, offering no explanation or justification. Was Barkla to be a stand-in for Moseley? Given the well-known experimentalist orientation of the Physics Committee, did Rutherford consider Barkla the man most likely to bring a prize and prestige to British physics? Was Rutherford keen to remind the world of Moseley's pivotal experiments? As likely as all these factors may have been, Rutherford did not leave a record of his intentions.

Rutherford could rely upon being noticed by Arrhenius, even if his candidate was less than stellar. For Arrhenius, what mattered was that Barkla was British, and was Rutherford's candidate. Carlheim-Gyllensköld took responsibility for producing the special report on Barkla. But to assume that they and others with a vote simply wanted a surrogate Moseley misses much of the background for what transpired in the fall of 1918.

Maybe the report was a mere formality to a foregone conclusion. Carlheim-Gyllensköld's account of Barkla's work was superficial and incomplete, based on the standards of the time; nothing is mentioned of his work and views after 1911.[34] Much of the report focused on the work of others, and in particular on Moseley's, and also, in briefer format, Arnold Sommerfeld's contributions to atomic theory, which although deemed as yet inconclusive, nevertheless indicated the value of X-ray spectroscopy for gaining insight into the inner parts of the atom. As for Moseley's work, Carlheim-Gyllensköld confessed that what he wrote in the report was not comprehensive or precise, yet it nevertheless could show that this work is of such a groundbreaking (*banebrytande*) nature that it might be permissible to assume that Mosely would have received a Nobel Prize had he not fallen in the war. Still it was important for Carlheim-Gyllensköld to underscore that Barkla's discovery was the result of the style of science preferred by the Academy's physicists: 'with simple experimental means and without cumbersome theoretical resources [he] managed to find a new and unpredicted phenomenon'.[35]

When, on 4 September 1918, the Physics Committee first met to discuss the year's evaluations, a tacit agreement was in place. Arrhenius had persuaded his colleagues that Barkla, and only Barkla, should receive a physics prize that year – the previously reserved 1917 Prize. Although a majority of the five members had several times declared Stark worthy of a prize, Arrhenius convinced his colleagues to keep him waiting. Awarding Stark at this point would only confirm to an anti-German world that Swedish neutrality could not be trusted. As long as there was a possibility that neutrals could work with the Allies to broker Central Power representation in post-war international scientific organisations, it was best to avoid provocation. Whether or not a willingness to give Barkla a prize marked a real desire to recognize the work of Moseley, the evidence seems to point to the importance of appeasing the Allies.

In the event, Arrhenius's vision was soon undermined. In 1919, the Academy's desire to support German science proved overwhelming. The two available awards in physics went to Planck and Stark, although both Carlheim-Gyllensköld and Arrhenius tried to substitute a non-German for

Stark. The Chemistry Committee was more outspoken; even before the war ended, it voted to award a prize in 1918 to Fritz Haber, which was rebuffed in the Academy. In 1919 the chemists again insisted on Haber and succeeded. German scientists were heartened. Anger and disbelief from abroad and locally came quickly. German scientists were elated. Many knew they could count on their racial brothers. Stark had already arranged an honorary doctorate for Arrhenius and noted that he was a true son of the Swedish '*Volk*', who through ancestry, religion, and cultural and political history had embraced the German '*Volk*' to their hearts.[36]

The first post-war Nobel ceremony was held in June 1920. The severe anti-German measures of the Versailles Treaty, and the exclusion of German science from the newly-created, pro-Allied International Research Council, fostered hostility among Swedish conservatives, both towards the League of Nations and towards international organizations for scientific cooperation.[37]

If Barkla was Rutherford's choice as a way of bringing honour to Moseley, he may well have experienced regret with the spectacle in Stockholm. Richards and the Braggs stayed at home. Barkla was the only non-German present. Although Barkla eagerly shook hands with the German laureates, this was far from the grand reconciliation that Arrhenius and others had envisioned. Barkla delivered a Nobel lecture that revealed little understanding of recent physics, and declared in his banquet speech that the Academy should itself receive a Nobel Prize for Peace.[38] But the boycott was in place, the Swedes were warned that they might soon be excluded, and the scorn over a clear pro-German bias, expressed through private letters and newspapers, could not easily be ignored.

At this first Nobel ceremony specially arranged to be held in June, it seemed appropriate that the weather was cheerlessly wet and the wind cold. The ceremony returned to December. The new chairman of the Nobel Foundation and politically-liberal leader of the Nobel Committee for Literature, Uppsala professor Henrick Schück, confessed in his opening speech that Alfred Nobel's dream was no more. Swedes' earlier pride over being entrusted with Nobel's donation and vision had been transformed in many quarters into dejection and shame. How might Nobel's legacy be brought back to life? He had no firm solution. In the new era just then emerging belief that the peaceful competition among civilized peoples leads to peace and social harmony – the basis of Nobel's vision – might have lost its hypnotic power, but much of the rhetoric from the pre-war age of illusion lived on. Through this lost dream and its language the prize was able to provide robes of idealism about humankind and the abundance of science more generally to cloak what was already emerging as the naked quest for furthering narrow professional and institutional interests, no less personal careerist gain.[39]

Appropriating Moseley

Neither in the speech presenting Barkla by Nobel Committee member Gustaf Granqvist, nor in Barkla's own Nobel Lecture was the name of Moseley ever mentioned. Although it is probably fair to say that had Moseley not fallen during the war, he would either have shared a prize, or stood alone on the podium, direct evidence is lacking. The one piece of evidence is ambiguous: the speech presenting Manne Siegbahn upon his receiving a Nobel Prize in 1925 (the previously reserved prize for 1924).

The work begun by Moseley and others on X-ray spectroscopic instrumentation and measurements that largely came to a stop during the war, continued to be developed in Lund in Sweden by Siegbahn together with his technical and scientific assistants. The ever improved techniques developed in his laboratory provided increasingly more precise measurements that enabled theorists to elaborate quantum atomic models. Siegbahn accepted the chair of physics in Uppsala in 1922 and a place on the Nobel committee. His acute need for additional funds and a rebuff from the Rockefeller Foundation for financial support, encouraged his Uppsala committee colleagues, Gullstrand and C. W. Oseen to advance his candidacy for a Nobel Prize in 1925. Already wary of the overly authoritative control over Swedish physics by 'the small popes in Uppsala, especially Siegbahn',[40] Arrhenius along with Carlheim-Gyllensköld opposed such an award. They pointed out that in spite of the great importance of the measurements coming from his laboratory, Siegbahn did not meet the statutory requirement of a discovery or invention. He perfected that which others had developed, drawing upon the instrumentation of Moseley, the Braggs, and Maurice de Broglie. None of the few nominators specified a discovery or invention. And although Oseen and Gullstrand recruited an outside X-ray radiologist as external expert who indicated that Siegbahn found evidence for a new series of spectral lines, the M lines, that can be construed as a discovery, this did not alleviate the challenge: none of the nominators had mentioned this achievement, nor was Siegbahn an author on the most important paper related to this discovery.[41]

Committee practice at the time was to exclude candidates for whom no specific discovery was mentioned by nominators. Oseen, who applied the statutes rigorously to exclude candidates, made a rare plea for tolerance and for a liberal interpretation in the case of Siegbahn. The Committee (minus Siegbahn) remained locked two against two, but Gullstrand, as chairman, could count his own vote twice. In the physics section Siegbahn prevailed six to three; the Academy backed this majority and voted to award the reserved 1924 prize to Siegbahn.[42] But disgruntlement toward committees was growing; some recognized a case of cronyism and disregard of established traditions for interpreting the statutes. In the following year the Academy rebelled, overturning both the Physics and Chemistry Committees' proposals and making awards to candidates who had been rejected (Jean Perrin, Richard Zsigmondy, and Theodor Svedberg).

No doubt aware that Siegbahn's prize was neither universally applauded nor strictly in conformity with the statutes, Gullstrand, as chairman of the Committee, designed his Nobel speech accordingly. In an unusually long presentation, Gullstrand depicted Siegbahn as part of a line of brilliant scientists who had developed X-ray spectroscopy into a highly valued resource for physics. Weaving into the narrative as many recent Nobel Prize winners as possible, and culminating with Siegbahn, Gullstrand was also effusive with comparable praise for Moseley.

> The second path through the newly discovered region of X-ray spectroscopy, namely the investigation of X-radiation in the different elements, was trodden with the greatest success by the young scientist Moseley, who was also an Englishman.... Moseley fell at the Dardanelles before he could be awarded the prize, but his researches had directed attention to the merits of Barkla, who consequently in 1918 was proposed for the Nobel Prize, which was awarded to him without delay.[43]

Such presentation lectures that offer capsulized histories of science for celebratory occasions rarely satisfy scholarly historical standards; simplification, omission, distortion, and playing to the audience are not uncommon. Gullstrand was a highly gifted academic diplomat; his skill in forging consensus at contentious Academic Senate meetings in Uppsala was legendary. In the early 1920s he assumed the delicate and trying task of negotiating for Sweden with the International Research Council in efforts to bring Germany into the organisation.[44]

Siegbahn was portrayed as a direct scientific descendent of Moseley. 1925 being the tenth anniversary of Moseley's death, Gullstrand implied that the prize to Siegbahn could in part be understood as well as a tribute to young

Moseley. Although Gullstrand invoked the names of numerous Nobel Prize winners in his praise of Siegbahn's achievements, he failed to mention those atomic physicists after Niels Bohr, who were making use of Siegbahn's precision measurements. Glaringly absent is Arnold Sommerfeld, who again had been passed over and who, along with experimental physicist Friedrich Paschen, would have, according to many, been worthy of recognition that year. In his own Nobel lecture, Siegbahn never once mentioned Moseley.[45]

No direct evidence has been found with which to assess what Gullstrand intended with his speech. Still the context suggests what could have motivated him to engage in historical exaggeration and to embellish the significance of Siegbahn's contributions. Siegbahn had faced strong opposition in the Nobel Committee. Although he received a prize in the end, among those in the know, his award was controversial. Other, possibly political considerations may well have been involved. Siegbahn was seeking major funding for his research in Uppsala, and this was an opportunity to win international support. The pro-German Gullstrand, as a leading representative of Swedish science, was keen to curry favour from British scientists who were not inclined to endorse the Allied boycott. We should be wary of any superficial reading innocent of the politics of the moment.

Sober reflection reveals that Henry Moseley and his contributions put to shame any form of Nobel infatuation that endows a special status to the winners as a unique population of researchers towering above all others. Today Henry Moseley stands as one of the great scientists of the past century; we do not need the award of a Nobel Prize to command our appreciation and respect.

Notes and References

[1] Center for History of Science, Royal Swedish Academy of Sciences (*KVA*), Stockholm, *KVA* protocol on Nobel matters for 1915. Letters of nomination for 1915 (hereafter *KVA/N* + year).

[2] Center for History of Science, Svante Arrhenius papers, Manuscripts, *Levnadsrön*; Elisabeth Crawford, *Arrhenius: From Ionic Theory to Greenhouse Effect* (Canton, MA: Science History, 1996), 1–108.

[3] Elisabeth Crawford and Robert Marc Friedman, 'The Prizes in Physics and Chemistry in the Context of Swedish Science', in Carl-Gustaf Bernhard, Elisabeth Crawford and

Per Sörbom (eds.), *Science, Technology and Society in the Time of Alfred Nobel* (Oxford: Pergamon for the Nobel Foundation, 1982), 311–331 (hereafter Crawford and Friedman, 'The Prizes'); Elisabeth Crawford, *The Beginnings of the Nobel Institution: The Science Prizes, 1901–1915* (Cambridge: Cambridge University Press, 1984), chap. 5–6 (hereafter Crawford, *Beginnings*); Robert Marc Friedman, *The Politics of Excellence: Behind the Nobel Prize in Science* (New York: Freeman & Times Books, Henry Holt & Co., 2001), chap. 2 (hereafter Friedman, *Politics of Excellence*).

[4] Minutes of meetings of the Nobel Committee for Chemistry of the Royal Swedish Academy of Sciences (hereafter, *Protokoll*, NKK and for the Physics Committee, NKF), 29 Sept 1915, Appendix 4, 'Öfversikt av H.G.J. Moseleys undersökningar angående det periodiska systemet'.

[5] *Protokoll*, NKK, 29 September 1915.

[6] *Protokoll*, NKF, 29 September 1915.

[7] Robert Marc Friedman, 'Nobel Physics Prize in Perspective', *Nature*, 292 (5826), (1981), 793–798 (hereafter Friedman, 'Nobel Physics'); Crawford, *Beginnings, op. cit.*; Friedman, *Politics of Excellence, op. cit.*

[8] *KVA/N*, 1915.

[9] NKF, 15 September 1914.

[10] Friedman, 'Nobel Physics' *op. cit.*; see also Crawford, *Beginnings, op. cit.* and Friedman, *Politics of Excellence, op. cit.*

[11] Elizabeth Crawford, 'Arrhenius, the Atomic Hypothesis, and the 1908 Nobel Prizes in Physics and Chemistry', *Isis*, 75 (3), (1984), 503–522; Crawford and Friedman, 'The Prizes', *op. cit.* 317–319; Center for History of Science, Svante Arrhenius Papers and Uppsala University Library, Oskar Widman Papers, correspondence between Arrhenius and Widman; Friedman, *Politics of Excellence, op. cit.* 36–39, 180–184; Diane Barkan, 'Simply a Matter of Chemistry? The Nobel Prize for 1920', *Perspectives on Science*, 2 (4), (1994), 357–395.

[12] Friedman, *Politics of Excellence, op. cit.* Chapter 5.

[13] *Aftonbladet*, 7 September 1914.

[14] Friedman, *Politics of Excellence, op. cit.* Chapter 5.

[15] *KVA*-Center, Christopher Aurivillius Papers, Olof Hammarsten to Aurivillius, 7 September 1914; Gustaf Retzius Papers, Hammarsten to Retzius, 8 September 1914.

[16] Friedman, *Politics of Excellence, op. cit.* 74.

[17] *KVA/N*, 1914.

[18] *KVA/N*, 1914. Letters of nomination for Planck came from, among others Wilhelm

Wien, H. A. Lorentz, Pieter Zeeman, J. D. van der Waals, Kamerling Onnes, and Vito Volterra.

[19] *NKF*, 15 September 1914, KU and Appendix 1. See also Paul Forman, 'The Discovery of the Diffraction of X-rays by Crystals: A Critique of the Myths', *Archive for the History of the Exact Sciences*, 6 (1), (1969), 38–71.

[20] Danish and Swedish translations of the manifesto *'Til hele Kulturverden!'* are found in the papers of many Academy members. Friedman, *Politics of Excellence*, 75–78; Lawrence Badash, 'British and American Views of the German Menace in World War I', *Notes and Records of the Royal Society of London*, 34 (1), (1979), 91–121; John Heilbron, *The Dilemmas of an Upright Man: Max Planck as Spokesman for German Science* (Berkeley: University of California Press, 1986), 69–72. See also Roy MacLeod, 'The Mobilisation of Minds and the Crisis in International Science: The *Krieg der Geister* and the Manifesto of the 93', *Journal of War and Culture Studies*, 10 (3), (2017), 1–21.

[21] For a recent assessment, see Roy MacLeod, 'The Great War and Modern Science: Lessons and Legacies', in Sarah Posman, Cedric van Dijck and Marysa Demoor (eds.), *The Intellectual Response to the First World War* (Brighton: Sussex Academic Press, 2017), 270–282.

[22] *Protokoll, KVA/N*, 28 Oktober 1914; *Dagens nyheter*, 26 October 1914, 'Nobelprisen'.

[23] *Svenska dagbladet*, 1 and 4 November 1914.

[24] The responses were published as *Världenskulturen och kriget: Huru återknyta de internationella förbindelserna?* (Stockholm: Albert Bonniers förlag, 1915); Friedman, *Politics of Excellence, op. cit.* 83–86.

[25] Rebecka Lettevall, Gerd Somsen, and Sven Widmalm, 'Introduction' in Lettevall, Somsen, & Widmalm (eds.), *Neutrality in Twentieth-Century Europe: Intersections of Science, Culture, and Politics after the First World War* (Routledge: New York & London, 2012), 3–4.

[26] Friedman, *Politics of Excellence, op. cit.* 87–91.

[27] *Protokoll, KVA/N*, 30 October 1915; National Archives (Helsinki), Edvard Hjelt Papers, Arrhenius to Hjelt, 25 October 1915.

[28] Royal Library (Stockholm), Gösta Mittag-Leffler Papers, diary entry, 11 November 1915.

[29] Arrhenius Papers, William Ramsay to Arrhenius, 23 December 1915.

[30] Friedman, *Politics of Excellence, op. cit.* 51–52.

[31] Library of Congress, Jacob Loeb Papers, Arrhenius to Loeb, 4 August 1918.

[32] *KVA/N*, 1918.

33 Paul Forman, 'Charles Glover Barkla', *Dictionary of Scientific Biography* (New York: Charles Scribner's Sons, 1970), vol. 1, 456–459.

34 *Protokoll, NKF*, 4 September1918, Appendix B.

35 *Ibid*.

36 Arrhenius papers, Johannes Stark to Arrhenius, 8 November 1919.

37 For further details, see Sven Widmalm, '"*A superior type of universal civilization*".Science as Politics in Sweden, 1917–1926', in Lettevall, Somsen, and Widmalm (eds.), *Neutrality in Twentieth-Century Europe*, 65–89. See also Roy MacLeod, 'The Great War and Modern Science: Lessons and Legacies', *Sartoniana*, 28 (2015), 13–32.

38 http://www.nobelprize.org/nobel_prizes/physics/laureates/1917/barkla-speech.html; http://www.nobelprize.org/nobel_prizes/physics/laureates/1917/barkla-lecture.pdf (accessed 29 November 2017).

39 Friedman, *Politics of Excellence, op. cit.* 113–115.

40 National Library (Oslo), Vilhelm Bjerknes papers, Arrhenius to Bjerknes, 22 February 1925.

41 I thank Russell Egdell for pointing out Siegbahn's absence on the most detailed paper from Lund dealing with M-type X-rays. See W. Stenström, 'Experimentelle Untersuchungen der Röntgenspektra. M-Reihe', *Annalen der Physik*, 362 (21), (1918), 347–376.

42 Friedman, *Politics of Excellence, op. cit.* 141–143, 154–158.

43 *Nobel Lectures, Physics 1922–1941* (Amsterdam: Elsevier Publishing Company, 1965).

44 Widmalm, *Superior type of universal civilization, op. cit.* 74–82.

45 http://www.nobelprize.org/nobel_prizes/physics/laureates/1924/siegbahn-lecture.pdf (accessed 3 December 2017).

CHAPTER SEVEN
MOSELEY AND THE MATTEUCCI MEDAL

*RUSSELL G EGDELL, FRANCESCO OFFI AND
GIANCARLO PANACCIONE*

Introduction

It must always remain a matter of speculation as to whether Henry Moseley would have won a Nobel Prize had he lived to be considered by the Royal Swedish Academy of Sciences in 1916 or beyond. However, Moseley was certainly awarded another major international physics prize, the Italian *Medaglia Matteucci* (Matteucci Medal). This *Medaglia* was first awarded in 1867, and so predates the Nobel Prize as an international prize in physics by over thirty years. The Medal was in the gift of the *Società Italiana delle Scienze detta dei XL* (Italian Society of Sciences, known as the Society of the Forty). Moseley won the *Medaglia* in 1919, four years after his death: the next recipient, in 1921, was Albert Einstein, for his work on relativity. Moseley remains the only person to have received the Matteucci Medal posthumously in the 150-year history of the award.

The Academy of the Forty

What eventually became the current *L'Accademia Nazionale delle Scienze detta dei XL* (The National Academy of Science, known as the Academy of the Forty) was founded in 1782 as the *Società Italiana* (Italian Society), following a proposal in 1766 for the establishment of an Italian Academy of Science by Antonio Maria Lorgna (1735–1796). Lorgna was a distinguished mathematician, hydraulic engineer and expert in ballistics.[1] His proposals were supported by Alessandro Volta (1745–1827), Lazzaro Spallanzani (1729–1799) and Ruggero Boscovich (1711–1787).[1] Part of Lorgna's vision was to promote a sense of unity amongst scientists from the different Italian states. The *Società* published a volume of *Memorie* with scientific papers reporting new results and discoveries. In his Preface to the first edition of the *Memorie*, published in 1782, Lorgna noted that 'the disadvantage of Italy is that its forces are divided [and there was a need] to start combining the knowledge and work of many famous separated Italians [in an association] that was not of any one state but of all of Italy'.[2] This desire to nurture a sense of national cohesion in the scientific community came almost a century before the political unification of Italy in 1861.

Over the next hundred years, the *Società Italiana* changed its name several times, first to *Società Italiana delle Scienze detta dei XL* (Italian Society of Sciences, known as the Society of the Forty – the name at the time of Moseley's award), then in 1949 to the *Accademia Nazionale dei XL* (National Academy of the Forty) and finally, in 1979, to the Accademia Nazionale delle Scienze detta dei XL (National Academy of the Sciences, known as the Academy of the Forty). The *Società Italiana* was initally based in Verona, but moved to Milan after Lorgna's death in 1796, and then to Modena. In 1875, the Society moved to Rome, the capital of recently unified Italy.[1] The *Accademia* currently occupies part of the *Villa Torlonia* complex in the *Quartiere V Nomentano* in the north of Rome.

On the foundation of the *Società Italiana* in 1782, a total of twenty-seven men were elected as Fellows, followed by a further seventeen in 1786. Subsequent elections kept the Fellowship at forty. Prominent early Fellows included Alessandro Volta (a foundation Fellow, elected 1782), Guiseppe Lodovico Lagrangia (1736–1813, known as Joseph-Louis Lagrange, elected 1786), Amedeo Avogadro (1776–1856, elected 1821), and Stanislao Cannizzaro (1826–1910, elected 1865).[1] Beginning in 1786, in common with academies throughout Europe and North America, the *Società* also elected up to twelve Foreign Members.[1] The earliest included Benjamin Franklin (1706–1790), Joseph Priestley (1733–1804), and Karl Wilhelm Scheele (1742–1786). The *Società* remained active and continued to elect both Fellows and Foreign Members in the years 1796–1814 when Italy was a client state of France. Foreign Members elected in 1919 – the year of Moseley's award – included Moseley's supervisor Ernest Rutherford (1871–1937, elected October 1919) and family friend Edwin Ray Lankester (1847–1929, elected May 1919).[3]

The Matteucci Medal

In 1867, the *Società* took a further step towards the recognition of international science in the establishment of the Matteucci Medal, thanks to Carlo Matteucci (1811–1868), a distinguished physicist and physiologist, internationally known for his work in bioelectricity (Fig.1).[4] Born to a physician's family in Forli, in today's Emilia-Romagna, Matteucci's early studies in mathematics at the University of Bologna between 1825 and 1829 led to three formative years at the *École Polytechnique* in Paris. These led in turn to studies in Bologna, Florence, Ravenna, and Pisa, and to a growing interest in the electrical nature

Fig. 1 Carlo Matteucci. (Source: Archive of L'Accademia Nazionale delle Scienze detta dei XL, Rome, 1860, by courtesy of L'Accademia Nazionale delle Scienze detta dei XL).

of muscle contraction and nerve conduction. Following in the tradition of Luigi Galvani (1737–1798), Matteucci's research reinforced the view that biological tissue can generate electrical current.[1,4] In Ravenna, he was appointed head of the city's hospital laboratory and taught at a local college. On the recommendation of his Parisian mentor, the distinguished physicist Francois Arago, he was appointed Professor of Physics at the University of Pisa in 1840. His contributions to electrophysiology were widely recognised. He was elected to the *Società Italiana* in 1840, and in 1844, received the distinguished Copley Medal of the Royal Society of London, 'for his various researches in animal electricity'.[5]

From 1847, Matteucci became active in politics, and in 1862 was appointed Minister of Education under Prime Minister Urbano Rattazzi (1808–1873). He was elected as President of the *Società* in 1866: to mark this event he established an international prize in physics, to be awarded annually by the *Società* to the author of the *Memoria* that reported the most important recent discovery in physics. This was to consist of a gold medal of the value of 200 lire (the equivalent of about £2000 today). He intended to make financial arrangements to fund the prize in perpetuity, but there is no evidence that there was a formal benefaction to the *Società* in 1866. Under his guidance the Ministry of Public Education was also persuaded to sponsor two other annual prizes, each consisting of gold medals valued at 400 lire, specifically for the best Italian *Memorie* in mathematics, and in the physical and natural sciences respectively.[6] These two prizes were ratified by a *Regio Decreto* (Royal Decree) on 13 October 1866, although the Matteucci Medal itself was not mentioned in the Decree.[7] Nonetheless three medals were awarded in 1867: the Matteucci Medal went to Charles Wheatstone (1802–1875), the Prize for Mathematical Sciences to Luigi Cremona (1830–1903), and the Prize for Physical and Natural Sciences to Giovanni Schiaparelli (1835–1910). The second Matteucci Medal was awarded to Hermann Helmholtz (1821–1894) in 1868. However Matteucci died in the same year, aged fifty-seven. His widow, Robinia Young-Matteucci, the daughter of the physicist Thomas Young (1773–1829), made a benefaction to the *Società* after his death. The status of the Matteucci Medal was later formalised in a *Regio Decreto* issued by Vittorio Emanuele II on 10 July 1870.[8]

Between 1869 and 1874, the Medal was not awarded. The *Società* was preoccupied with moving its headquarters from Modena to Rome (which became capital of Italy in 1871) and with discussions of a possible merger with the recently re-established *Accademia Reale dei Lincei*.[1] Normal business was resumed only with the appointment of Arcangelo Scacchi (1810–1893) as president of the *Società* in 1875 and after the move to Rome had been completed.

The first Medal to be awarded after the decree of 1870 was to Henri Victor Regnault (1810–1878) in 1875, followed by William Thomson, 1st Baron Kelvin (1824–1907) in 1876, and Gustav Kirchhoff (1824–1887) in 1877.[9] Since then, the Medal has been awarded in a somewhat sporadic fashion, with three gaps of five years in the latter half of the nineteenth century. Only one Medal was awarded between 1933 and 1975, that to Wolfgang Pauli in 1956. Twenty-three Medals have been awarded since 1975. The latest went in May 2017 to Marco Tavani for his work on gamma ray astrophysics.

In the 150 years of the award, there have been seventy-two Medalists, seventeen of whom (23 per cent) have been Italians. This suggests an understandable national bias. Nonetheless, the Medal has retained a distinctly international flavour, in keeping with its intention to reward 'Italian and foreign physicists for fundamental contributions to science'.

The Nomination Process

Recommendations for the Medal in the early twentieth century were based on the deliberations of a *Commissione* (committee) of three Fellows. The *Commissione* included a *Presidente* (who was distinct from the President of the *Società*) and a *Relatore* (spokesman), who communicated the choice of the *Commissione* to the President of the Society on behalf of the President of the Committee. The formal proposal was supported by a short document, enitled *Relazióne per il conferimento della medaglia Matteucci*, which outlined the significance of the work that had led to the nomination. The archives of the academy in Rome for the years 1910–1920 convey no sense of political or other motives (if any) lying behind the decisions of the *Commissione*, but simply preserve the letters of recommendation and the *Relazione*. Neither is there any record of where suggestions by the committee came from – or any hint of the sense of self-importance that surrounded the Nobel committees.

Fig. 2 The *Commissione* who nominated Moseley for the Matteucci Medal in 1919. Left to right: Orso Mario Corbino, Augusto Righi and Antonio Roiti. (Source: Archive of *L'Accademia Nazionale delle Scienze detta dei XL*, Rome, undated, by courtesy of *L'Accademia Nazionale delle Scienze detta dei XL*).

The *Commissione* members in 1919 were Antonio Roiti (1843–1921, *Presidente*), Augusto Righi (1850–1920), and Orso Mario Corbino (1876–1937, *Relatore*). They had been elected to the *Società* in 1892, 1891 and 1911 respectively (Fig. 2). They were a colourful trio, influential in both science and politics in Italy and beyond. Roiti studied mathematics at the University of Pisa, but in 1866 interrupted his academic career to fight under Giuseppe Garibaldi (1807–1882) in the third Italian War of Independence.[10] He was taken prisoner by the Austrians but escaped to continue fighting. He was appointed professor of first chemistry and then physics in the University of Pisa in 1866 and 1868 respectively, and published on a diverse range of topics, including the motion of liquids, the speed of sound and electrical resistance. Eventually, he was appointed director of the *Museo degli strumenti antichi di astronomia e di fisica di Firenze*, and became Councillor for Education of the City of Florence.

Righi graduated as a civil engineer from the University of Bologna in 1872 and went on to teach physics at the Technical Institute of Bologna until 1880.[11] He taught at the University of Palermo between 1880 and 1885, then moved to Padua between 1885 and 1889, and returned to Bologna in 1889 until his death. Righi's work touched on the photoelectric effect, the Zeeman effect, magnetic hysteresis, the Michelson-Morley experiment, and X-rays. In addition, he is credited as being the first to generate electromagnetic radiation with wavelengths of less than 10 cm using oscillator circuits, thus opening up the new field of microwave science. His pre-eminence in this area attracted a young Guglielmo Marconi (1874–1937)[12] to attend Righi's lectures and to work for a couple of years in Righi's laboratory in Bologna, starting in 1894. Marconi was not formally enrolled as a student of the university, but started his experimental work on radio waves in his country house *Villa Griffone* in Pontecchio, not far from Bologna. There was clearly mutual respect, especially after Marconi won the Nobel Prize for Physics in 1909 and enjoyed growing international recognition as an inventor and entrepreneur. Righi was awarded the Matteucci Medal in 1882, the Hughes Medal of the Royal Society of London in 1905 and received forty nominations for the Nobel Prize in Physics between 1905 and 1920 – almost three times the fifteen nominations received by Marconi.[13] Righi also served as a Senator of the Kingdom of Italy.

In 1919, Corbino was the third and junior member of the *Commissione*, but he too became a major figure in Italian science and public life.[14] Born and educated in Sicily, he studied in Catania and Palermo, and became Professor of Physics at the University of Messina in 1905. In 1909, he moved to the Institute of Physics in Rome to work under Pietro Blaserna (1839–1918), whom he succeeded as director in 1918. His most significant work, involving the study of the motion of electrons in metals under the influence of a magnetic field, began in Palermo, and led to the award of the Medal in 1909.

Corbino was appointed Minister for Education in the Bonomi government in 1921 and also joined the boards of directors of several companies in the Italian electrical industry. In 1923, he accepted responsibility for the national economy as a minister in the first Mussolini government, although he was personally at odds with many Fascist policies. He is probably best remembered as Director of the Institute of Physics in Rome at the time when *i ragazzi di Via*

Panisperna (the Via Panisperna boys), including Enrico Fermi (1901–1954) and Emilio Segrè (1905–1989), were at the forefront of research in nuclear physics. Corbino had been the driving force behind the establishment of Italy's first chair in theoretical physics at his Institute, to which Fermi was appointed professor in 1926, aged only twenty-four. Fermi won the 1937 Nobel Prize in Physics for his work on radioactivity induced by neutron bombardment,[13] but left Italy for the USA after attending the Nobel ceremony in 1938.[15] Segrè is known for the discovery of technetium and astatine, and the subatomic anti-proton. He also left Italy for the USA in 1938, to avoid Mussolini's anti-semitic policies, and won the Nobel Prize in Physics in 1959.[13]

In 1919, Roiti and Righi were long-standing members of the *Commissione*, but Corbino had replaced his colleague and mentor Blaserna in 1916 – the year in which the Braggs were awarded the Medal. Moseley's *Relazione* occupies just over two typed pages, and in translation reads as follows:[16]

REPORT ON THE AWARD OF THE MATTEUCCI MEDAL FOR 1919

After the discovery by Bragg of the selective reflection of X-rays on a crystalline sheet the young English physicist H. G. J. Moseley started a deservedly renowned line of research, thus establishing a new area of spectroscopy: the ultraspectra of the elements.

In two consecutive papers published in 1914 the results of these studies were presented, including the high frequency spectra of the vast majority of the known elements as emitted under bombardment by cathodic rays. Each element showed itself to be characterised by the emission of a double line, which was similar for all elements, with frequency increasing regularly as a function of the order number of the element as defined by the ordered series of increasing atomic weights. The frequency of any double line is linked by a very simple algebraic expression to the number of the abovementioned element, while no simple relationship is found with atomic weight.

The theoretical importance of such memorable research grows with each passing day. First of all it was possible to establish that only three more hitherto unknown chemical elements can exist, and the place of each element in the periodic table has been found. Moreover, since the number of each element corresponds to the number of elemental positive charges in the atomic nucleus, as in the Rutherford and Bohr model, it was also possible to demonstrate that the only parameter that determines all the physical and chemical properties of an element, with the exception of the atomic weight, is the atomic number and not the atomic weight.

These laws also apply to radioactive elements undergoing atomic disintegration. Due to the fact that they may change their nuclear charge, and hence atomic number, independently from a change in their atomic weight, it may happen that elements with the same atomic weight have different atomic numbers and conversely that elements with different atomic weights may have the same atomic number. In the latter case, no physical or chemical property will allow one to distinguish one from the other. In particular they will show the same emission spectrum, both in the high frequency X-ray region, and in the visible region. Such expectations clearly arising from Moseley's research were verified by Rutherford.

From these short comments one can appreciate the importance that should be attributed to the work of Moseley, work that was interrupted when he took up front-line service in the English army at the outbreak of war. Moseley was struck down by a shot through the forehead in the attack on the Dardenelles, in the bloom of his youth, aged only twenty-six [sic].

In deciding that the Matteucci Medal for the year 1919 should be awarded in his memory, the committee feels it is paying appropriate tribute to the author of a number of papers that will remain among the most important of the past few years: while also expressing the huge regret of the Society at his premature death which constitutes a truly great misfortune for Science throughout the world.

Rome, 12 July 1919

The Committee:
Prof. Roiti, President
Augusto Righi
O.M. Corbino, Spokesman

The *Relazione* links Moseley's contributions to that of the Braggs. Likewise, the *Relazione* for the Braggs compiled in 1916 shows that the importance of Moseley's work was thoroughly recognised at that time.[17] There is an emphasis on the relationship of Moseley's discoveries to other work on radioactive decay, but little is said about the emerging Bohr Model of the atom. Finally, the *Relazione* specifically emphasises that the award should be viewed as a tribute to Moseley's memory. A short note from Corbino to Vito Volterra (1860–1940), President of the *Società*, reiterates that the award is posthumous, although it gives the wrong date in August (the 8th) and the wrong year (1916) for Moseley's death.[18]

Presentation of the Medal

Although it has been recent practice for the *Medaglia* to be presented at a closed meeting of the Academy held in early spring,[19] there is no surviving record of such ceremonies for the period 1910–1920. Instead, the award process seems to have involved a letter from the *Presidente* (of the *Società*) informing a recipient of the selection, with the actual Medal being dispatched at a later date. Self-evidently in Moseley's case it was necessary to convey the decision to someone else. In principle, the news could have been relayed to Rutherford or to the Royal Society, but the *Società* chose instead to write to Moseley's mother (who had remarried and was now Mrs. Sollas). There was difficulty in finding her address, and it is possible that family friend Sir Edwin Ray Lankester, who in 1919 had been elected a Foreign Member of the *Società*, was asked for details.[20] Even so, *Il Presidente*'s letter was sent to the wrong place, and in a gracious reply, Mrs. Sollas ends by giving her correct address (Fig. 3).[21]

The records of the *Accademia* contain several indications of a backlog in the minting of gold medals during and immediately after the Great War. There is an undated scribbled note listing all the awards for the Mathematics Prize, the Science Prize and the Matteucci Medal for the years 1914–1919, with the implication that by 1919 or later, all these Medals had to be struck.[22] A second document gives detailed instructions to the Royal Mint in Rome for medals to be struck for Stark, Wood, Moseley and Einstein.[23] Another undated document gives a list of '*Medaglie coniate e spedite*' (medals struck and dispatched) between 1915 and 1921,[24] with a note at the end that Stark's medal for 1915 had to be remade in 1922, which suggests a delay of up to seven years. The same document tells us that the mathematics and science medals were minted in gold in 1915, 1916 and 1917, but between 1918 and 1921, silver was used instead (Fig. 4). However, the Matteucci Medal remained *d'oro* – perhaps to save face on the international scene. The archives contain two acknowledgments of receipt relevant to our story. The first is a letter from William Henry Bragg, FRS (1862–1942), recording thanks for a 'Gold Medal awarded to my son and myself by the *Società Italiana delle Scienze*'.[25] Bragg's letter is dated 28 June 1922, six years after his award and three years after Moseley's. Dated just under a month later, 14 July 1922, we find an acknowledgement form signed by Amabel Nevill Sollas.[26] A photograph of the Medal is shown in Fig. 5: analysis by X-ray spectroscopy reveals that it contains 99 per cent gold by weight, along with traces of copper and silver to harden the metal.[27]

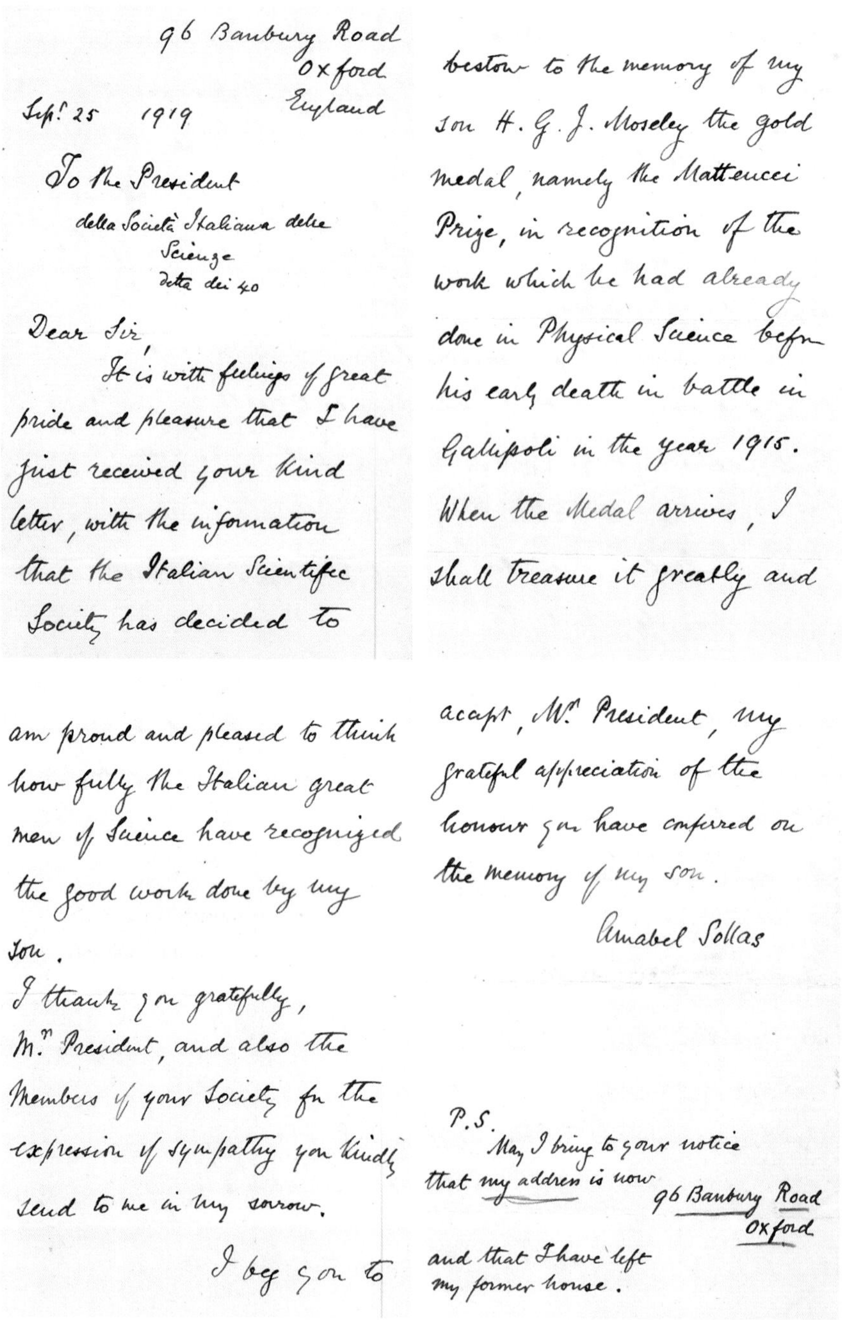

> 96 Banbury Road
> Oxford
> England
> Sep.t 25 1919
>
> To the President
>   della Società Italiana delle
>   Scienze
>   detta dei 40
>
> Dear Sir,
>   It is with feelings of great pride and pleasure that I have just received your kind letter, with the information that the Italian Scientific Society has decided to bestow to the memory of my son H. G. J. Moseley the gold medal, namely the Matteucci Prize, in recognition of the work which he had already done in Physical Science before his early death in battle in Gallipoli in the year 1915. When the Medal arrives, I shall treasure it greatly and am proud and pleased to think how fully the Italian great men of Science have recognized the good work done by my son.
>   I thank you gratefully, Mr. President, and also the Members of your Society for the expression of sympathy you kindly send to me in my sorrow.
>   I beg you to accept, Mr. President, my grateful appreciation of the honour you have conferred on the memory of my son.
>
>   Amabel Sollas
>
> P.S. May I bring to your notice that my address is now 96 Banbury Road Oxford and that I have left my former house.

Fig. 3 Letter from Amabel Sollas acknowledging Moseley's nomination for the Matteucci Medal for 1919. (Source: Archive of *L'Accademia Nazionale delle Scienze detta dei XL* Rome, 25 September 1919, by courtesy of *L'Accademia Nazionale delle Scienze detta dei XL*).

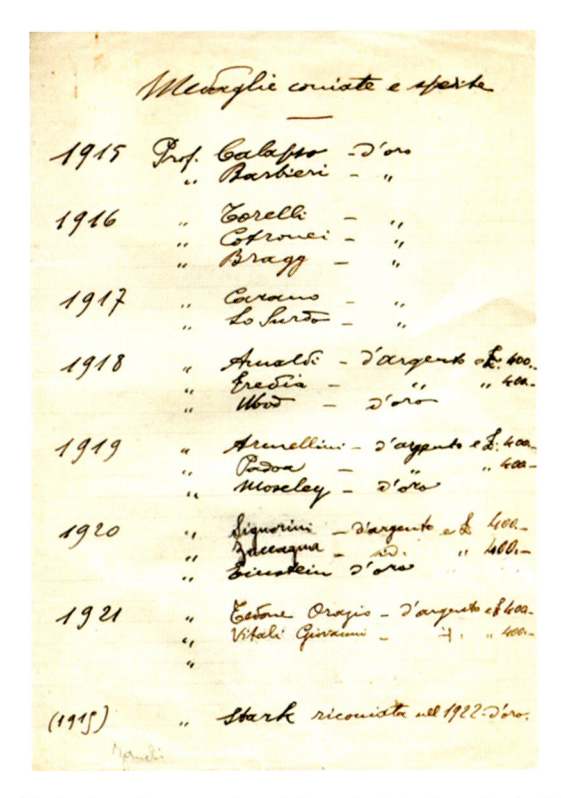

Fig. 4 List of medals that have been struck and dispatched, dealing with the big backlog in medals that developed in the First World War. (Source: Archive of *L'Accademia Nazionale delle Scienze detta dei XL* Rome, Rome, undated but probably 1922, by courtesy of *L'Accademia Nazionale delle Scienze detta dei XL*).

Fig. 5 Moseley's Matteucci Medal. (Source: Clare Hopkins, 2017, by courtesy of William le Fleming).

The Matteucci Medal and the Nobel Prize

There has always been speculation as to whether Moseley would have won a Nobel Prize had he survived the Great War.[28] It is, of course, impossible to know, but it is interesting to compare names and dates of winners and nominees for the Prizes in Physics and Chemistry with winners of the Matteucci Medal.

Table 1 gives details of the 'Nobel Status' of winners of the *Medaglia Matteucci* for the period 1896–1930.[9,13] In 1896, the Medal went to Konrad Röntgen, who in 1901 became the first winner of the Nobel Prize in Physics for his discovery of X-rays. No Matteucci Medals were awarded between 1897–1900. However, direct comparisons from 1901 onwards are possible. By 1930 most of those prize winners connected with Moseley's discoveries and their impact have been included, so the table does not extend beyond this year. The period between 1901 and 1930 also represents the three decades of greatest continuity in the award of the Medal. However, no Medals were awarded in 1903, 1920 and 1922, and there were joint winners only in 1904 (the Curies) and 1916 (the Braggs). By contrast, the only year for which no Nobel Prize in Physics was awarded in this period was 1916, and there were six joint or multiple awards. Consequently there were twenty-nine Matteucci Medalists between 1901 and 1930 but thirty-six Nobel Laureates in Physics. Thus Matteucci winners are the more exclusive group, or at least the smaller.

There is significant overlap between winners of the Matteucci Medal and Nobel Prize winners in Physics, but in no sense did the Matteucci *Commissione* simply follow in the footsteps of the Swedish Academy. In any case, this would have been impossible before 1900. Indeed Table 1 identifies ten scientists who first won the Matteucci Medal and went on to win a Nobel Prize at a later date. Max von Laue and Albert Einstein won the Prize and the Medal for the same year, 1914 and 1921, respectively. However, Einstein's Nobel Prize for 1921 was awarded as a deferred Prize in 1922, so the Matteucci Medal came a year earlier.

Einstein's *Relazione* has not been preserved in the Rome archive. However, there is a letter from Antonio Garbasso to an *Onorevole Senatore* (Honourable Senator), presumably Emanuele Paternò (1847–1935), which tells us that Einstein was nominated for his work on the theory of relativity.[29] Garbasso had assumed the role of *Relatore* following the death of Augusto Righi on 8 June 1920. This letter is dated 2 November 1921, just before the death on 8 November 1921 of Antonio Roiti, a second member of Moseley's *Commissione*. This left Corbino as the senior active member of the *Commissione* at the time of Einstein's nomination. Under his influence the Italians were much less conservative than their Nobel counterparts in recognising the importance of theoretical physics:[29] the Swedish Academy felt unable to award a prize for relativity, and instead fudged matters by giving Einstein a prize for his work on the photoelectric effect.[13] Garbasso's letter also suggests that a Medal should be awarded to Francis Aston for his work on positively charged ions. This Medal was never awarded, but Aston did win the Nobel Prize in Chemistry in 1922 for his work on mass spectroscopy.

Only eight Matteucci Medal winners in the table had previously won a Nobel Prize in Physics, and this includes the two Curies and the two Braggs, so only in six years in this period did the Medal go to an individual who had already won been recognised by the Academy in Stockholm. In addition, William Ramsay and Ernest Rutherford were awarded the Medal after receiving a Nobel Prize in Chemistry. Rutherford's received his Medal in 1913, after the Geiger-Marsden experiment and introduction of the idea of the nuclear atom. However, his *Relazione* makes no mention of this discovery,[30] and the Medal was awarded for his work on radioactivity, as for the earlier Nobel Prize.[28,31,32] This provides support for a statement by Niels Bohr in 1960: '… you see actually the Rutherford work [the nuclear atom] was not taken seriously. We cannot understand today, but it was not taken seriously at all. There was no mention of it any place. The great change came from Moseley'.[33]

Rutherford also appears in a group of six who were nominated for a Nobel Prize in Physics, but did not receive the award. This group includes Moseley, whose single nomination is discussed in Chapter 6. Henry Poincaré, Robert Wood and Arnold Sommerfeld are also in this group, each having received many Nobel nominations (fifty-one, thirty-nine and eighty-four, respectively) without winning a Prize. Only three have won a Matteucci Medal without receiving a single Nobel nomination: Antonio Garbasso, Orso Corbino and Antonino lo Surdo were all Italians.

The last and biggest group (listed separately after the table) consists of eighteen who were awarded Nobel Prizes in Physics but who did not win a Matteucci Medal. Members of this group include Hendrik Lorentz, Antoine Becquerel, J. J. Thompson, Johannes van der Waals, Max Planck and Louis de Broglie. Less surprising, in view of discussion elsewhere in this volume,[34] Charles Barkla and Manne Siegbahn never won a Matteucci Medal for their work on X-rays.

Concluding Remarks

The Matteucci Medal predates the Nobel Prize as an international award in physics by over thirty years. However, the award of this Medal has never aroused the level of public interest that has surrounded the Nobel awards since their inception in 1901. This is hardly surprising, as the gold Matteucci Medal is not accompanied by a cash prize, and the presentation of the Medal has never involved a high profile ceremony involving royalty. The Medal has also lacked the continuity associated with the Nobel Prize. Nonetheless, the Matteucci Medal stands as an important indicator of esteem within the international physics community, especially in the decades between 1900 and 1930. In this context, Moseley occupies a special place in the history of the Medal as the only posthumous recipient in 150 years.

The Royal Swedish Academy of Sciences recognised the field of characteristic X-rays and X-ray spectroscopy with the award of two Nobel Prizes in Physics: those to Charles Barkla for 1917 and to Manne Siegbahn for 1924. Both Prizes were deferred, and the awards were made in 1918 and 1925, respectively. Both awards were controversial and Barkla in particular was a surprising choice. He had received only one nomination (from Rutherford), and that arrived too late for formal consideration for the 1917 Prize.[13,28,31] In 1919, when the *Commissione* for the *Medaglia Matteucci* deliberated on the Medal for that year, both Moseley and Barkla were possible candidates. Moseley was preferred. Indeed, it is possible that Moseley's nomination was prompted by the Nobel award to Barkla the previous year. Manne Siegbahn was never awarded a Matteucci Medal, so in Italy at least Moseley was clearly regarded as the most significant figure in the foundation of X-ray spectroscopy. The sequence of Medals awarded – to Moseley in 1919, Einstein in 1921 and Bohr in 1923 – provides a snapshot of the huge importance accorded Moseley's discoveries in the early part of the twentieth century.

Table 1. A comparison of winners of the Matteucci Medal between 1896 and 1930, with nominations and awards of Nobel Prizes. The year of the Matteucci award is given in the first column.

Year	Name	Nobel Nominations	Nobel Prize
1896	Konrad Röntgen (1845–1923)	21 nominations for physics between 1901 and 1922. Nominations between 1906 and 1922 were for medicine	1901 Physics
1896	Philipp Lenard (1862–1947)	15 nominations for physics between 1901 and 1925	1905 Physics
1901	Guglielmo Marconi (1874–1937)	15 nominations for physics between 1902 and 1933	1909 Physics
1903	Albert A. Michelson (1852–1931)	4 nominations for physics between 1904 and 1907	1907 Physics
1904	Pierre Curie (1859–1906) Marie Curie-Sklodowska (1867–1934)	Pierre Curie 3 nominations for physics in 1902 and 5 in 1903 Marie Curie 2 nominations for physics in 1902 and 1 1903. 2 nominations for chemistry in 1911	1903 Physics (Marie and Pierre Curie, shared with Becquerel) 1911 Chemistry (Marie Curie)
1905	Henry Poincaré (1854–1912)	51 nominations for physics between 1904 and 1912	–
1906	James Dewar (1842–1923)	5 nominations for physics between 1905 and 1913 and 3 for chemistry in 1910 and 1911	–
1907	William Ramsay (1852–1916)	1 nomination for physics in 1904 and 30 nominations for chemistry between 1902 and 1904	1904 Chemistry
1908	Antonio Garbasso (1871–1933)	No nominations	–
1909	Orso Mario Corbino (1876–1931)	No nominations	–

Year	Name	Nobel Nominations	Nobel Prize
1910	H. Kammerlingh Onnes (1853–1926)	22 nominations for physics between 1909 and 1913	Physics 1913
1911	Jean Perrin (1870–1942)	36 nominations for physics between 1913 and 1926 and 11 nominations for chemistry between 1915 and 1926	Physics 1926
1912	Pieter Zeeman (1865–1943)	2 nominations for physics in 1901 and 1 for physics in 1902	Physics 1902
1913	Ernest Rutherford (1871–1937)	20 nominations for physics between 1907 and 1937. 1 nomination for chemistry in 1907 and 3 in 1908	Chemistry 1908
1914	Max von Laue (1879–1960)	2 nominations for physics in 1914 and 3 in 1915	Physics 1914
1915	Johannes Stark (1874–1957)	11 nominations for physics between 1914 and 1919	Physics 1919
1916	W. Henry Bragg (1862–1942) W. Lawrence Bragg (1890–1971)	W. Lawrence Bragg 1 nomination for physics in 1914 and 4 in 1915 1 nomination for chemistry in 1915 W. Lawrence Bragg 2 nominations for physics in 1915	Physics 1915 (W. L. Bragg and W. H. Bragg)
1917	Antonino lo Surdo (1880–1949)	No nominations	–
1918	Robert W. Wood (1868–1955)	39 nominations for physics between 1915 and 1950	–
1919	Henry G. J. Moseley (1887–1915)	1 nomination for physics and 1 for chemistry in 1915	–
1921	Albert Einstein (1871–1955)	62 nominations for physics between 1910 and 1922	Physics 1921
1923	Niels Bohr (1885–1962)	20 nominations for physics between 1917 and 1922 and 1 for chemistry in each of 1920 and 1929	Physics 1922

Year	Name	Nobel Nominations	Nobel Prize
1924	Arnold Sommerfeld (1868–1951)	84 nominations for physics between 1917 and 1951	–
1925	Robert Millikan (1868–1953)	19 nominations for physics between 1916 and 1926	Physics 1923
1926	Enrico Fermi (1901–1954)	35 nominations for physics between 1935 and 1948 and 3 for chemistry between 1935 and 1937	Physics 1938
1927	Erwin Schrödinger (1887–1961)	41 nominations for physics between 1929 and 1933	Physics 1933
1928	Venkata Raman (1888–1970)	2 nominations for physics in 1929 and 10 for physics in 1930	Physics 1930
1929	Werner Heisenberg (1901–1976)	29 nominations for physics between 1929 and 1933	Physics 1932
1930	H. Arthur Compton (1892–1962)	15 nominations for physics between 1925 and 1927	Physics 1927

Nobel Laureates in Physics who did not win the Matteucci Medal
(year of birth and death, followed by year of Nobel Prize)
Hendrik Lorentz (1853–1928, 1902)
Antoine Becquerel (1852–1908, 1904)
Joseph J. Thomson (1856–1940, 1906)
Gabriel Lippmann (1845–1921, 1908)
Karl Braun (1850–1918, 1909)
Johannes van der Waals (1837–1923, 1910)
Wilhelm Wien (1864–1928, 1911)
Gustaf Nils Dalén (1869–1937, 1912)
Charles Barkla (1877–1944, 1917)
Max Planck (1858–1947, 1918)
Charles Guillame (1861–1938, 1920)
Manne Siegbahn (1886–1978, 1924)
James Franck (1882–1964, 1925)
Gustav Hertz (1887–1975, 1925)
Charles Wilson (1869–1959, 1927)
Owen Richardson (1879–1959, 1928)
Louis de Broglie (1892–1987, 1929)

Notes and References

Documents in the archive of *L'Accademia Nazionale delle Scienze detta dei XL* (hereafter *ANSXL*) are identified by a *fascicolo* (dossier) number and a *busta* (envelope) number.

[1] G. Penso, *Scienziati Italiani e Unità d'Italia – Storia dell'Accademia Nazionale dei XL* (Roma: Bardi Editóre, 1978).

[2] A.M. Lorgna, *preface to Memorie di Matematica e Fisica della Società Italiana, Tomo I* (Verona: Ramanzini, 1782).

[3] *Memorie di Matematica e di Scienze Fisiche Naturali della Società Italiana delle Scienze (detta dei XL), Tomo XXL*, (Roma: Tipografia della reale Accademia dei Lincei, 1920).

[4] G. Moruzzi, 'The Electrophysiological Work of Carlo Matteucci', *Brain Research Bulletin*, 40 (2), (1996), 69-71. This paper is an English translation of a paper published in Italian in 1964 and includes a foreward by M. Piccolino. See also M. Bresadola, 'Carlo Matteucci and the Legacy of Luigi Galvani', *Archives Italiennes de Biologie*, 149 (Supp.), (2001), 3-9.

[5] *Electrical Magazine* conducted by Charles Walker (London: Simkin, Marshall and Co., 1845), vol. 1, 610.

[6] *ANSXL, Busta 25, Fascicolo 6*, Letter sent on behalf of Carlo Matteucci to '*Caro Professore*' [presumed to be Pietro Marianini] in Modena, 7 September 1866, received 8 September 1866. The letter is on headed notepaper of *Ministero della Marina, Bolletino Meterlogico, Firenze* although the headings are crossed out. The letter discusses arrangements for award of two gold medal each valued at 400 Lire to be funded by the Ministry of Education.

[7] *Raccolta Ufficiale delle Leggi e Decreti del Regno, n. 3288 del 13 ottobre 1866,* published in the *Gazzetta Ufficiale del Regno d'Italia n. 305 di martedì 6 novembre 1866.*

[8] *Raccolta Ufficiale delle Leggi e Decreti del Regno, n. 5762 del 10 luglio 1870,* published in the *Gazzetta Ufficiale del Regno d'Italia n. 219 di giovedì 11 agosto 1870.*

[9] A complete list of winners of the medal appears on the website of the *L'Accademia Nazionale delle Scienze detta dei XL:* http://www.accademiaxl.it/medaglia-matteucci/.

[10] G. Diaz de Santillana, entry on 'Antonio Roiti', in *Enciclopedia Italiana di Scienze, Lettere ed Arti,* (Roma: Istituto dell'Enciclopedia Italiana, 1949), vol. 29, 583.

[11] 'Righi, Augusto', in *Complete Dictionary of Scientific Biography* (Detroit: Charles Scribner's Sons, 2008), vol. 11, 460–461.

[12] Maria Grazia Ianniello, entry on 'Marconi, Guglielmo', *in Dizionario biografico degli italiani* (Roma: Istituto dell'Enciclopedia Italiana, 2007), vol. 69, 793–797.

[13] Nominations for Nobel Prizes may be searched on the Official Website of the Nobel Prize: https://www.nobelprize.org/nomination/archive/ (accessed on 29 November 2017).

[14] Edoardo Amaldi and Luciano Segreto, entry on 'Corbino, Orso Mario', in *Dizionario biografico degli italiani* (Roma: Istituto dell'Enciclopedia Italiana, 1983), vol. 28, 760-766.

[15] Emilio Segrè, *Enrico Fermi, Physicist* (Chicago: University of Chicago Press, 1970).

[16] ANSXL, Busta 28, Fascicolo 17, *Relazióne per il conferimento della medaglia Matteucci per il 1919*, 12 July 1919. The *Relazióne* is incorrect in stating that Moseley's two key papers were both published in 1914.

[17] ANSXL, Busta 28, Fascicolo 17, *Relazióne sul conferimento della medaglia Matteucci per il 1916*, undated but probably July 1916.

[18] ANSXL, Busta 28, Fascicolo 17, *Orso Corbino to Caro Presidente*, 7 July 1919.

[19] A. Grandolini, personal communication.

[20] ANSXL, *Busta* 27, *Fascicolo* 4, Letter giving Lankester's address, undated but must be 1919.

[21] ANSXL, *Busta* 27, *Fascicolo* 4, Amabel Sollas to the The President *Società Italiana delle Scienze detta dei XL*, 25 September 1919.

[22] ANSXL, *Busta* 27, *Fascicolo* 4, Handwritten list of medals for the years 1914–1919 [presumed to be ones that were to be minted], undated but probably 1919.

[23] ANSXL, *Busta* 29, *Fascicolo* 29, Typed letter instructing the Royal Mint to strike four gold medals using stamp 3894, undated but probably 1922.

[24] ANSXL, *Busta* 34, *Fascicolo* 25, Handwritten list of *medaglie coniate e spedite* (medals that have been struck and dispatched), undated but probably 1922. The year of Einstein's Medal is given incorrectly as 1920 rather than 1921 and there is no evidence that this mistake was corrected on the medal itself.

[25] ANSXL, *Busta* 27, *Fascicolo* 4, W. H. Bragg to *Cassiere Economo, R. Scuola Ingegneri, via delle Sette Sale 11B*, Rome of acknowledging receipt of the Matteucci Medal, 28 June 1922.

[26] ANSXL, *Busta* 29, *Fascicolo* 29, Receipt for *una medaglia d'oro* signed by Amabel Nevill Sollas, 14 July 1922. We are grateful to Will le Fleming (great grandson of Henry Moseley's sister Margery) for locating the medal within the family archive.

[27] Analysis of the Medal by energy dispersive X-ray spectroscopy in a low-voltage scanning electron microscope was conducted by Dr. Alison Crossley and Dr. Chris Salter in the Materials Department of Oxford University on 23 June 2017. The weight percentage of gold in the Medal (99%) is higher than found in typical 22 karat costume jewllry, which contains only 92% gold.

[28] Robert Marc Friedman, *The Politics of Excellence: Behind the Nobel Prize in Science* (New York: Henry Holt and Company, 2001).

[29] *ANSXL, Busta 29, Fascicolo 37*, Antonio Garbasso to *Onorevole Senatore* [presumably Emanuele Paternò President of the *Società* between 1921 and 1932], 2 November 1921.

[30] *ANSXL, Busta 28, Fascicolo 17, Relazione sul conferimento della Medaglia Matteucci (anno 1913)*, undated but probably July 1913.

[31] Celia Jarlskog, 'Lord Rutherford of Nelson, His 1908 Nobel Prize in Chemistry, and Why He Didn't Get a Second Prize'. Extended version of a talk at the conference *Neutrino 2008*, Christchurch, New Zealand, 25–31 May 2008.

[32] K. B. Hasselberg, 'Award Ceremony Speech, 1908', in *Nobel Lectures, Chemistry, 1901–1921* (Amsterdam: Elsevier, 1966).

[33] American Institute of Physics, Oral History Interviews. Niels Bohr interviewed by Thomas S. Kuhn, Leon Rosenfeld, Aage Petersen, and Erik Rudinger, 14 November 1962, Session I https://photos.aip.org/history-programs/niels-bohr-library/oral-histories/4517-1 (accessed on 29 November 2017). Quoted in Richard Rhodes, *The Making of the Atomic Bomb* (London: Simon and Schuster, 1986), 85.

[34] R. M. Friedman, Chapter 6 this volume.

CHAPTER EIGHT

X-RAY SPECTROSCOPY AND THE DISCOVERY OF NEW ELEMENTS

RUSSELL G EGDELL

Introduction

In his first paper on the High-Frequency spectra of the elements, Moseley realised that X-ray spectroscopy could play a role in the verification of claims for the discovery of new elements:

> Its [X-ray spectroscopy's] advantage over ordinary spectroscopic methods lies in the simplicity of the spectra and the impossibility of one substance masking the radiation from another. It may even lead to the discovery of new elements, as it will be possible to predict the position of their characteristic line.[1]

This conclusion was prescient, if not entirely accurate: X-ray spectra turned out to be considerably more complicated than Moseley had anticipated,[2] and interferences did prove problematic when attempting to analyse for trace elements. As discussed in Chapter 3, Moseley's final, but unpublished, plot of atomic numbers against the square roots of X-ray frequencies left room for four new elements – those with atomic numbers (or Z values) 43 (technetium), 61 (promethium), 72 (hafnium), and 75 (rhenium). Moseley's compilation of X-ray spectra reached only as far as gold, with $Z = 79$. In 1916, the range was extended by the Swedish physicist Manne Siegbahn (1886–1978) to include all remaining elements where measurements were feasible up to $Z = 92$ (uranium).[3] His paper included a plot along the same lines as Moseley's diagram (but with the axes switched round), and left vacant slots for elements 84-89 (Fig. 1). By 1916, four of these elements had already been discovered and were known as polonium, radon, radium and actinium. However they were all highly radioactive and Siegbahn could not handle them in his spectrometer. Nonetheless it was fairly obvious from their chemical properties that they should occupy positions 84, 86, 88 and 89 respectively within the periodic table. So by 1916 only elements 85 and 87 remained unknown from within Siegbahn's 'gap of six'.

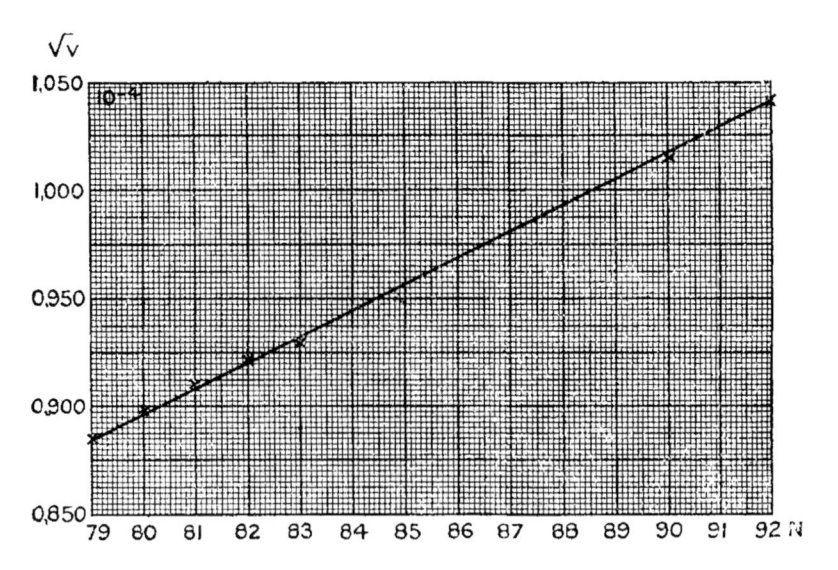

Fig. 1 Manne Siegbahn's plot of X-ray frequencies for elements 79-92. Unlike Moseley, Siegbahn does not include error bars in his graph and he plots $\sqrt{\upsilon}$ against N, instead of treating N as a function of υ, as in Moseley's papers. (Source: *Philosophical Magazine Series* 6, 31 (184), (1916), 403–406).

Siegbahn also left a vacant slot for element 91. In fact an isotope of element 91 had already been identified by Kasimir Fajans and Oswald Göhring in 1913.[4] Due to the very short half-life of only 6.7 hours they proposed the name brevium.[5] In 1917 the longer lived $^{231}_{91}91$ isotope was identified by Lise Meitner and Otto Hahn, using radiological techniques. With a half-life of over 30,000 years this isotope was found to undergo alpha decay to $^{227}_{89}Ac$.[6] For this reason they suggested the name proto-actinium, which was eventually abbreviated to protactinium in 1949.

X-ray spectroscopy made no contribution to the discovery of element 91, although an X-ray spectrum was eventually measured in 1930.[7] However, just as Moseley had predicted, X-ray spectroscopy was central to the quest to find the remaining 'missing' elements, those with atomic numbers 43, 61, 72, 75, 85 and 87. The full but often convoluted history of the completion of the periodic table has been charted in two recent books.[8] This chapter retraces the narrative with a focus on the role of X-ray spectroscopy in the discovery of the new elements. The stories of elements 43 and 75 (technetium and rhenium) and of elements 85 and 87 (astatine and francium) are intertwined, so these two pairs are each treated together. The artificial elements so far discovered beyond uranium are outside the scope of this chapter, but a postscript looks at the hunt for primordial (i.e. naturally occurring) superheavy elements in the 1970s. Here X-ray spectroscopy was yet again the key technique, and Moseley's legacy continued to be influential in the hunt for new elements.

Element 72: Hafnium (Hf).

Moseley's only direct involvement in the search for missing elements was concerned with element 72. As discussed in Chapter 3, he was able to show that Urbain's sample of celtium was in fact a mixture of elements 70 and 71– neoytterbium (now known as ytterbium) and lutecium (now lutetium). Although Urbain immediately recognised the potential of X-ray spectroscopy as a method for identifying elements, he was unwilling to accept that he had not succeeded in isolating celtium. In 1922, eight years after Urbain's visit to Oxford, Alexandre Dauvillier re-examined the X-ray spectrum of the same sample that Moseley had studied and was able to identify '*deux lignes extrêmement faibles démontrent l'existence d'une trace de celtium et lui assignent le nombre atomique 72* (two extremely weak lines demonstrating the existence of a trace of celtium and assigning atomic number 72 to it)'.[9]

This result appeared in a paper in the '*Comptes rendus, présentée par M. G. Urbain*', followed by a short commentary by Urbain himself, which ended with the conclusion that '*le résultat négatif relatif au celtium, par la méthode de Moseley, tenait seulement à son défaut de sensibilité* (the negative result for celtium by Moseley's method was due only to his lack of sensitivity)'.[10] No spectra were presented in Dauvillier's paper and it seems that not everyone could see the two weak X-ray lines on the original plate. Manne Siegbahn famously went so far as to state that 'I did not see any celtium lines on the photographic plate that Dauvillier showed me. I think they are probably only visible to Frenchmen'.[11]

By the early 1920s, it was becoming obvious based on chemical reasoning that element 72 should not be considered a rare earth, but rather occupied a place below zirconium in group IV of the standard periodic table.[12] In 1923, Dirk Coster and Georg von Hevesy, working at Bohr's institute in Copenhagen, examined several minerals containing zirconium. They were immediately able to identify strong L emission lines of element 72, and proposed the name 'hafnium' after the Latin name of Copenhagen, where their work was done.[13] Their first publication was a report of X-ray wavelengths, without supporting original data, but shortly afterwards thoroughly convincing X-ray spectra were published by Coster (Fig. 2).[14] Chemical isolation of hafnium followed shortly afterwards.[15]

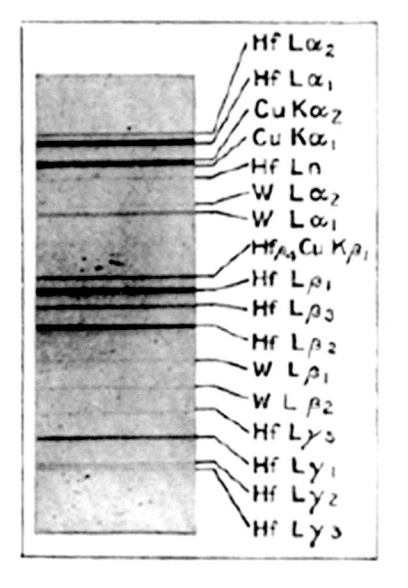

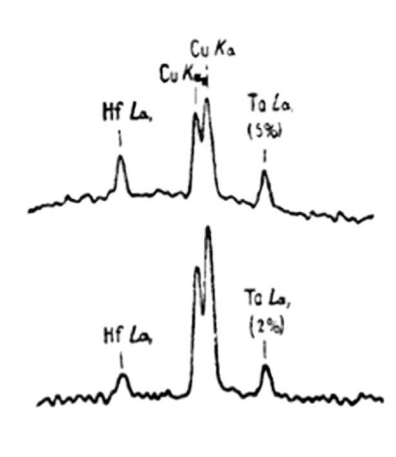

Hf L-SPECTRUM.

Fig. 2 One of Dirk Coster's X-ray plates of the L emission spectrum of element 72, along with a photometer trace derived from the photographic plate. (Source: *Philosophical Magazine Series 6*, 46 (275), (1923), 956–963).

The naming of element 72 aroused a political controversy that lasted for many years.[16] In an editorial for *Chemical News* in 1923, W. P. Wynne wrote:

> We adhere to the original word celtium given to it by Urbain as a representative of the great French nation which was loyal to us throughout the war. We do not accept the name which was given it by the Danes who only pocketed the spoils of war.[17]

Neither Coster (Dutch) nor von Hevesy (Hungarian) were, in fact, Danish. The International Committee on Atomic Weights were asked to arbitrate on the matter. The Committee initially prevaricated by suggesting that both names could be used.[18] They eventually found in favour of hafnium, but only after Urbain's death in 1938.[19]

Elements 43 and 75: Technetium (Tc) and Rhenium (Re).

In 1871, Mendeleev had left room for an element under manganese – *eka-manganese* – in the position occupied by what is now known as technetium.

Refinement of his table led to a second element in the same group below *eka*-manganese, which was dubbed *dwi*-manganese (the prefixes *eka* and *dwi* were derived from Sanskrit for one and two). Moseley's second X-ray paper assigned atomic numbers 43 and 75 to these two missing elements.

Reports of the isolation of elements in the manganese group actually predate both Mendeleev and Moseley. However, the first plausible claim for the discovery of *eka*-manganese was based on work beginning in 1904, by the Japanese chemist Masataka Ogawa. This was conducted under the supervision of Sir William Ramsay, the Nobel-Prize winning professor of chemistry, at University College London. Ogawa's work was never substantiated, but there have been recent claims that he may have isolated *dwi*-manganese instead.[20]

The first concerted effort to use X-ray spectroscopy to identify element 43 was undertaken by Claude H. Bosanquet and T. C. Keeley. Bosanquet had worked with William Henry Bragg, and was an expert in X-ray techniques. Together the pair measured X-ray spectra of fourteen manganese-rich minerals, including manganite and pyrolusite, in the hope of identifying the Kα lines of element 43, which were expected to be shifted to short wavelength of those of molybdenum: with their apparatus, they were able to detect 0.2 per cent molybdenum in a sample. Despite their best efforts, they concluded in 1924 that 'the results, so far, have been negative'.[21] Nonetheless, there was a widespread sense of anticipation that the discovery of the element was imminent. In 1925, Richard Hamer of the University of Pittsburgh went so far as to submit a letter to *Science*, proposing the name moseleyum be given the element 43 when it was eventually found,[22] a proposal welcomed by Sir Richard Gregory, the editor of *Nature*.[23]

However, the suggestion of moseleyum was ignored by the German team of Walter Noddack, Ida Tacke (later Ida Tacke-Noddack), and Otto Berg, who in 1925 published a paper in *Naturwissenschaften* reporting the measured positions of K-series X-ray lines from element 43 and L lines from element 75.[24] The paper had two sections, the first under the names of Noddack and Tacke dealing with chemical procedures for the enrichment of samples of the new elements. The second, *Röntgenspektroskopischer*, section was by Tacke and the spectroscopist Berg.

Noddack and Tacke reasoned that the best chance of finding elements 43 and 75 was in the ores of elements in the middle of the second and third transition series, rather than in samples rich in first row manganese. This strategy appeared to meet with success. The reported L spectrum of element 75 contained five lines, some of which overlapped with lines of tungsten and zinc. They proposed the name rhenium (after the Rhineland – Ida Tacke was born in Wesel in North Rhine-Westphalia), which was said to be present at a level of 5 per cent in samples of Norwegian columbite. The data for *dwi*-manganese included a schematic of the spectrum with 'estimated intensities' rather than the original X-ray photograph (Fig. 3).

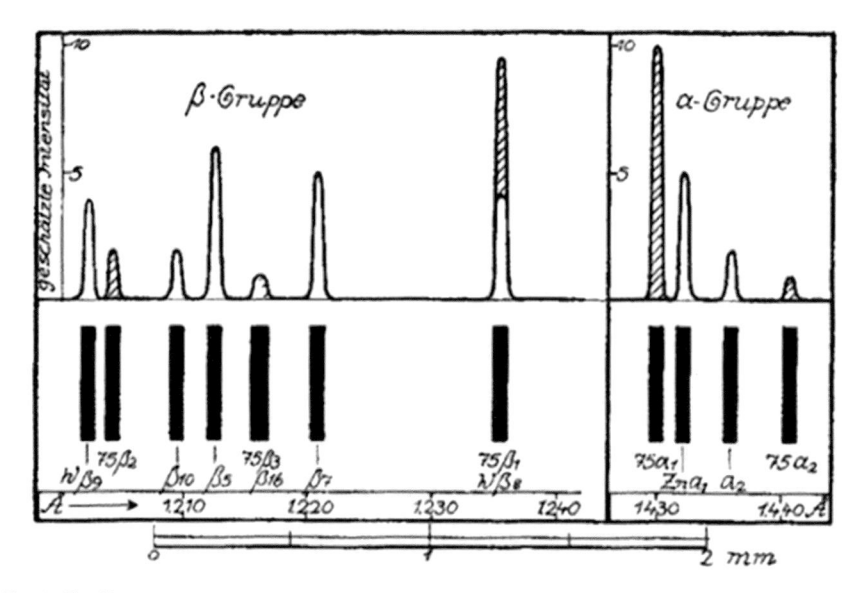

Fig. 3 The X-ray spectrum upon which the discovery of element 75 (rhenium) was based. (Source: *Naturwissenschaften*, 13 (5), (1925), 567–574).

In addition, the positions of three K emission lines from element 43 were tabulated and compared with the expected positions. The concentration of *eka*-manganese in columbite was stated be ten times lower than that of *dwi*-manganese. For this new element, the positions of the X-ray lines were simply given in a table, and no attempt was made to give a representation of the original spectra. Noddack and Tacke proposed that element 43 should be known as masurium, a controversial choice, evoking memories of a major German victory over the Russian army in the Masurian district of East Prussia during the Great War. However, this was Walter Noddack's birthplace, so at least there was consistency in the two suggested names.

The Noddacks married in 1926, and set out to confirm their claims for the discovery of element 75. By 1929, they had succeeded in preparing a one gram sample of rhenium.[25] However, element 43 proved more elusive. This is not surprising, as the eventual verifiable discovery of element 43 in 1937 was based not on the analysis a geological sample, but on a piece of molybdenum that had been bombarded by neutrons and deuterons in the Berkeley cyclotron of Ernest Lawrence. The irradiated molybdenum deflection plate had made its way to Palermo, where it was studied by Emilio Segrè (one of the *via Panisperna* boys who had gone to work with Lawrence) and Carlo Perrier. Unusual radioactivity in the irradiated sample was linked to element 43 using chemical separation procedures and radiological detection (that is measurement of the characteristics of the radioactivity).[26] In 1940, Segrè and Wu went further, and showed that element 43 could also be found amongst the fission products of uranium.[27] It was proposed (but not by Segrè) that the element should be known as panormium, after the Latin Panormus for Palermo. The name technetium eventually adopted in 1947 derives from the Greek τεχνητός (artificial), and recognises that element 43 'is the first [to be] artificially made'.[28]

Segrè met with the Walter Noddack on two occasions in 1937 – first, when he visited his laboratory in Freiburg and second, a few weeks later in Palermo. Segrè recalls asking to see the original X-ray plates which had led to the wavelengths given for masurium in the 1925 paper. Walter Noddack claimed that the plates had all been broken, and that he was unable to produce a sample of masurium because all his material had been sent to F.W. Aston for isotopic analysis. When they met in Palermo, Segrè showed Noddack his radioactive sample of element 43.[29]

Noddack's evasiveness is suspicious. In 1927, Berg had reported his observation of the X-ray lines of element 43 in no less than twenty-eight photographic plates from amongst the 1000 measurements he had made.[30] Moreover, in 1925, a reproduction of an X-ray plate was in fact published in the *Sitzungsberichte der Preussisschen Akademie der Wissenschaften phys.-math* (Fig. 4).[31] Diffuse intensity was just about discernible on the X-ray photograph at the position expected for the strongest $K\alpha_1$ emission line, but the $K\beta_1$ and $K\alpha_2$ lines '*sind in der Wiedergrabe schwer erkennbar*' (are hardly seen in the reproduction). A digitised version of the spectrum, obtained by counting developed grains in the original plate, is more convincing in showing all three X-ray lines for element 43 as weak but distinct features. The integrity of the counting procedure has however been called into question as there are discrepancies between the grain count plot and the plate from which it was purportedly derived.[32] Segrè was unambiguous in doubting that the Noddacks had ever found evidence for the existence of element 43 in their samples, and in an interview with C. Weiner conducted in 1937 he said 'They had done

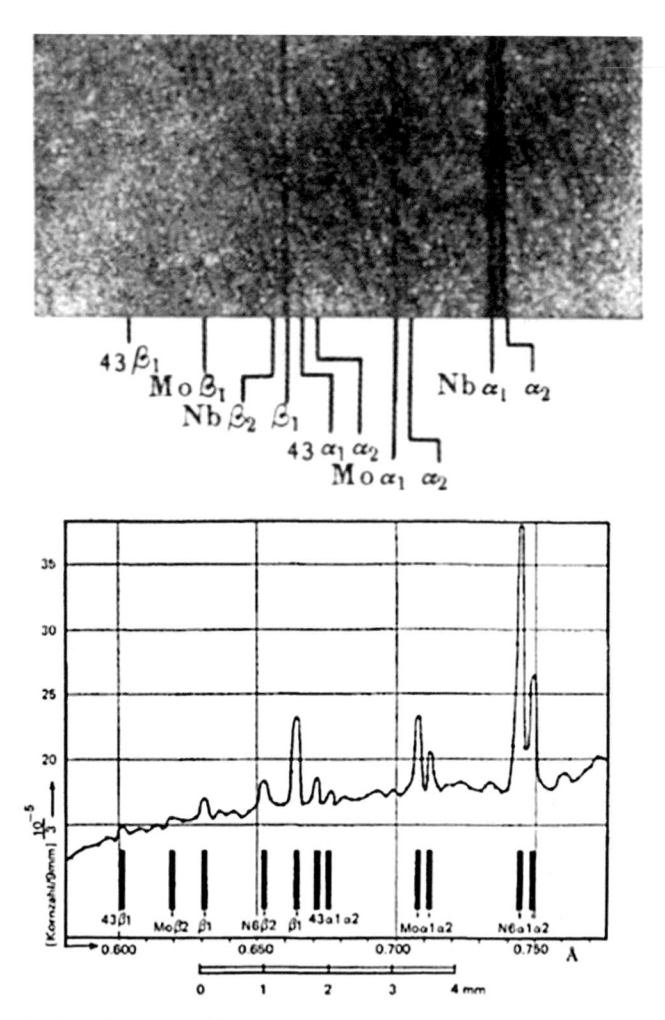

Fig. 4 Reproduction of an original X-ray plate used to support claims for discovery of element 43, along with a plot derived by counting grains in the photograph. The strongest line in the plot corresponds to twelve million grains having been counted. (Source: *Sitzungsberichte der Preussisschen Akademie der Wissenschaften phys.-math*, Klasse XIX (1925), 400).

excellent work on rhenium but they had been plain dishonest – there's no other way of saying it – on technetium 43 or what they called masurium'.[33]

Although identifiable isotopes of technetium were indeed first produced by artificial means and all of its isotopes are radioactive, its presence amongst the decay products of uranium means that it can be found in geological samples, although only in trace amounts. This led Pieter van Assche (in 1988) to attempt to rehabilitate the Noddacks 1925 claim.[34] Making a series of *ad hoc* assumptions, he arrived at the conclusion that the detection limit in Berg's spectroscopic measurements should have been about one part in 10^{-12} and not 10^{-9}, as was stated in the 1925 paper. He further argued that an abundance of element 43 at levels of 10^{-13} was not unreasonable if the mineral contained uranium. There was therefore a strong possibility that element 43 may have been detected in 1925.

In 1989, these arguments were contested by P. K. Kuroda, who had personal experience of X-ray spectroscopy in the 1930s.[35] In particular, he highlighted errors in van Assche's assumptions about the possible concentrations of technetium in the samples studied by the Noddacks, and further noted that Berg had been claimed that element 43 was to be found in samples of the mineral sperrylite ($PtAs_2$), which does not contain uranium and therefore cannot contain technetium. For years, this controversy rumbled on, and as recently as 2005, Roberto Zingales has reiterated the arguments for reinstating the Noddacks' claims.[36] His paper has in turn been persuasively criticised by Fathi Habashi,[37] who has drawn attention to Kuruda's rebuttal of van Assche. Some closure on the matter seems to have been reached with a retraction from Zingales, in response to Habashi's comments.[38]

Elements 85 and 87: Astatine and Francium

In Mendeleev's periodic table of 1871, there was a possibility of there being elements below iodine and caesium, *eka*-iodine, and *eka*-caesium. After the discovery of radon and radium, there remained obvious gaps to be filled in to the left of these two elements, at the bottom of the halogen and alkali metal groups, respectively. It is now known that all isotopes of both element 85 (now astatine, with chemical symbol At) and element 87 (now francium, with chemical symbol Fr) are radioactive, with short half-lives: the most stable isotopes are $^{210}_{85}At$ with a half-life of 8.1 hours and $^{232}_{87}Fr$, which has an even shorter half-life of 22 minutes. These isotopes are found in the decay sequences of uranium, thorium, radon and actinium. Whilst both these elements are in a strict sense naturally occurring, their terrestrial abundances are exceedingly low: there is estimated to be at most 15 gm of francium and 0.05 gm of astatine in the top kilometre of the earth's crust.[39] As with technetium, this means that attempts to establish their existence by measuring X-ray spectra from samples derived from naturally occurring minerals were doomed to failure.

Nonetheless, there have been two episodes where claims were made for the discovery of both elements based on X-ray spectroscopy. The first involved a collaboration between F. H. Loring and J. G. F. Druce, who reported in

Chemical News observation of 'a strange line in the X-ray spectrum' of a sample of crude rhenium oxide prepared by Druce.[40] This had a wavelength intermediate between the expected positions of $L\alpha_1$ and $L\alpha_2$ lines of element 87. However, their failure to observe the corresponding $L\beta$ lines was ignored. A week after the discovery of element 87, Loring and Druce also announced that they had also found $L\alpha$ and $L\beta$ emission lines characteristic of element 85.[41] They did, however, realise that the evidence for element 85 was flimsy, and cautioned:

> ...we feel compelled nevertheless, to regard these lines as uncertain indications of this new element being present in the sample .... We fully realise that much more work is necessary to establish the two elements, here mentioned on a firm basis.

Their provisional claims were never substantiated.

The second episode involved Yvette Cauchois (1908–1999) and Horia Hulubei (1896–1972), under the patronage Jean Baptiste Perrin (1870–1942), who won the Nobel Prize in Physics in 1926. For her doctorate at the Sorbonne, Cauchois designed and built a high performance spectrometer that used curved crystals to both focus and monochromate X-rays or soft γ-rays. This gave her instrument unparalleled sensitivity and resolution.[42] Hulubei was Romanian, but interrupted his studies at the University of Jassy to join the *Aeronautique Militaire* when Romania entered the Great War in 1916. After the War he completed his degree at Jassy, and returned to France in 1926.

Element 87 is first mentioned explicitly in a paper by Hulubei as sole author in 1936.[43] In this study, Hulubei worked with mineral samples rich in caesium, the element above number 87 in the periodic table. He reported emission lines at '1032 u.x. and 1043 u.x'., which he attributed to $L\alpha_1$ and $L\alpha_2$ transitions of element 87, but without giving any images of the X-ray plates to support his claims. In a second paper, Hulubei listed wavelengths for weaker $L\beta$ and $L\gamma$ emission lines.[44] In homage to his homeland, he proposed the name moldavium in the first paper; in the second, he introduced the chemical symbol Ml for the element in his title. Hulubei's claims were roundly criticised from Cornell University by F.R. Hirsh – first in 1937[45] and again in 1943.[46] Despite careful searches, Hirsh could find no evidence of emission lines from element 87 in samples of $CsHSO_4$ derived from the mineral lepidolite. Moreover, Hirsh dismissed Hulubei's emission lines as arising from mercury.[45]

The experiments performed by Hulubei and Cauchois in some cases exploited the 'internal' excitation produced by the radioactive disintegration of elements within the decay sequence of radon: a tube filled with radon gas then replaced the X-ray tube normally used.[47] This approach was inspired by work of Ernest Rutherford, reported in the *Proceedings of the Cambridge Philosophical Society* in 1925.[48] Incidentally, Rutherford's X-ray spectrometer incorporated the same sodium chloride crystal that Moseley had used in his first X-ray paper. With W. A. Wooster, Rutherford had shown that L-emission lines characteristic of element 83 (bismuth) could be observed during the decay of radium B as it was then known. Radium B is in fact $^{214}_{82}Pb$, which undergoes β-decay to $^{214}_{83}Bi$. Rutherford and Wooster concluded that:

> At the moment of the expulsion of the disintegration electron from radium B, the internal atom structure of radium B corresponds to an element of number 82, but an instant later, when the electron has escaped the nucleus, the charge on the latter is 83 and there must be a reorganisation of the external electrons.

This 'reorganisation' could lead to the emission of characteristic X-rays.

A decay sequence that could lead to element 85 from radon envisaged α-decay of $^{222}_{86}Rn$ to $^{218}_{84}Po$, followed by β decay to $^{218}_{85}85$ and emission of X-rays characteristic of the final decay product. The of detection of element 85 in this way did not depend on natural abundances, but rather on the branching ratios for the decay sequences available to radon, and the efficiency of the 'internal reorganisation' processes in promoting X-ray emission.[49]

The discovery of element 85 in a radon tube was announced in *Comptes rendus* by Hulubei and Cauchois in 1939, just before the outbreak of the Second World War. The positions of Lα₁, Lβ₁ and Kα₁ emissions were listed,[50] but yet again no original X-ray plates were reproduced. The authors were piqued when, the following year, W. Minder presented radiological evidence for element 85 in *Helvetica Chimica Acta*, without mentioning the French paper.[51] In 1940, Hulubei and Cauchois therefore published a short letter in *Comptes rendus* in which they reiterated their prior claim for element 85, and stated that they hoped to present a complete spectral analysis with the reproduction of photographic evidence.[52] The paper ends with the sad mitigation that '*malheureusement nos travaux ont été interrompus par la guerre* (sadly our work was interrupted by the war)'. Their research was in fact continued at the Instituto Superiore di Sanita in Rome by Manuel Valdares, who had studied under Cauchois. Valdares' experiments employed a larger sample of radon than had been available in Paris, and the characteristic X-ray lines of element 87 again revealed themselves. Most of this work was published in the proceedings of his institute.[53]

However, this war-time X-ray work was soon superseded by the now-accepted discovery of element 87 by Marguerite Perey, who had joined *l'Institut de Radium* in Paris as a technician in 1929. By 1939, she had realised that although $^{227}_{89}Ac$ decays mainly via β-emission, a small fraction (about 1%) undergoes α decay to element 87, which in turn undergoes β-decay with an extremely short half-life of twenty-one minutes. This is very much less than the half-life for β-decay of actinium, thus allowing the two decay processes to be differentiated.[54] Ironically, her claim for the discovery of element 87 appeared in a paper presented by Jean Perrin, who had earlier promoted Cauchois and Hulubei. Perey took some time to persuade herself that her observations could not be interpreted in terms of a new isotope of actinium. However, after the war her claims for element 87 gained general acceptance in the international scientific community. After flirting with the name catium,[55] she opted in 1951 to name her new element francium, after the country in which her research had been conducted.[56]

The eventual discovery of element 85 also involved radiological detection, coupled with chemical separation procedures that eliminated other known elements. In 1940, Corson, Mackenzie and Segrè announced that an artificial α-particle emitter with a half-life of 7.5 hours was produced when $^{209}_{83}Bi$ was bombarded with high energy α-particles.[57] They concluded that $^{211}_{85}85$ had been formed in an (α,2n) reaction. This new element had some characteristics of a metal, but could be co-distilled with iodine and was found to accumulate in the thyroid glands of guinea pigs (thus establishing physiological activity similar to that of iodine). These unfortunate animals died forty-five minutes after the injection of a preparation containing the new element.[58] Naming the element was not seen as a priority during the war, but in 1947, the trio proposed the name astatine, from the Greek for unstable.[59] Their short note to *Nature* with this proposal appeared immediately after a similar note naming element 43 technetium.[28]

Hulubei continued to believe in the priority of his claim for discovery of element 85, and continued to argue his case in the scientific literature.[60] In 1941, he moved back to Romania as professor of physics and Rector of Bucharest University, which suffered Allied bombardment in 1944, resulting in the destruction of his equipment. In a submission to the Romanian Royal Academy of Science, he pleaded that:

> We would like to call this element Dor (Do). It was identified during a period of terrible suffering for humanity. The name would, by its meaning in Romanian [hope or longing], recall a longing for a time when peace will bring an end to the most hateful war history has ever known.[61]

The periodic table would have been enriched by a name violating all the rules now imposed by the International Union of Pure and Applied Chemistry (IUPAC), but 'Dor' was not accepted. Hulubei also continued to promote his priority for element 87, and in 1947 produced a robust defence against the criticisms of Hirsh: his long delay in responding to Hirsh's 1943 paper was attributed to delays in American literature reaching Romania.[62] However, by this point, Perey's work had gained general acceptance, and moldavium was also consigned to history.

Element 61: Promethium (Pm)

In 1924, Moseley's second 'High-Frequency' paper was clear in demonstrating that there must be a missing element between neodymium and samarium, but he was not the first to make this suggestion. In 1902, the Bohemian chemist Bohuslav Brauner realised that the large difference in atomic weight between Nd and Sm pointed towards the existence of an unknown element.[63] However, his conclusions appeared only the *Proceedings of the Bohemian Academy of Sciences*, and did not attract international attention.[64]

The 'non-discovery' of element 61 by X-ray spectroscopy involves an even more convoluted sequence of events than for the elements discussed earlier, described by Clarence J. Murphy as a 'tangled web'.[65] Working in Florence in the early 1920s, Luigi Rolla and Lorenzo Fernandez performed multiple fractional recrystallisations of rare earth mixtures extracted from Brazilian monazite sand, in the hope of separating element 61. They measured L-shell X-ray spectra of their extracts, but were not entirely certain that they could see characteristic lines on their own X-ray plates. This work lay dormant between 1924 and 1926, with the X-ray plates and the samples deposited in a sealed package in the *Accademia dei Lincei* in Rome. Rolla and Fernandez were prompted to publish their findings in 1926 in response to two competing claims from the USA and after measurement of rather more convincing K-shell X-ray spectra by their colleague Professora Rita Brunetti.[66] They suggested their element should be known as florentinium. The American claims arose from collaborations led, respectively, by Charles James and B. Smith Hopkins. James had begun working on element 61 as early as 1912, following a letter from Sir William Ramsay that highlighted the large jump in atomic weights between neodymium and samarium. James was fully aware of Moseley's work, and in 1914 had been in correspondence with him about rare earth samples. In the same year, James claimed that weak visible region spectroscopic bands could be attributed to element 61.[67]

These claims were never substantiated, but beginning in 1923, James stepped up his efforts to isolate element 61 from Brazilian monazite sand, again using fractional crystallisation techniques. By this stage, he had moved from Britain to the University of New Hampshire, where he worked with a graduate student, James Cork. Whilst awaiting X-ray spectroscopic measurements to be made by H. C. Fogg in the Physics Department of the University of Michigan, James was asked to comment on two papers from the University of Illinois just submitted to the *Journal of the American Chemical Society*. This request was dated 26 April 1926, and the papers were: 'Observations on the Rare Earths XXIII. Element Number 61. Part One. Concentration and Isolation in Impure State' by J. Allen Harris and B. Smith Hopkins; and 'Observations on the Rare Earths XXIII. Number 61. Part Two. X-ray Analysis' by J. Allen Harris with L. F. Yntema and B. Smith Hopkins. The two papers were published on 5 June 1926.[68]

Prior publication gave Hopkins precedence in naming the element, and in a Special Article in *Science* published the day before (June 4), he proposed the name Illinium.[69] This article recounted the story of element 61, and concluded:

> It is true that later arrangements of the Periodic Table indicated that eka-caesium, eka and dwi-manganese and eka-iodine were missing, but there were no theoretical grounds for supposing that eka-neodymium might exist, until Moseley's rule showed that element number 61 was still to be identified.

This statement was repeated in a letter to *Nature*, also appearing on 4 June.[70] Hopkins' letter prompted Brauner to write to *Nature* from Prague disputing the statement, and drawing attention to his own earlier prediction of a missing element between neodymium and samarium in a paper read before the Bohemian Academy of Science in 1902.[71] Finally in November, 1926, Cork, James and Fogg submitted their own paper describing the concentration and identification of element 61 to the *Proceedings of the National Academy of Science*, and this was published later that year.[72]

Looking at the two X-ray papers from the USA, it is striking that the contribution by Harris, Yntema and Hopkins tabulates values for wavelengths of two L series X-ray lines of element 61, but does not present any of the six original X-ray plates mentioned in the paper. Instead, we have: 'In all cases our lines were faint, indicating that our material is not in a very concentrated form. Our plates have in all cases been verified by disinterested observers'. However, there is no indication as to who these observers actually were. Moreover, the paper compares wavelengths for the X-rays of element 61 with values interpolated from measurements on adjacent elements, so it is hardly surprising that the lines are reported to appear exactly where they should.

By contrast, Cork, James and Fogg did present an L shell X-ray spectrum, with the positions of seven lines of element 61 marked on the photographic plate (Fig 5). They then asserted that:

> The estimated amount of oxalate of element 61 was from 1 to 1.5 percent. While there are more than 20 L series lines for each element only about seven of these are fairly strong.... all seven of these lines for element 61 are observable on the original plates, although some of them are rather faint in the reproduction.[72]

Today, it requires more than an eye of faith to see any of the lines which

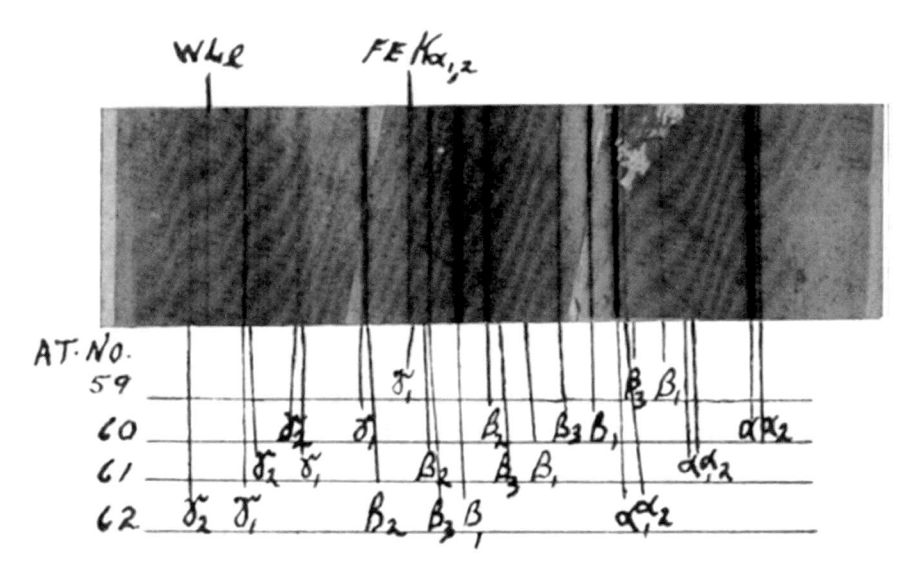

Fig. 5 An X-ray plate upon which Cork, James and Fogg based their claims for discovery of element 61. (Source: *Proceedings of the National Academy of Sciences of the United States of America*, 12 (12), (1926), 696-699).

are identified in the published photograph, even the strongest $L\alpha_{1,2}$ lines. In retrospect, attempts to 'discover' element 61 by X-ray spectroscopy could never have met with success in the 1920s, as yet again it is now known that all isotopes of the element are highly radioactive, and that the terrestrial abundance of the element is far too low for it to be detected by X-ray techniques using material derived from naturally occurring samples.

Element 61 was finally produced at the Clinton Laboratories of America's wartime Manhattan Project at Oak Ridge, Tennessee: the work was completed in 1945, but not published until 1947.[73] In collaboration with Jacob Marinsky and Lawrence Glendenin at the Massachusetts Institute of Technology, the atomic scientist Charles Coryell reported that samples of the fission products of uranium fuel from a graphite reactor had been analysed by ion exchange chromatography. This technique had been developed in the Manhattan Project as a means of separating rare earths and actinides in solution. An elution peak with a retention time on the chromatographic column slightly different to that of neodymium ($Z = 60$) was identified radiologically, thus proving the existence of element 61.

The Americans first considered that the new element should be called clintonium, after the laboratory where it was discovered. However, Grace Mary Coryell, the wife of Charles Coryell, came up with the more appealing prometheum, after Prometheus who stole fire from Mount Olympus.[74] This spelling was modified later to promethium by IUPAC. In 1949, W.F. Burkhart, W. F. Peed and E. J. Spitzer of the Carbide and Carbon Chemicals Corporation (also based at Oak Ridge) published K-shell and L-shell X-ray emission spectra of promethium, using samples from what had now become the Oak Ridge National Laboratory.[75] Remarkably, the energies of the L shell lines almost perfectly matched those reported by Cork, James and Fogg. The 1949 paper makes no comment on this issue, but the striking coincidence between the X-ray spectra led Clarence Murphy to speculate that James and co-workers may have indeed isolated and characterised promethium in 1926.[65] However, the terrestrial abundance of promethium is so low that it must have been impossible to measure true X-ray spectra of element 61 at the time.

Primordial Superheavy Elements

Looking further forward than Moseley could have imagined, recent research on elements beyond $Z = 100$ has involved the synthesis of new isotopes by nuclear fusion. Elements produced in this way invariably have very short half-lives and are obtained in almost vanishingly small amounts: in extreme cases, formal ratification of a new element by IUPAC may involve only a handful of atoms. Here, the established techniques of X-ray spectroscopy are not applicable. However, in the 1970s, there was widespread speculation that an 'island' of nuclear stability might be found amongst so-called 'superheavy' elements.

Radiation damage from radioactive isotopes, which undergo α-particle decay within transparent minerals such as mica, leads to the appearance of coloured 'spherical halos'. The radii of the halos provide a measure of the energy of the α-particles responsible for introducing the crystal defects that produce the colouration. In 1976, R.V. Gentry from the Oak Ridge National Laboratory, with co-workers from Florida and California, reported investigations of 'giant halos' in samples of monazite from Madagascar. The halo diameters pointed to α energies in excess of 14 MeV, which could only arise from hitherto unknown elements.

Using focused beams of protons with energies of the order of 5 MeV to excite the X-ray spectra, these scientists examined the energy window between the L-emission lines of thorium and uranium and the K-emission lines of the lanthanides. Very weak X-ray emission lines in this region, with energies around 27 keV, were attributed to the L emissions of superheavy elements with Z values of 116, 124, 126 and 127. The results were published in the most prestigious physics journal, *Physical Review Letters*.[76] Unusually, this paper was not made subject to peer review. Publication provoked a small flurry of submissions to *Physical Review Letters*, using exotic models of nuclear structure to rationalise the apparent stability of the new elements, with predictions of half-lives of many millions of years in the oasis of stability.[77]

Another paper from the University of Pittsburgh put forward the bizarre suggestion that the elusive element 116 (which falls in the same group as selenium and polonium) might be concentrated in the livers of lobsters, shrimps and crabs, and that physicists should harvest these creatures as a source of superheavy elements.[78] This paper was not embraced by the physics community. Meanwhile, the Florida group realised that at least some of the purported superheavy X-ray lines could arise from internal nuclear transitions resulting from excited isotopes produced by the proton bombardment: properly speaking, soft γ-rays rather than X-rays were being measured.[79] Indeed, this possibility had been raised in their original paper. In subsequent papers, the Oak Ridge group showed that no evidence for superheavy elements could be found when synchrotron radiation was used to excite the X-ray spectra.[80]

The lid on the coffin of speculation about primordial superheavy elements was sealed by extremely sensitive nuclear fission experiments, which also yielded negative results, thus ending this chapter in the history of new elements and X-ray lines.[81] However, almost forty years later, two short-lived isotopes of element 116 with atomic weights of 290 and 291 were made by nuclear fusion of $^{48}_{20}Ca$ and $^{245}_{96}Cm$.[82] The element was given the name livermorium, in recognition of the Lawrence Livermore Laboratory.[83]

Conclusion

Moseley's two 'High-Frequency' papers, published before the Great War, both defined how missing elements were to be identified and provided a method for their discovery. In their paper of 1923, announcing that element 72 was to be found in zirconium minerals, Coster and Hevesy wrote that:

> Since Moseley's discovery of the fundamental laws of X-ray emission it has become clear that the most simple and conclusive characteristic of a chemical element is given by its X-ray spectrum. In addition, Moseley's laws allow us to calculate very accurately the wave-lengths of the X-ray spectral lines for any element in the periodic table if those of the elements in its neighbourhood are known. Taking into account that the presence of a very small proportion of a definite element in any chemical substance suffices to give a good X-ray spectrum of this element it is quite evident that for the eventual discovery of any unknown element X-ray spectroscopy, especially as it has been develop by Siegbahn, represents the most effective method.[14]

As we have seen, this message was embraced by a generation of element hunters and, after hafnium, led to the discovery of rhenium. However, it was not foreseen that the four elements remaining after these two discoveries were all highly radioactive, and had exceedingly small terrestrial abundances. Three of these elements were 'discovered' only after synthesis in a nuclear reactor, and all required radiological detection (that is measurement of the characteristics of their radioactivity). To the modern historian of science, the story of elements 43, 61, 85 and 87 was muddied by reports of X-ray wavelengths based on data that would not convince a contemporary referee. Knowledge of where the X-ray lines should be found was more of a hindrance than a help, and led to what, in retrospect, must be judged as episodes of self-deception. Moreover, the willingness of journals to publish values for X-ray wavelengths of hitherto unknown elements without supporting data, does not speak well of earlier scientific conventions, or conform to current norms. Even after radiological techniques had emerged as the best way to characterise new radioactive elements, X-rays reared their head again in the 1970s with claims for discovery of primordial superheavy elements based on marginally plausible X-ray spectra. Although this proved to be a blind avenue of research, it is clear that Moseley's papers of 1913 and 1914 were still shaping scientific practice more than sixty years after his death in Gallipoli.

Notes and References

[1] H. G. J. Moseley, 'The High-Frequency Spectra of the Elements'. *Philosophical Magazine Series 6*, 26 (156), (1913), 1024–1034.

[2] M. Siegbahn, *Spektroskopie der Röntgenstrahlen* (Berlin: Springer-Verlag, 1931).

[3] M. Siegbahn and E. Friman, 'On the High-Frequency Spectra of the Elements Gold-Uranium', *Philosophical Magazine Series 6*, 31 (184), (1916), 403–406.

[4] K. Fajans and O. Göhring, 'Über die Komplexe Natur des Ur X'. *Naturwissenschaften*, 1 (14), (1913), 339.; K. Fajans and O. Göhring, 'Über das Uran X2 – Das Neue Element der Uranreihe', *Physikalische Zeitschrift*, 14 (1913), 877–884

[5] K. Fajans and P. Beer, 'Das Verhalten der Radio-Elemente Fallungsreaktionen', *Berichte der deutschen chemischen Gesellschaft*, 46 (3), (1913), 348–349.

[6] O. Hahn and L. Meitner, 'Die Muttersubstanz des Actiniums, Ein Neues Radioaktives Element von Lebendauer', *Physikalische Zeitschrift*, 19 (1918), 208–218.

[7] H. Beuthe and A. von Grosse, 'Die L-Serie des Elements 91 – Protactinium', *Zeitschrift für Physik*, 61 (3-4), (1930), 170–173.

[8] E. Scerri, *A Tale of Seven Elements* (Oxford: Oxford University Press, 2013). Hereafter, Scerri, *Seven Elements*; M. Fontani, M. Costa and M. V. Orna, *The Lost Elements: The Periodic Table's Shadow Side* (Oxford: Oxford University Press, 2015). Hereafter, Fontani, *Periodic Table*.

[9] A. Dauvillier, 'Sur le Series L du Lutecium et de l'Ytterbium et sur l'Identification du Celtium avec l'Element de Nombre 72', *Comptes rendus de l'Académie des sciences*, 174 (1922), 1347–1349.

[10] G. Urbain, 'Les Numeros Atomique du Neo-Ytterbium, du Lutecium et du Celtium', *Comptes rendus de l'Académie des sciences*, 174 (1922), 1349–1351.

[11] Fontani, *Periodic Table, op. cit.* 237–238.

[12] Scerri, *Seven Elements, op. cit.* 95–98.

[13] D. Coster and G. Hevesy, 'On the Missing Element of Atomic Number 72', *Nature*, 111 (2777), (1923), 79.

[14] D. Coster, 'On the X-Ray Spectra of Hafnium and Thulium', *Philosophical Magazine Series 6*, 46 (275), (1923), 956–963.

[15] G. Hevesy, 'The Discovery and Properties of Hafnium', *Chemical Reviews*, 2 (1), (1925), 1–41.

[16] H. Kragh, 'Anatomy of a Priority Conflict – the Case of Element-72', *Centaurus*, 23 (4), (1980), 275–301.

[17] Scerri, *Seven Elements, op. cit.* 91.

[18] Fontani, *Periodic Table, op. cit.* 241.

[19] *Ibid*. 117.

[20] Scerri, *Seven Elements, op. cit.* 102–104.

[21] C. H. Bosanquet and T. C. Keeley, 'Note on a Search for the Missing Element No. 43', *Philosophical Magazine Series 6*, 48 (283), (1924), 145–147.

[22] R. Hamer, 'Moseleyum', *Science*, 61 (1573), (1925), 208–209.

[23] Editorial, 'Current Topics and Events', *Nature*, 115 (2893), (1925), 545.

[24] W. Noddack, I. Tacke and O. Berg, 'Die Ekamangane', *Naturwissenschaften*, 13 (5), (1925), 567–574.

[25] J. Noddack and W. Noddack, 'Die Herstellung von Einem Gramm Rhenium', *Zeitschrift für anorganischen und allgemeine chemie*, 183 (1), (1929), 353-375.

[26] C. Perrier and E. Segrè, 'Some Chemical Properties of Element 43', *Journal of Chemical Physics*, 5 (9), (1937), 712–717.

[27] E. Segrè and C. S. Wu, 'Some Fission Products of Uranium', *Physical Review*, 57 (6), (1940), 552.

[28] C. Perrier and E. Segrè, 'Technetium – the Element of Atomic Number-43', *Nature*, 159 (4027), (1947), 24.

[29] Fontani, *Periodic Table, op. cit.* 313–314.

[30] O. Berg, 'Über der Röntgenspektroskopischen Nachweis der Ekamangane', *Zeitschrift fur angewandte chemie*, 40 (9), (1927), 254–256.

[31] W. Noddack, I. Tacke and O. Berg, 'Ekamangane', *Sitzungsberichte der Preussisschen Akademie der Wissenschaften phys.-math, Klasse XIX* (1925), 400.

[32] G. Herrmann, 'Technetium or Masurium – a Comment on the History of Element-43', *Nuclear Physics A*, 505 (2), (1989), 352–360.

[33] Scerri, *Seven Elements, op. cit.* 136–137

[34] P. H. M. van Assche, 'The Ignored Discovery of Element Z = 43', *Nuclear Physics A*, 480 (2), (1988), 205–214.

[35] P. K. Kuroda, 'A Note on the Discovery of Technetium', *Nuclear Physics A*, 503 (1), (1989), 178–182.

[36] R. Zingales, 'From Masurium to Trinacrium: The Troubled Story of Element 43', *Journal of Chemical Education*, 82 (2), (2005), 221–227.

[37] F. Habashi, 'The History of Element 43 – Technetium', *Journal of Chemical Education*, 83 (2), (2006), 213.

[38] R. Zingales, 'The History of Element 43 – Technetium (the Author Replies)', *Journal of Chemical Education*, 83 (2), (2006), 213.

[39] N.N. Greenwood and A. Earnshaw, *Chemistry of the Elements* (Oxford, Pergamon Press, 1989), 76; *Ibid*. 928.

[40] F. H. Loring and J. G. F. Druce, 'Eka-Caesium', *The Chemical News and Journal of Industrial Science*, 131 (6 November), (1925), 289–289.

[41] F. H. Loring and J. G. F. Druce, 'Eka-Caesium and Eka-Iodine', *The Chemical News and Journal of Industrial Science*, 131 (13 November), (1925), 305.

[42] Y. Cauchois, 'Spectrographe Lumineux par Transmission de Rayons X non Canalises a Travers un Mica Courbe', *Comptes rendus de l'Académie des sciences*, 194 (1932), 362–365; Y. Cauchois, 'Spectrographie Des Rayons X Par Transmission d'un Faisceau non Canalise a Travers un Cristal Courbe – I', *Journal de Physique et Le Radium*, 3 (1932), 320–336; Y. Cauchois, 'Spectrographie des Rayons X par Transmission d'un Faiseau non Canalise a Travers un Cristal Courbe II', *Comptes rendus de l'Académie des sciences*, 194, (1932), 1479–1482.

[43] H. Hulubei, 'Recherches Relatives a l'Element 87', *Comptes rendus de l'Académie des sciences*, 202 (1936), 1927–1929.

[44] H. Hulubei, 'Nouvelle Recherche sur l'Element 87 (Ml)', *Comptes rendus de l'Académie des sciences*, 205 (1937), 854–856.

[45] F. R. Hirsh, 'A Note on the Search for Element 87', *Physical Review*, 51 (7), (1937), 584–587.

[46] F. R. Hirsh, 'The Search for Element 87', *Physical Review*, 63 (3-4), (1943), 93–96.

[47] H. Hulubei and Y. Cauchois, 'Nouvelle Technique dans la Spectrographie Cristalline Des Rayons Gamma'. *Comptes rendus de l'Académie des sciences*, 199 (1934), 857–859.

[48] E. Rutherford and W. A. Wooster, 'The Natural X-Ray Spectrum of Radium B', *Proceedings of the Cambridge Philosophical Society*, 22 (6), (1925), 834–837.

[49] B. F. Thornton, 'Finding Eka-Iodine: Discovery Priority in Modern Times', *Bulletin for the History of Chemistry*, 35 (2), (2010), 86–96.

[50] H. Hulubei and Y. Cauchois, *Comptes rendus de l'Académie des sciences*, 209 (1939), 39. This paper is cited in subsequent papers (without the title being mentioned) but it is not available in the *Comptes rendus* archive.

[51] W. Minder, 'Über die Beta-Strahlung des Ra A und die Bildung des Elementes mit der Kernladungszahl 85', *Helvetica Chimica Acta*, 13 (2), (1940), 144–152.

[52] H. Hulubei and Y. Cauchois, 'Sur la Presence de l'Element 85 parmi les Produits de Disintigration du Radon', *Comptes rendus de l'Académie des sciences*, 210 (1940), 696–697.

[53] Fontani, *Periodic Table op. cit.* 331–334.

[54] M. Perey, 'Sur un Element 87, Derive de l'Actinium', *Comptes rendus de l'Académie des sciences*, 208 (1939), 97–98.

[55] Scerri, *Seven Elements, op. cit.* 156–161.

[56] M. Perey, 'Le Francium – Element-87', *Bulletin de la Societe Chimique de France*, 18 (11–1), (1951), 779–785.

[57] D. R. Corson, K. R. Mackenzie and E. Segrè, 'Possible Production of Radioactive Isotopes of Element 85', *Physical Review*, 57 (5), (1940), 459.

[58] D. R. Corson, K. R. Mackenzie and E. Segre, 'Artificially Radioactive Element 85', *Physical Review*, 58 (8), (1940), 672–678.

[59] D. R. Corson, K. R. Mackenzie and E. Segre, 'Astatine – the Element of Atomic Number-85', *Nature*, 159 (4027), (1947), 24.

[60] Scerri, *Seven Elements, op. cit.* 168.

[61] Fontani, *Periodic Table, op. cit.* 333.

[62] H. Hulubei, 'Search for Element 87', *Physical Review*, 71 (10), (1947), 740–741.

[63] Fontani, *Periodic Table, op. cit.* 290

[64] Scerri, *Seven Elements, op. cit.* 176–180

[65] C. J. Murphy, 'Charles James, B. Smith Hopkins and the Tangled Web of Element 61', *Bulletin for the History of Chemistry*, 31 (1), (2006), 9–18.

[66] Fontani, *Periodic Table, op. cit.* 291–295

[67] Scerri, *Seven Elements, op. cit.* 182–184

[68] J. A. Harris and B. S. Hopkins, 'Observations on the Rare Earths XXIII. Element Number 61. Part One. Concentration and Isolation in Impure State', *Journal of the American Chemical Society*, 48 (6), (1926), 1585–1594.; J. A. Harris, L. F. Yntema and B. S.

Hopkins, 'Observations on the Rare Earths XXIV. Element Number 61. Part Two. X-Ray Analysis', *Journal of the American Chemical Society*, 48 (6), (1926), 1594–1598.

[69] J. A. Harris, L. F. Yntema and B. S. Hopkins, 'Illinium', *Science*, 63 (1640), (1926), 575–576.

[70] J. A. Harris, L. F. Yntema and B. S. Hopkins, 'The Element of Atomic Number 61: Illinium', *Nature*, 117 (2953), (1926), 792–793.

[71] B. Brauner, 'The New Element of Atomic Number 61: Illinium', *Nature*, 118 (2959), (1926), 84–85.

[72] J. M. Cork, C. James and H. C. Fogg, 'The Concentration and Identification of the Element of Atomic Number 61', *Proceedings of the National Academy of Sciences of the United States of America*, 12 (12), (1926), 696–699.

[73] J. A. Marinsky, L. E. Glendenin and C. D. Coryell, 'The Chemical Identification of Radioisotopes of Neodymium and of Element-61', *Journal of the American Chemical Society*, 69 (11), (1947), 2781–2785.

[74] Fontani, *Periodic Table, op. cit.* 304–306

[75] L. E. Burkhart, W. F. Peed and E. J. Spitzer, 'The K-Spectra of Element-61', *Physical Review*, 75 (1), (1949), 86–89.; W. F. Peed, E. J. Spitzer and L. E. Burkhart, 'The L-Spectrum of Element-61', *Physical Review*, 76 (1), (1949), 143–144.

[76] R. V. Gentry, T. A. Cahill, N. R. Fletcher, H. C. Kaufmann, L. R. Medsker, J. W. Nelson and R. G. Flocchini, 'Evidence for Primordial Superheavy Elements', *Physical Review Letters*, 37 (1), (1976), 11–15.

[77] F. Petrovich, R. J. Philpott, D. Robson, J. J. Bevelacqua, M. Golin and D. Stanley, 'Comments on Primordial Superheavy Elements', *Physical Review Letters*, 37 (9), (1976), 558–561.; C. Y. Wong, 'Primordial Superheavy Element-126', *Physical Review Letters*, 37 (11), (1976), 664–666.

[78] R. L. Wolke, 'Proposed Experiment to Corroborate Existence of Superheavy Element 116 in Nature', *Physical Review Letters*, 37 (16), (1976), 1098–1100.

[79] J. D. Fox, W. J. Courtney, K. W. Kemper, A. H. Lumpkin, N. R. Fletcher and L. R. Medsker, 'Comment on Evidence for Primordial Superheavy Elements', *Physical Review Letters*, 37 (10), (1976), 629–631.

[80] C. J. Sparks, S. Raman, E. Ricci, R. V. Gentry and M. O. Krause, 'Evidence against Superheavy Elements in Giant-Halo Inclusions Reexamined with Synchrotron Radiation', *Physical Review Letters*, 40 (8), (1978), 507–511.

[81] R. V. Gentry, J. H. Halperin, B. H. Ketelle, G. D. Okelley, R. W. Stoughton and J. A. S. Adams, 'Reinvestigation of Alpha-Activity of Conway Granite', *Nature*, 273 (5659),

(1978), 217–218.; C. Stephan, M. Epherre, E. Cieslak, M. Sowinski and J. Tys, 'Search for Superheavy Elements in Monazite Ore from Madagascar'. *Physical Review Letters*, 37 (23), (1976), 1534–1536.

[82] Y. S. Oganessian, V. K. Utyonkov, Y. V. Lobanov, F. S. Abdullin, A. N. Polyakov, I. V. Shirokovsky, Y. S. Tsyganov, G. G. Gulbekian, S. L. Bogomolov, B. N. Gikal, A. N. Mezentsev, S. Iliev, V. G. Subbotin, A. M. Sukhov, A. A. Voinov, G. V. Buklanov, K. Subotic, V. I. Zagrebaev, M. G. Itkis, J. B. Patin, K. J. Moody, J. F. Wild, M. A. Stoyer, N. J. Stoyer, D. A. Shaughnessy, J. M. Kenneally and R. W. Lougheed, 'Measurements of Cross Sections for the Fusion-Evaporation Reactions $^{244}Pu\,(^{48}Ca,xn)\,^{292-x}114$ and $^{245}Cm\,(^{48}Ca,xn)\,^{293-x}116$', *Physical Review C*, 69 (5), (2004), 054607/1–9

[83] R. D. Loss and J. Corish, 'Names and Symbols of the Elements with Atomic Numbers 114 and 116 (IUPAC Recommendations 2012)', *Pure and Applied Chemistry*, 84 (7), (2012), 1669–1672.

CHAPTER NINE

X-RAY SPECTROSCOPY 100 YEARS ON

RUSSELL G EGDELL

Introduction

In the 100 years since Moseley first measured 'High-Frequency' spectra of the elements, X-ray spectroscopy has evolved to span an enormous and diverse range of activity in both fundamental and applied research. For example, X-ray emission spectroscopy has established itself as an everyday tool in elemental analysis, used in such different fields as scrap metal dealing and mineralogical surveying. X-ray spectrometers have found their way into space – not least in NASA's Mars Rover, *Curiosity*. The technique is now applied with atomic scale spatial resolution within electron microscopes, and on timescales in the femtosecond regime using powerful X-ray lasers. In parallel with these developments, other techniques have emerged based on the characteristic energies of atomic core levels: these include X-ray absorption spectroscopy, X-ray photoelectron spectroscopy, and Auger electron spectroscopy. Along with X-ray emission, these spectroscopies are now used to probe fundamental aspects of the electronic structure of solids and their surfaces. Looking across the very broad spectrum of contemporary research effort in the field of X-ray spectroscopy, we can well appreciate the ongoing impact of the work conducted by Moseley before the Great War.

Background

X-ray emission spectroscopy (XES) developed rapidly after Moseley's two pioneering papers.[1] Key figures, including Manne Siegbahn (1886–1978), Alexandre Dauvillier (1882–1975), Maurice de Broglie (1875–1960) and Dirk Coster (1889–1950) have been introduced in earlier chapters. The Swedish physicist Manne Siegbahn played a central role in consolidating the technique, first as a professor in Lund, and after 1923, in Uppsala. His initial contribution was to extend measurement of X-ray spectra beyond gold, which had been Moseley's limit, to include all available elements up to uranium.[2] He proved to be a fastidious builder of increasingly precise instruments, embracing a wide range of energies, from the softest to the hardest of X-rays.[3] The soft M-type rays that Moseley had predicted were first observed in Lund[4] after development of improved high vacuum techniques.[5] Siegbahn also introduced the xu unit, which was widely embraced when reporting wavelengths in X-ray spectra.[6] In the presentation speech at the award of the deferred 1924 Nobel Prize in Physics, the Chairman of the Physics Committee, Allvar Gullstrand (1862–1930), warmly praised his countryman:

> The high level to which he [Siegbahn] has brought X-ray spectroscopy can perhaps be best defined by the statement that the exactitude with which wavelengths can now be measured by his methods is a thousand time greater than attained by Moseley…. Siegbahn's work attains the character that is required for the award of a Nobel Prize not only because his methods of measurement provide an implement of hitherto undreamt-of exactitude, apt to further new scientific advances, or because he himself has used them to make a number of new discoveries, but above all owing to the importance for atomic physics that his methods of measurement and discovery have.[7]

Although Siegbahn himself did not contribute directly to the development of atomic theory, his precision measurements were undoubtedly of great importance to Niels Bohr and Arnold Sommerfeld, with whom he collaborated. In 1934, the words chosen in his nomination for the Hughes Medal of the Royal Society of London for his 'work as a physicist and technician on long-wave X-rays'[8] were surely intended as a compliment.

In parallel with X-ray emission spectroscopy (XES), came the field of X-ray absorption spectroscopy (XAS). The first indication of discrete and characteristic absorption edges were found in experiments performed in 1913 by the French physicist Maurice de Broglie, elder brother of Louis-Victor (who later became well-known for his ideas about the wave properties of electrons). Maurice de Broglie perfected a means of measuring emission spectra using a rotating crystal mounted on the drum of a recording barometer, coupled with photographic detection of the X-rays.[9] He also pioneered excitation of emission spectra using X-ray rather than electron bombardment, thus opening the technique to non-metallic elements.[10] Whilst recording an emission spectrum of tungsten, he found discontinuities on the photographic plate at energies corresponding to the K-edges of silver (Ag) and bromine (Br),[10] arising from the AgBr in the photographic emulsion.[11] De Broglie went on to make systematic measurements of absorption edges of elements between bromine and thorium using thin sample layers.[12] X-ray absorption spectroscopy was further developed by H. Fricke, who measured K shell edges for all elements between magnesium and chromium using one of Siegbahn's vacuum spectrographs;[13] and by G. Hertz, who concentrated on lower-energy L-shell absorptions in heavier elements between caesium and neodymium.[14]

To a first approximation, the energies of X-ray absorption edges and emission lines are essentially characteristic of an element, but there are small shifts dependent on the chemical state of the element – particularly, the oxidation state. Reviewing progress in the field in 1930, Harold Roper Robinson (1889–1955) noted that so-called 'chemical shifts' in absorption edges were first identified in 1920, four years before discovery of similar shifts in emission spectra.[15] In Siegbahn's 1931 edition of *Spektroskopie der Röntgenstrahlen*, the topic of *absorption und chemische bindung* (absorption and chemical binding) merited a whole chapter.[6] A further important landmark in 1931 was the observation by J. D. Hanawalt that oscillatory structure above the absorption edge found for solid Zn was absent for Zn vapour.[16] This laid the foundations for development of the analysis of the oscillations (so-called extended X-ray absorption fine structure or EXAFS) as a structural tool in the 1970s.

Following the success of XES and XAS, X-ray photoelectron spectroscopy (XPS) became the third important field of core level spectroscopy to emerge in the past 100 years. Although Moseley is not generally credited with this technique, he did in fact have an influence on its development, by virtue of the bequest to the Royal Society made in his will. Summarising established knowledge about X-rays in 1906, Charles Barkla noted that when X-rays fell on a sample, corpuscular radiation (which is made up of what we would now call photoelectrons) was emitted as well as the characteristic X-rays.[17] The first attempt to measure the kinetic energy of these electrons was probably due to P. D. Innes.[18] However, the development of XPS was largely due to the work of Harold Roper Robinson, FRS (1889–1955).[19] Robinson joined the physics department in Manchester in 1908, took a first class degree in 1911 and served his apprenticeship in radioactivity. He worked alongside Moseley, with whom he published one paper on the number of ions produced by the β- and γ-radiations from radium.[20]

Robinson's most significant work in Manchester involved measurement of the kinetic energies of β-rays (also called secondary corpuscular rays) that are emitted when characteristic X-rays fall on a metal target. His initial experiment involved a nickel anticathode and magnetic deflection of the electrons emitted from iron and lead targets onto a photographic plate,[21] but this work was interrupted by the outbreak of the Great War. Robinson was commissioned as a second lieutenant in the Royal Garrison Artillery in February 1915 and, like Lawrence Bragg and Charles G. Darwin, worked on techniques of sound ranging to pinpoint enemy guns. Robinson returned to Manchester after the war, but in the summer of 1919, Rutherford moved to the Cavendish in Cambridge. Robinson followed him in 1921 as the first Moseley Research Student appointed by the Royal Society. This Studentship derived from Moseley's bequest to the Royal Society, 'to be applied to the furtherance of experimental research in Pathology, Physics, Physiology, Chemistry or other branches of science'.[22]

In Cambridge, Robinson continued his work on corpuscular rays with an improved copper-Kα X-ray source to excite the spectra. The frequencies of the core levels derived by application of the Einstein equation for the photoelectric effect were compared with X-ray emission frequencies for a series of seventeen elements, including gold, lead and bismuth, thus laying the foundations of the field of X-ray photoelectron spectroscopy. In his seminal paper, Robinson acknowledges that the 'Moseley Studentship … made it possible for me to resume scientific work'.[23]

Robinson continued in this field after successive moves to Edinburgh, Cardiff and what became Queen Mary College in London. In 1930, he was the first to demonstrate a chemical shift in X-ray photoelectron spectra.[15] He was also the first to provide a convincing quantitative demonstration that the energies of so-called Auger electrons that are emitted under X-ray excitation have energies independent of the energy of the exciting radiation.[24] As the name suggests, Pierre Auger (1899–1993) is generally credited with being first to establish that a second family of characteristic electron is emitted along with photoelectrons when an atom suffers core excitation.[25] These were identified in cloud chambers, where two distinct electron tracks were seen to emerge from core-ionised molecules or atoms in the gas phase. Initially, the Auger electrons were thought to arise from internal reabsorption of a photon produced by X-ray fluorescence, and in Robinson's paper from 1926 they were described as 'tertiary cathode rays'. It is now known that Auger electron emission is a synchronous process which competes with X-ray emission as a channel for decay of core-hole excited states and does not involve emission and reabsorption, although energetically the two amount to the same thing. Despite the best efforts of Robinson and others, electron spectroscopy remained an obscure and unpopular technique for several decades. Its eventual emergence as a routine spectroscopic method in the 1950s required a series of technical innovations, most of them introduced by Kai Siegbahn (1918–2007), the son of Manne Siegbahn.

The rapid development of X-ray spectroscopy as an academic field in its first two decades may be gauged by the number of papers referenced in the 1931 edition of Manne Siegbahn's monograph.[5] The number of citations increased from 41 in 1916 to 187 in 1930. Following the Second World War, the growth in publications has been explosive: a simple online search of the literature throws up almost 550,000 papers on X-ray spectroscopies broadly defined since 1945.[26] This includes about 170,000 on XPS, which has come to play the biggest role in academic publications. In 2016 alone, 13,000 papers on XPS were published. This is a big field.

X-Ray Emission Spectroscopy as a Routine Analytical Technique

In the first of his two 'high frequency' papers, Moseley anticipated that '[X-ray spectroscopy] may prove a powerful method of chemical analysis'.[27] X-ray emission spectroscopy (XES) – or X-ray fluorescence (XRF), as it is often called – has indeed developed to become a routine analytical technique of enormous importance. The market in X-ray spectrometers of various sorts must now be counted in hundreds of millions of pounds per annum. It is applicable to solids, powders and liquids, and is used in a wide range of industries and technologies.[28]

Commercial instruments enable XRF to be performed in two main ways: these are energy-dispersive X-ray fluorescence (EDXRF) and wavelength-dispersive X-ray fluorescence (WDXRF). A typical bench-top EDXRF instrument includes an X-ray tube (usually with a rhodium or silver anode) producing radiation hard enough to ionise all elements up to uranium. Detectors are based on a silicon diode or a strip of ultrahigh purity silicon. X-rays incident on the detector can excite valence electrons to levels far above the band gap, leaving holes in the valence band. Highly excited electrons and holes quickly lose energy to give electrons and holes sitting at the band edges. The number of electron-hole pairs produced in this way depends on the energy of the incident X-ray, and this in turn determines the magnitude of a charge pulse registered by the detection system.

This instrument configuration is amenable to miniaturisation to give portable hand held X-ray spectrometers which are now used in the field in a diverse range of areas. These include mining and geological surveying; environmental monitoring; jewellery and pigment analysis (for example in paintings or in glazes); and scrap metal dealing. Fig. 1 shows a portable instrument being used in a metal yard, where it is obviously important to know the chemical composition of the scrap. A handheld EDXRF instrument

Fig. 1 Portable energy-dispersive X-ray fluorescence instrument in use in a scrap metal yard. (Source: Christelle Petiot, 2016, by courtesy of Oxford Instruments).

can cost as little as £15,000 and weigh around two kilograms. A detection limit of about 50 ppm can be achieved with measuring times of a minute in optimal cases (for example when analysing for nickel in steel), and all elements from magnesium onward are routinely amenable to analysis.

Although EDXRF instruments are cheap and convenient to use, better energy resolution and improved sensitivity are obtained in bench-top WDXRF instruments. These are based on crystal monochromators similar in principle to those used in Moseley's experiments, and excitation again involves an X-ray tube. Detection of X-rays from the lightest elements typically uses gas filled counters not dissimilar to those used by Moseley and Darwin, whereas harder X-rays are detected by scintillation counters. These instruments can analyse for all elements between beryllium and uranium.

XES in Space

Since the 1960s, XES has become well established as a technique for elemental analysis in space missions within our solar system. One early application of XES in the Apollo 15 and 16 missions in 1972 mapped the distribution of magnesium, aluminium and silicon across the lunar surface using characteristic X-rays excited by the X-ray component of the solar radiation.[29]

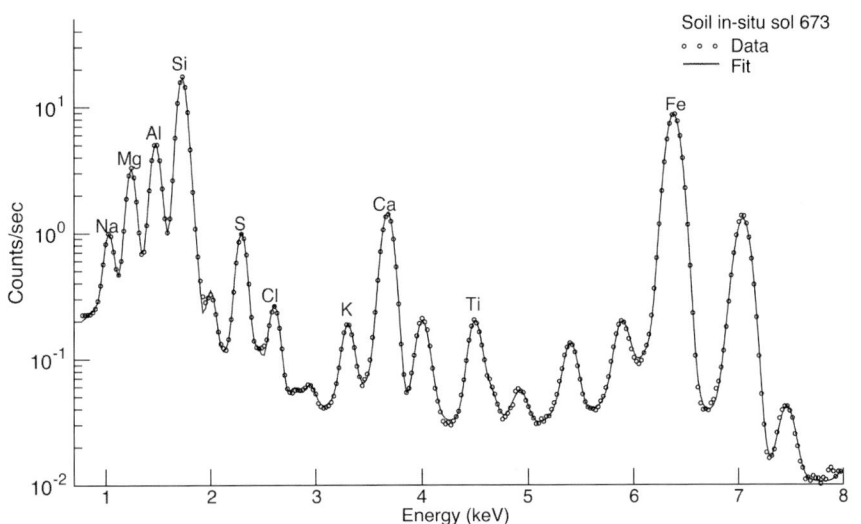

Fig. 2 An image of the alpha particle X-ray spectrometer deployed at the end of a robotic arm examining rocks on the Martian surface (taken on 24 January 2017, Sol 1589), along with an X-ray spectrum of Martian soil adapted from *Geophysical Research Letters*, 43 (1), (2016), 67–75. (Source: J. A. Berger, 2017, by courtesy of NASA/JPL-Caltech).

The lunar X-rays were detected by an array of proportional counters mounted on the orbiting spacecraft.[30] A detection system based on similar counters with X-ray filter windows[31] was also used on the two landers of the Viking mission to Mars.[32] This ultra-miniature spectrometer employed $^{55}_{26}Fe$ and $^{109}_{48}Cd$

radioisotope sources to excite the spectra, allowing abundances of elements ranging between magnesium and iron to be measured at both landing sites.[33]

In several recent missions, NASA has used a configuration that combines three techniques (which all trace their origins to Rutherford's group) into one instrument. This so-called Alpha Particle X-ray spectrometer (APXS) contains a radioactive $^{242}_{96}Cm$ curium source which emits both mono-energetic α-particles and X-rays derived from plutonium, a daughter product of curium. The α-particles may be backscattered through large angles as in the Geiger-Marsden experiment, and lose energy depending on the mass of the nuclei which cause the scattering: the energy loss is determined by conservation of momentum in 'billiard-ball' collisions. Thus measurement of the energy of the scattered α-particles allows elemental analysis based on atomic mass. Alternatively, the α-particles (or the X-rays) may cause inner-shell ionisation with subsequent emission of characteristic X-rays. These are energy-analysed with a silicon detector. The X-ray spectrometer was optimised for heavier elements and achieved sensitivity approaching the ppm level, while Rutherford backscattering was better in analysing for lighter elements.[34] Finally, the α-particles may induce (α,p) nuclear reactions of elements between fluorine and silicon between fluorine and silicon, with emission of protons at characteristic energies.

These three techniques were first used together in the Mars *Pathfinder* mission in 1996, following the success of the backscattering experiment on the Surveyor V, VI and VII missions to the Moon in 1967 and 1968.[35] The X-ray spectrometer was able to detect and quantify elements between sodium and iron in the Martian soil, as well as argon in the atmosphere.[36] More recently, the rover *Curiosity*, which landed on the surface of Mars in 2012, incorporates the latest version of the triple-technique spectrometer (Fig. 2).[37] *Curiosity* also carried the so-called ChemMin (Chemistry and Mineralogy) experiment, which combines the techniques of X-ray diffraction and X-ray spectroscopy using a conventional cobalt X-ray tube and a large area charge coupled device detector. The *Rosetta* mission to the comet 67P/Churuyumov-Gerasimenko also incorporated an APXS instrument,[38] but battery power was lost two days after touchdown on the surface of the comet on 20 November 2014, so it is unlikely that X-ray spectra from the mission will ever be published.

The coronae of most stars emit characteristic X-rays, and the measurement of X-ray spectra of stars is a further area of X-ray astronomy: these measurements cannot be performed on earth due to absorption of X-rays in the atmosphere. The X-ray Multi Mirror mission probe *XMM-Newton* incorporates a reflecting grating spectrometer,[39] while the *Chandra* space probe has two X-ray spectrometers. These are the Low-Energy Transmission Grating Spectrometer (LETGS)[40] and the High-Energy Transmission Grating Spectrometer operating in a higher energy regime extending to 12 keV (i.e. 1Å).[41]

The spectra measured with these instruments typically involve highly ionised states of light elements within the coronal plasma of a star.[42] No doubt Moseley would have been intrigued to know that the energies of $K\alpha_1$ X-ray lines of ions such as C^{5+}, O^{7+} and Fe^{25+}, which each contain only one electron, follow an N^2 dependence (historically the symbol N was used for atomic number, although the symbol Z is now more popular). Moseley realised that $N–1$ rather than N appears in his famous formula for the frequency of K-type X-rays because 'the repulsion of the other electrons cannot be neglected compared with the attraction of the nucleus, and then N must be replaced by $N-\sigma_n$'.[43] The repulsion is obviously absent in ions with only one electron so that $\sigma_n = 0$. The X-ray spectroscopic work that is now being conducted on stars is part of the broader field of X-ray astronomy that was recognised in the award of a half-share in the Nobel Prize in Physics for 2002 to Riccardo Giacconi (1931–) 'for pioneering contributions to astrophysics, which have led to the discovery of cosmic X-ray sources'.

X-Ray Spectroscopy in the Electron Microscope

All microscopes give magnified images of objects, but electron microscopes can now achieve a spatial resolution of less than one Ångstrom (10^{-10} meters), permitting one to visualise individual atoms. This is possible because the de Broglie wavelength of high energy electrons is very much smaller than the wavelength of visible light.

German physicist Ernst Ruska (1906–1988) and electrical engineer Max Knoll (1897–1969) built a prototype electron microscope in 1931, and the first commercial instrument was manufactured by Siemens in 1938. Sixty years later, in 1998, Ruska was awarded a half-share of the Nobel Prize in Physics for this work. Today there are three main variants of the electron microscope. In scanning electron microscopy (SEM) a beam of electrons with energies typically between 400 eV and 40,000 eV is focused down to a spot size that may be as small as 0.4 nm, and the beam is steered across the surface in a series of parallel side to side scans (this is known as raster scanning). Interaction with atoms within the sample produces high energy back-scattered electrons, but also low energy secondary electrons. Characteristic

X-rays and Auger electrons are also emitted as a result of core ionisation by the incident electron beam. Imaging most commonly involves collection of secondary electrons, but chemical analysis can be performed concurrently by monitoring the characteristic X-rays using either energy or wavelength dispersive spectrometers (Fig. 3).[44] Analysis without imaging is also possible in electron microprobe instruments. The resolution in SEM cannot be equated to the beam size, but is determined by the volume within which the probe beam interacts with the sample: this increases with increasing beam energy and optimal spatial analytical resolution is obtained with energies of 5 keV or lower. The effective sampling range with a 1 keV electron beam can be as small as 10 nm or lower.

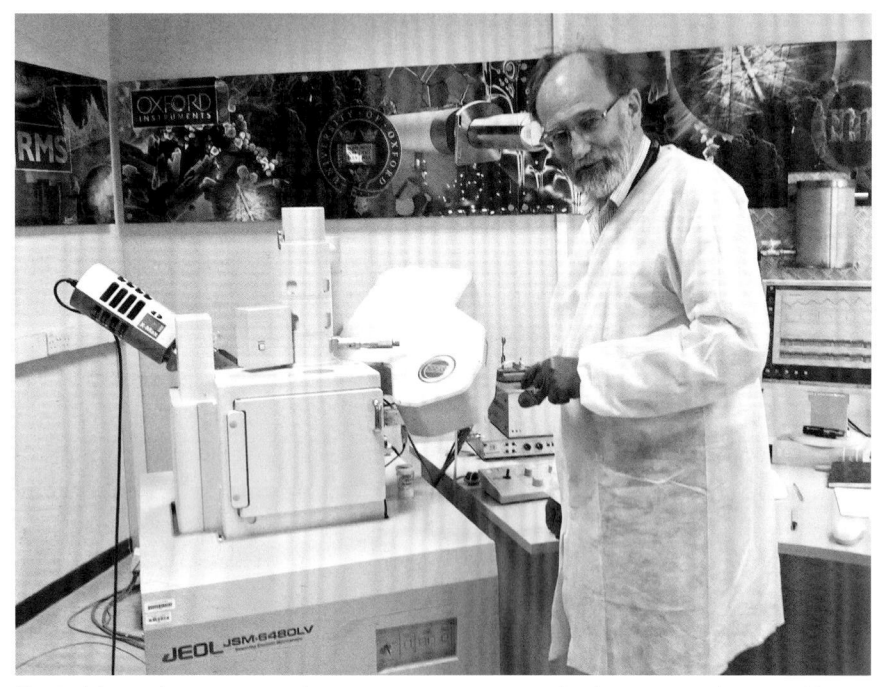

Fig. 3 A low-voltage scanning electron microscope equipped with an energy-dispersive X-ray analyser, housed in the Department of Materials Science at the University of Oxford. This instrument was used to analyse Moseley's Matteucci Medal. (Source: Clare Hopkins, 2017, by courtesy of Department of Materials, University of Oxford).

Higher electron beam energies of up to 400 keV are used in transmission electron microscopy (TEM), which requires ultrathin samples to allow the electron beam to pass through. The transmitted beam contains information about the structure of the specimen and is magnified by an electron optical lens system. Scanning transmission electron microscopy (STEM) combines raster scanning of the incident beam as in SEM with detection of electrons that have passed through a thin sample. Historically, the spatial resolution in both TEM and STEM was limited by aberrations, which occur before the specimen in STEM but after it in TEM. New-generation microscopes are able to correct for these aberrations. Elemental analysis with energy or wavelength dispersive spectrometers is possible with both techniques,[45] but is simpler in STEM. Measurement of what are in effect X-ray absorption spectra by core edge electron energy loss spectroscopy (EELS) is also becoming increasingly popular.[46]

Within the past ten years, it has become possible to achieve the ultimate goal in materials characterisation: chemical analysis that combines atomic-level spatial resolution with single-atom sensitivity in aberration-corrected STEM. Atomic scale chemical analysis was first reported in the EELS mode in 2008,[47] but more recently (2010) atomic scale EDX mapping has also been implemented.[48]

Synchrotron Based XES [49]

Synchrotrons are large scale national facilities where bunches of electrons moving close to the speed of light and with energies typically in the giga-electron volt (GeV) range are steered around a large 'storage ring' by bending magnets. Radiation is emitted whenever an electron is accelerated, and the bending involves acceleration. Synchrotron 'light', as it is often called, extends across a wide spectral range, from the infrared into the hard X-ray region. First-generation light sources performed experiments with synchrotron radiation parasitic upon the particle physics experiments for which these facilities were principally built, and would typically include only a handful of beamlines. In second-generation machines, the whole facility was geared toward the production of synchrotron radiation, which could be tapped off at each of the bending magnets on the storage ring.

Current third-generation synchrotrons include linear 'insertion devices' within which the electron beam is made to 'wiggle' or 'undulate' by a periodic array of magnets with alternating polarity along the straight section. The radiation derived from wigglers and undulators is much more intense (by a factor of around 10^4) than from the bending magnets, and the brightness of the X-rays from a modern undulator may be around 10^{13} times higher than that from a conventional rotating-anode X-ray source – this is the most intense laboratory-based source now available. A typical synchrotron facility thus has a series of beamlines, each drawing radiation from a bending magnet or an insertion device and selecting a spectral range for a specific experiment using crystal or grating monochromators. The beamline is completed with an experimental end-station. Some end-stations perform diffraction experiments, but typically up to about half will be set up for X-ray spectroscopies of various sorts.[49]

Worldwide, there are now upward of fifty synchrotron light sources in operation or under construction. The largest is the Spring 8 facility in Japan, where the storage ring has a circumference of just under 1.5 km and operates at 8 GeV. The UK Diamond Light Source (Fig. 4) has a smaller 0.5 km storage ring operating at 3 GeV. This facility cost around £400 million in its first two stages of construction, with phase three now underway. The machine will eventually incorporate forty beam-lines, and has annual running costs upward of £40 million.

Fig. 4 An aerial view of the Diamond Light Source, Harwell, Oxfordshire. (Source: Diamond Communications Team, 2017, by courtesy of Diamond Light Source Ltd).

X-ray emission is intrinsically a weak process, especially for lighter elements where core-hole decay is dominated by the Auger mechanism.[50] Thus there is much to be gained by using high intensity radiation from an electron synchrotron to measure X-ray emission spectra. Most research on X-ray emission beam-lines focuses on emission involving valence electrons – for example, the decay of K-shell hole states in elements such as nitrogen, oxygen or fluorine by transitions from occupied valence $2p$ states; or by transitions from occupied 3*d* states in transition metal compounds into L shell (2*p*) hole states.

These spectra provide information about electronic structure and bonding complementary to that derived from X-ray photoelectron spectra. In the simplest picture, the emission spectrum should be independent of the energy of the radiation used to excite the spectra. However, with synchrotron sources it is possible to tune the energy of incident X-rays to lie close to the threshold for excitation of the core level of interest. Straightforward X-ray emission then morphs into resonant inelastic X-ray scattering (RIXS), in which the emitted photon energy distribution is shifted by a constant energy loss from the incident photon energy. Electronic excitations are being measured by the energy loss, with elemental specificity.

Some of the most exciting developments in X-ray emission in the past few years have involved the use of X-ray free-electron lasers, part of the growing family of 'fourth-generation' light sources.[49] The world's first X-ray laser is the Stanford Linac Coherent Light Source (LCLS), which first produced light in 2009. Electrons generated by a pulse from an ultraviolet laser striking a copper surface are accelerated to energies in the range 3.5–15 GeV in a linear accelerator over a distance of about 1 km before entering a straight undulator section having a length of 132 m. This contains an array of over 3,000 magnets with alternating polarity (Fig. 5).[51] The radiation that is emitted is both coherent and polarised and is confined to short pulses whose duration is counted in femtoseconds. The brightness of the light in these pulses may be a billion times higher than can be obtained from conventional synchrotron undulators.

In one remarkable recent experiment, X-ray emission spectra of aluminium were measured as a function of the incident photon energy.[52] The results (Fig. 6) bear a striking resemblance to Moseley's staircase, where the progressive changes in X-ray frequency are due to a change in the atomic number N (or Z). Although the physics is somewhat different in the recent experiment, it is still governed by a formula that Moseley would have understood. As the photon energy increases, increasing numbers of electrons can be stripped off the aluminium in the dense hot plasma that follows the X-ray pulse. The results shown in Fig. 6 simply reflect a decrease in the σ_n value (discussed above) as increasing number of electrons are removed from the atom and repulsion from the 'other electrons' decreases.

X-Ray Absorption Spectroscopy[50]

Although X-ray absorption started as a laboratory-based technique, it does require a continuous source of X-rays. The continuum emitted from

Fig. 5 The undulator hall of the X-ray free electron laser at the Linac Coherent Light Source (LCLS) at Stanford in California, USA. (Source: P. Emma, 2017, by courtesy of P. Emma and LCLS).

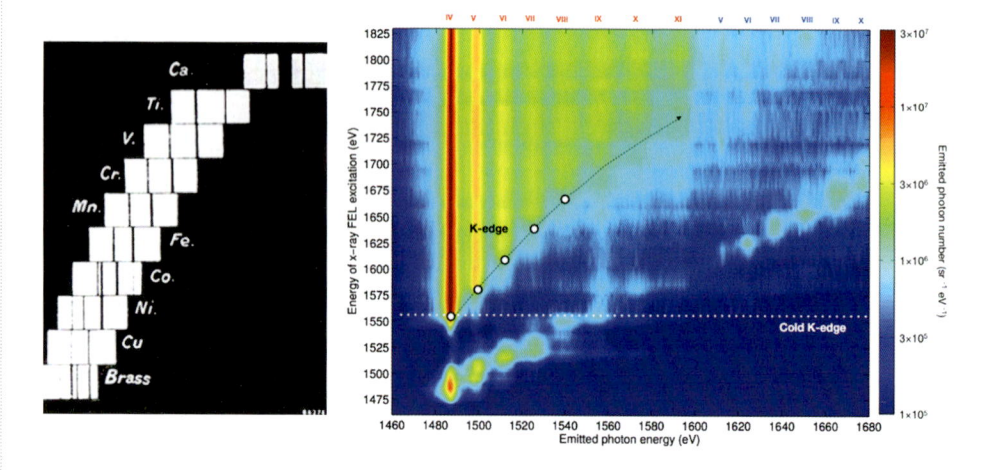

Fig. 6 Moseley's staircase as published in 1913 compared with a staircase in X-ray emission spectra of aluminium in different charge states measured in 2012 (adapted from Nature, 482 (7383), (2012), 59-63). (Source: *Philosophical Magazine Series 6*, 26 (156), (1913), 1024-1034; and J. S. Wark, 2017, by courtesy of J. S. Wark).

conventional X-ray tubes that was used in early work is not ideal, as the sharp characteristic X-ray lines are superimposed on the so-called 'Bremstrahlung' background. Increasingly X-ray absorption spectroscopy (XAS) experiments are carried out at synchrotron facilities, and indeed there are probably more XAS beam-lines worldwide than for any other spectroscopic technique.[50] When dealing with relatively hard X-rays, spectra can be measured in a transmission mode by measuring the intensity in the X-ray beam before and after passage through a sample, as in the pioneering experiments discussed earlier. Alternatively, the absorption may be monitored by measuring the yield of emitted characteristic X-rays or the yield of low-energy secondary electrons. This approach is a necessity for the softest X-rays, which are absorbed too strongly to allow transmission through most samples. Electron yield XAS is particularly easy to implement for samples that are electrical conductors and simply involves the measurement of the current passing through the sample with a sensitive ammeter. Alternatively, the yield of Auger electrons may be measured: this requires an electron energy analyser, but confers greater surface sensitivity on the experiment since the inelastic mean free path for typical Auger electrons is much shorter than the path of the low energy electrons that dominate the secondary electron yield.

In addition to techniques using photons, measurement of the energy loss experienced by fast electrons losing energy to promote core level excitation mimics X-ray absorption spectra. Electron energy loss spectroscopy (EELS) may be carried out in an electron microscope, although better energy resolution can be achieved in dedicated transmission EELS spectrometers without aiming for good spatial resolution. Historically, what were in effect absorption spectra measured by EELS played an important role in exploring the electronic structure of the high temperature oxide superconductors discovered in the 1980s.[53]

The near-edge region in X-ray absorption spectra permits elemental analysis, and also provides information about the chemical state of the absorbing atom in terms of oxidation states and local coordination environment.[54] In addition, the polarisation dependence in so-called NEXAFS spectra provides information about the molecular orientation of species adsorbed on well-ordered surfaces.[55]

Fine structure in X-ray absorption spectra extending well beyond the absorption edge arises from backscattering of electrons by atoms surrounding the absorption site.[55] The backscattered wave interferes with the outgoing wave to produce oscillations as a function of the energy (and hence wavelength) of the outgoing electron. Mathematical analysis of the extended X-ray absorption fine structure (EXAFS) thus provides information about the distances of scattering atoms from the absorption site. This technique is applicable to all solids and to liquids, and unlike classical diffraction experiments it does not require long range order.

X-Ray Photoelectron Spectroscopy and Auger Electron Spectroscopy

The modern era in X-ray photoelectron spectroscopy began with the work of Kai Siegbahn (1918–2007), who, like his father Manne, was based in Uppsala. Until the mid-1950s, the younger Siegbahn had worked mainly on analysis of the energy of β-particles found in radioactive decay. In particular, he measured the conversion electrons produced when the energy of an internal nuclear transition in the γ-region excites an external electron.[56]

This work involved high performance magnetic deflection spectrometers with very much better energy resolution than ever achieved by Robinson. These were adapted for work on photoelectron spectra, initially using fairly hard X-rays such as Cu-Kα radiation to excite the spectra.[57] A series of innovations within Siegbahn's group[58] included the use of softer X-rays derived from Mg or Al anodes to excite the spectra, along with introduction of rotating anode X-ray sources coupled with bent crystal monochromator to narrow the X-ray output and remove the background continuum.[59] Siegbahn also introduced high resolution electron energy analysers using electrostatic rather than magnetic deflection and improved means for detection of the photoelectrons.

The culmination of this work was a high performance XPS instrument based around a massive electrostatic analyser incorporating two hemispheres with a mean radius of 300 mm.[60] This instrument became available commercially in the 1980s (Fig. 7).[61] In 1958, the Uppsala group rediscovered that oxidation of a metal (Cu in their case) leads to a chemical shift in core binding energies,[62] and further developed the concept of chemical shift with work on simple organic and inorganic compounds.[63] $Na_2S_2O_3$ is one of the most striking early cases that still features in textbooks.[64] It should also be noted that the Uppsala group was also responsible for ongoing developments in X-ray emission, particularly in the design of grating monochromators for use in the soft X-ray regime.[65] Kai Siegbahn received a half-share of the Nobel Prize in Physics in 1981 for 'for his contribution to the development of high-resolution electron spectroscopy', continuing the Swedish dynasty in X-ray spectroscopy.

A particular aspect of XPS is that the spectra excited by soft Mg or Al X-rays

Fig. 7 An ESCA 300 X-ray photoelectron spectrometer developed by Kai Siegbahn's group in Uppsala. The instrument in this photograph was at the UK Daresbury National Laboratory, near Warrington. (Source: The author, 2005, by courtesy of National Centre for Electron Spectroscopy and Surface Analysis, Daresbury Laboratory).

are dominated by electrons that originate within the outer few nanometres of a sample: photoelectrons from deeper layers suffer inelastic collisions and do not reach the surface. Ultrahigh vacuum is needed to produce clean surfaces of the sort needed for academic study, and the technique has played a major role in development of the broader field of surface science.[54] Although XPS has excellent depth resolution, lateral resolution cannot match that in electron microscopes. This is because the incident X-rays cannot be focused to the same extent as an electron beam, and limited spatial resolution in the range of hundreds of nanometres is achieved only in the electron optics used to collect the photoelectrons.

However, Auger lines are invariably present in X-ray photoelectron spectra, and may be identified as such by changing the photon energy used to excite the spectra: the kinetic energy of the Auger lines is unchanged, but photoelectron lines shift in energy in accordance with the Einstein equation. Auger electron spectra may also be excited by electron bombardment. If a finely focused electron beam is used for this purpose, surface analysis with excellent spatial resolution becomes possible in the technique of scanning Auger microscopy (SAM).

XPS is not a cheap technique. An entry level commercial instrument would typically cost around £200,000 and high performance ultrahigh vacuum systems incorporating XPS along with other surface science techniques can easily reach £2,000,000. Nonetheless the world market in XPS instruments was estimated by a market research company based in the USA to be in excess of $400 million in 2015 and is expected to increase to about $750 million by 2023.[66] Its applications include forensic analysis; the analysis of problems in technologies dealing with adhesion and corrosion; the monitoring of surface contamination in the semiconductor and healthcare sectors; and quality control in preparation of industrial heterogeneous catalysts.[67]

XPS is also widely performed on synchrotron sources, especially in academic work. The tunability of the radiation allows a number of effects to be exploited, including resonance emission when the photon energy is tuned to match that of a core level.[68] One interesting recent trend has been toward so-called hard X-ray photoelectron spectroscopy (HAXPES), where the higher energy of the outgoing photoelectrons reduces the surface sensitivity.[69] This is advantageous when trying to study the bulk electronic structure of materials. In some ways HAXPES takes the technique back to where it started with Robinson's experiments in Manchester.

Conclusion

Since Moseley's day, X-ray spectroscopy has grown to encompass a wide range of disciplines and approaches. It is today one of the most important of all analytical techniques, but is also of growing importance in fundamental investigations of the structure of matter in all conditions, from superconductors at low temperatures to plasmas in the coronae of stars. Achievements in the field of X-ray spectroscopy, broadly defined, have so far led to the award of four Nobel Prizes in Physics – to Charles Barkla, Manne Siegbahn, Kai Siegbahn and Riccardo Giacconi. The impact of Moseley's two 'High-Frequency' papers published just before the Great War continues to be felt over 100 years on from his first experiments with X-rays.

Notes and References

[1] H. G. J. Moseley, 'The High-Frequency Spectra of the Elements', *Philosophical Magazine Series 6*, 26 (156), (1913), 1024-1034 (hereafter Moseley, 'High-Frequency I'); *idem*, 'The High-Frequency Spectra of the Elements. Part II', *Philosophical Magazine Series 6*, 27 (160), (1914), 703-713 (hereafter Moseley, 'High-Frequency II').

[2] M. Siegbahn and E. Friman, 'On the High-Frequency Spectra of the Elements Gold-Uranium', *Philosophical Magazine Series 6*, 31 (184), (1916), 403–406.

[3] M. Siegbahn, 'Precision Measurements in the X-Ray Spectra', *Philosophical Magazine Series 6*, 37 (222), (1919), 601–612.

[4] W. Stenström, 'Experimentelle Untersuchungen der Röntgenspektra. M-Reihe', *Annalen der physik*, 362 (21), (1918), 347–376. The 'discovery' of M-type X-rays was central to the award of Siegbahn's Nobel Prize for Physics in 1925, so it is ironic that he is not on the most thorough paper related to the discovery.

[5] M. Siegbahn, *Spektroskopie der Röntgenstrahlen* (Berlin: Springer-Verlag, 1931).

[6] The xu unit was intended to be 10^{-13} m, but the value was defined assuming a constant lattice spacing in calcite crystals. Because of variations between different crystals, this assumption introduced systematic errors into the literature which persisted for many years. See for example J. A. Bearden, 'X-Ray Wavelengths', *Reviews of Modern Physics*, 39 (1), (1967), 78–124.

[7] A. Gullstrand, Award Ceremony Speech for the Nobel Prize in Physics 1924 Manne Siegbahn. In *Nobel Lectures, Physics 1922–1941* (Amsterdam: Elsevier, 1965). Available online at https://www.nobelprize.org/nobel_prizes/physics/laureates/1924/press.html (accessed 27 November 2017).

[8] See https://Royalsociety.Org/Grants-Schemes-Awards/Awards/Hughes-Medal (accessed 27 November 2017) for a list of medal winners.

[9] M. de Broglie, 'Sur un Nouveau Procede Permettant d'Obtenir la Photographie de Raies de Rayons Röntgen', *Comptes rendus de l'Académie des sciences*, 157 (1913), 924–926.

[10] M. de Broglie, 'Sur la Spectroscopie de Rayons Secondaire Emis hors des Tubes a Rayons de Röntgen, et les Spectres d'Absorption', *Comptes rendus de l'Académie des sciences*, 158 (1914), 1493–1495.

[11] W. H. Bragg and M. Siegbahn are credited with suggesting this assignment.

[12] M. de Broglie, 'Sur la Bande d'Absorption K des Elements pour les Rayons X, Suivie du Brome au Bismuth, et l'Emission d'un Tube Coolidge vers les Tres CourtesLongeurs d'Onde', *Comptes rendus de l'Académie des sciences*, 163 (1916), 87–89.

[13] H. Fricke, 'The K-Characteristic Absorption Frequencies for the Chemical Elements Magnesium to Chromium', *Physical Review*, 16 (3), (1920), 202–216.

[14] G. Hertz, 'Über die Absorptionsgrenzen in Der L-Serie', *Zeitschrift für Physik*, 3 (1), (1920), 19–25.

[15] H. R. Robinson and C. L. Young, 'The Influence of Chemical State on Critical X-Ray Absorption Frequencies', *The London, Edinburgh and Dublin Philosophical Magazine and Journal of Science*, 10 (62), (1930), 71–75.

[16] J. D. Hanawalt, 'The Dependence of X-Ray Absorption Spectra upon Chemical and Physical State', *Physical Review*, 37 (6), (1931), 715–726.

[17] C. Barkla, 'Secondary Röntgen Radiation', *Philosophical Magazine Series 6*, 11 (66), (1906), 812–828.

[18] P. D. Innes, 'On the Velocity of Cathode Particles Emitted by Various Metals under the Influence of Röntgen Rays, and its Bearing on the Theory of Atomic Disintegration', *Proceedings of the Royal Society A*, 79 (532), (1907), 442–462.

[19] E. N. da C. Andrade, 'Harold Roper Robinson', *Biographical Memoirs of Fellows of the Royal Society*, 3 (1957), 161–172.

[20] H. G. J. Moseley and H. Robinson, 'The Number of Ions Produced by the Beta and Gamma Radiations from Radium', *Philosophical Magazine Series 6*, 28 (165), (1914), 327–337. This was Moseley's last published paper.

[21] H. R. Robinson and W. F. Rawlinson, 'The Magnetic Spectrum of the Beta Rays Excited in Metals by Soft X-rays', *Philosophical Magazine Series 6*, 28 (164), (1914), 277–281.

[22] Moseley to his mother, 27 June 1915, Letter 134 in J. L. Heilbron, *H. G. J. Moseley: The Life and Letters of an English Physicist, 1887–1915* (Berkeley: University of California Press, 1974), 271–272. Moseley starts this letter with the words 'This is the last will and testament of me Henry Gwyn Jeffrey [sic] Moseley Second Lieutenant Royal Engineers…'. Moseley's bequest was supplemented by the sale of Pick's Hill by his mother.

[23] H. Robinson, 'The Secondary Corpuscular Rays Produced by Homogenous X-Rays', *Proceedings of the Royal Society A*, 104 (727), (1923), 455–479.

[24] H. Robinson and A. M. Cassie, 'The Secondary and Tertiary Cathode Rays Produced by External and Internal Absorption of Homogenous X-Rays', *Proceedings of the Royal Society A*, 113 (764), (1926), 282–310.

[25] P. Auger, 'Sur les Rayons Beta Secondaire Produits dans un Gas par des Rayons X', *Comptes rendus de l'Académie des sciences*, 177 (1923), 169–171. Lise Meitner should probably take credit for discovering the Auger effect, as she published on the topic a year before Auger. See L. Meitner, 'Über die Entstehhung der Beta-Strahl-Spektren Radioaktiver Substansen', *Zeitschrift für Physik*, 9 (1), (1922), 131–144.

²⁶ These numbers come from an automated topic search with the Web of Science Search Engine using the following terms: X-ray and spectroscopy, X-ray and absorption, X-ray and emission, X-ray and photoelectron, X-ray and photoemission, Auger and spectroscopy, XES, XAS, XPS, ESCA, HAXPES, XRF, RIXS, EDXRF, WDXRF, EDX, EDAX, EXAFS.

²⁷ Moseley, 'High-Frequency I', *op. cit.* 1029–1030 discusses detection of Fe and Ni in Co and Cu in Mn and V.

²⁸ B. Beckhoff, B. Kanngiesser, N. Langhoff, R. Wedell and H. Wolff, *Handbook of Practical X-Ray Fluoresence Analysis* (Berlin Heidelberg: Springer, 2006).

²⁹ I. Adler, P. Bjorkholm, H. Blodget, P. Gorenstein, R. Schmadebeck, J. Trombka, L. Yin, P. Lowman, E. Eller, J. Gerard and R. Lamothe, 'Apollo-15 Geochemical X-Ray-Fluorescence Experiment – Preliminary Report', *Science*, 175 (4020), (1972), 436–440.

³⁰ I. Adler, P. Bjorkholm, H. Blodget, R. Schmadebeck, J. Trombka, L. Yin, P. Lowman, P. Gorenstein, H. Gursky, B. Harris, R. Lamothe, J. Gerard, E. Eller and G. Osswald, 'Apollo-16 Geochemical X-Ray-Fluorescence Experiment – Preliminary Report', *Science*, 177 (4045), (1972), 256–259.

³¹ B. C. Clark and A. K. Baird, 'Ultraminiature X-Ray-Fluorescence Spectrometer for in-situ Geochemical Analysis on Mars', *Earth and Planetary Science Letters*, 19 (3), (1973), 359–368.

³² B. C. Clark, A. K. Baird, H. J. Rose, P. Toulmin, K. Keil, A. J. Castro, W. C. Kelliher, C. D. Rowe and P. H. Evans, 'Inorganic Analyses of Martian Surface Samples at Viking Landing Sites', *Science*, 194 (4271), (1976), 1283–1288.

³³ B. C. Clark, A. K. Baird, R. J. Weldon, D. M. Tsusaki, L. Schnabel and M. P. Candelaria, 'Chemical-Composition of Martian Fines', *Journal of Geophysical Research*, 87 (NB12), (1982), 59–67.

³⁴ R. Rieder, H. Wanke, T. Economou and A. Turkevich, 'Determination of the Chemical Composition of Martian Soil and Rocks: The Alpha Proton X-ray Spectrometer', *Journal of Geophysical Research-Planets*, 102 (E2), (1997), 4027–4044.

³⁵ J. H. Patterson, E. J. Franzgrote, A. L. Turkevich, W. A. Anderson, T. E. Economou, H. E. Griffin, S. L. Grotch and K. P. Sowinski, 'Alpha-Scattering Experiment on Surveyor 7 – Comparison with Surveyors 5 and 6', *Journal of Geophysical Research*, 74 (25), (1969), 6120–6125.

³⁶ R. Rieder, T. Economou, H. Wanke, A. Turkevich, J. Crisp, J. Bruckner, G. Dreibus and H. Y. McSween, 'The Chemical Composition of Martian Soil and Rocks Returned by the Mobile Alpha Proton X-Ray Spectrometer: Preliminary Results from the X-Ray Mode', *Science*, 278 (5344), (1997), 1771–1774.

37 J. A. Berger, M. E. Schmidt, R. Gellert, J. L. Campbell, P. L. King, R. L. Flemming, D. W. Ming, B. C. Clark, I. Pradler, S. J. V. VanBommel, M. E. Minitti, A. G. Fairen, N. I. Boyd, L. M. Thompson, G. M. Perrett, B. E. Elliott and E. Desouza, 'A Global Mars Dust Composition Refined by the Alpha-Particle X-Ray Spectrometer in Gale Crater', *Geophysical Research Letters*, 43 (1), (2016), 67–75.

38 A. H. Treiman, D. L. Bish, D. T. Vaniman, S. J. Chipera, D. F. Blake, D. W. Ming, R. V. Morris, T. F. Bristow, S. M. Morrison, M. B. Baker, E. B. Rampe, R. T. Downs, J. Filiberto, A. F. Glazner, R. Gellert, L. M. Thompson, M. E. Schmidt, L. Le Deit, R. C. Wiens, A. C. McAdam, C. N. Achilles, K. S. Edgett, J. D. Farmer, K. V. Fendrich, J. P. Grotzinger, S. Gupta, J. M. Morookian, M. E. Newcombe, M. S. Rice, J. G. Spray, E. M. Stolper, D. Y. Sumner, A. R. Vasavada and A. S. Yen, 'Mineralogy, Provenance, and Diagenesis of a Potassic Basaltic Sandstone on Mars: Chemin X-Ray Diffraction of the Windjana Sample (Kimberley Area, Gale Crater)', *Journal of Geophysical Research-Planets*, 121 (1), (2016), 75–106.

39 J. W. den Herder, A. C. Brinkman, S. M. Kahn, G. Branduardi-Raymont, K. Thomsen, H. Aarts, M. Audard, J. V. Bixler, A. J. den Boggende, J. Cottam, T. Decker, L. Dubbeldam, C. Erd, H. Goulooze, M. Gudel, P. Guttridge, C. J. Hailey, K. Al Janabi, J. S. Kaastra, P. A. J. de Korte, B. J. van Leeuwen, C. Mauche, A. J. McCalden, R. Mewe, A. Naber, F. B. Paerels, J. R. Peterson, A. P. Rasmussen, K. Rees, I. Sakelliou, M. Sako, J. Spodek, M. Stern, T. Tamura, J. Tandy, C. P. de Vries, S. Welch and A. Zehnder, 'The Reflection Grating Spectrometer on Board XMM-Newton', *Astronomy & Astrophysics,* 365 (1), (2001), L7–L17.

40 A. C. Brinkman, C. J. T. Gunsing, J. S. Kaastra, R. L. J. van der Meer, R. Mewe, F. Paerels, A. J. J. Raassen, J. J. van Rooijen, H. Brauninger, W. Burkert, V. Burwitz, G. Hartner, P. Predehl, J. U. Ness, J. Schmitt, J. J. Drake, O. Johnson, M. Juda, V. Kashyap, S. S. Murray, D. Pease, P. Ratzlaff and B. J. Wargelin, 'First Light Measurements of Capella with the Low-Energy Transmission Grating Spectrometer aboard the Chandra X-Ray Observatory', *Astrophysical Journal*, 530 (2), (2000), L111–L114.

41 M. F. Corcoran, J. H. Swank, R. Petre, K. Ishibashi, K. Davidson, L. Townsley, R. Smith, S. White, R. Viotti and A. Damineli, 'The Chandra HETGS X-Ray Grating Spectrum of Eta Carinae', *Astrophysical Journal*, 562 (2), (2001), 1031–1037.

42 M. Gudel and Y. Naze, 'X-Ray Spectroscopy of Stars', *Astronomy and Astrophysics Review*, 17 (3), (2009), 309–408.

43 H. Moseley, 'Atomic Models and X-Ray Spectra', *Nature*, 92 (2307), (1914), 554.

44 J. Goldstein, D. E. Newbury, D. C. Joy, C. E. Lyman, P. Echlin, E. Lifshin, L. Sawyer and J. R. Michael, *Scanning Electron Microscopy and X-Ray Microanalysis* (New York: Springer US, 3rd edn 2003).

45 C. B. Carter and D. B. Williams, *Transmission Electron Microscopy: Diffraction, Imaging, and Spectrometry* (Cham: Springer International Publishing, 2016).

46 R. F. Egerton, *Electron Energy Loss Spectroscopy in the Electron Microscope* (New York: Springer US, 2011).

47 D. A. Muller, L. F. Kourkoutis, M. Murfitt, J. H. Song, H. Y. Hwang, J. Silcox, N. Dellby and O. L. Krivanek, 'Atomic-Scale Chemical Imaging of Composition and Bonding by Aberration-Corrected Microscopy', *Science*, 319 (5866), (2008), 1073–1076.

48 A. J. D'Alfonso, B. Freitag, D. Klenov and L. J. Allen, 'Atomic-Resolution Chemical Mapping Using Energy-Dispersive X-Ray Spectroscopy', *Physical Review B*, 81 (10), (2010), 100101/1–4.

49 P. Willmott, *An Introduction to Synchrotron Radiation* (Hoboken: John Wiley, 2011).

50 J. A. Van Bokhoven and C. Lamberti, *X-Ray Absorption and X-Ray Emission Spectroscopy: Theory and Applications* (Chichester: John Wiley, 2016).

51 P. Emma, R. Akre, J. Arthur, R. Bionta, C. Bostedt, J. Bozek, A. Brachmann, P. Bucksbaum, R. Coffee, F. J. Decker, Y. Ding, D. Dowell, S. Edstrom, A. Fisher, J. Frisch, S. Gilevich, J. Hastings, G. Hays, P. Hering, Z. Huang, R. Iverson, H. Loos, M. Messerschmidt, A. Miahnahri, S. Moeller, H. D. Nuhn, G. Pile, D. Ratner, J. Rzepiela, D. Schultz, T. Smith, P. Stefan, H. Tompkins, J. Turner, J. Welch, W. White, J. Wu, G. Yocky and J. Galayda, 'First Lasing and Operation of an Ångstrom-Wavelength Free-Electron Laser', *Nature Photonics*, 4 (9), (2010), 641–647.

52 S. M. Vinko, O. Ciricosta, B. I. Cho, K. Engelhorn, H. K. Chung, C. R. D. Brown, T. Burian, J. Chalupsky, R. W. Falcone, C. Graves, V. Hajkova, A. Higginbotham, L. Juha, J. Krzywinski, H. J. Lee, M. Messerschmidt, C. D. Murphy, Y. Ping, A. Scherz, W. Schlotter, S. Toleikis, J. J. Turner, L. Vysin, T. Wang, B. Wu, U. Zastrau, D. Zhu, R. W. Lee, P. A. Heimann, B. Nagler and J. S. Wark, 'Creation and Diagnosis of a Solid-Density Plasma with an X-Ray Free-Electron Laser', *Nature*, 482 (7383), (2012), 59–63.

53 N. Nücker, J. Fink, J. C. Fuggle, P. J. Durham and W. M. Temmerman, 'Evidence for Holes on Oxygen Sites in the High-Tc Superconductors $La_{2-x}Sr_xCuO_4$ and $YBa_2Cu_3O_{7-x}$', *Physical Review B*, 37 (10), (1988), 5158–5163.

54 D. P. Woodruff and T. A. Delchar, *Modern Techniques of Surface Science* (Cambridge: Cambridge University Press, 2nd edn, 2010).

55 B. K. Agarwal, *X-Ray Spectroscopy* (Berlin: Springer-Verlag, 2nd edn, 1991).

56 A. Hedgran, K. Siegbahn and N. Svartholm, 'A Large Beta-Spectrometer with 2-Directional Focusing for Precise Measurements of Nuclear Radiation', *Proceedings of the Physical Society of London A*, 63 (369), (1950), 960–986.

[57] E. Sokolowski, C. Nordling and K. Siegbahn, 'Magnetic Analysis of X-Ray Produced Photo and Auger Electrons', *Arkiv for Fysik*, 12 (4), (1957), 301–318.

[58] K. Siegbahn, 'From X-Ray to Electron Spectroscopy', Nishina Memorial Foundation (eds.), *Nishina Memorial Lectures: Creators of Modern Physics* (Heidelberg: Springer, 2008), 137–228; B. Wannberg, U. Gelius and K. Siegbahn, 'Design Principles in Electron-Spectroscopy', *Journal of Physics E-Scientific Instruments*, 7 (3), (1974), 149–159; H. Fellnerfeldegg, U. Gelius, B. Wannberg, A. G. Nilsson, E. Basilier and K. Siegbahn, 'New Developments in ESCA-Instrumentation', *Journal of Electron Spectroscopy and Related Phenomena*, 5 (NOV-D), (1974), 643–689.

[59] A. Fahlman and K. Siegbahn, 'ESCA Method Using Monochromatic X-Rays and a Permanent Magnet Spectrograph', *Arkiv for Fysik*, 32 (2–3), (1966), 111–118.

[60] U. Gelius, L. Asplund, E. Basilier, S. Hedman, K. Helenelund and K. Siegbahn, 'A High-Resolution Multipurpose ESCA Instrument with X-Ray Monochromator', *Nuclear Instruments & Methods in Physics Research Section B-Beam Interactions with Materials and Atoms*, 229 (1), (1984), 85–117; U. Gelius, B. Wannberg, P. Baltzer, H. Fellnerfeldegg, G. Carlsson, C. G. Johansson, J. Larsson, P. Munger and G. Vegerfors, 'A New ESCA Instrument with Improved Surface Sensitivity, Fast Imaging Properties and Excellent Energy Resolution', *Journal of Electron Spectroscopy and Related Phenomena*, 52 (1990), 747–758.

[61] G. Beamson, D. Briggs, S. F. Davies, I. W. Fletcher, D. T. Clark, J. Howard, U. Gelius, B. Wannberg and P. Balzer, 'Performance and Application of the Scienta ESCA 300 Spectrometer', *Surface and Interface Analysis*, 15 (9), (1990), 541–549.

[62] E. Sokolowski, C. Nordling and K. Siegbahn, 'Chemical Shift Effect in Inner Electronic Levels of Cu Due to Oxidation', *Physical Review*, 110 (3), (1958), 776.

[63] K. Hamrin, G. Johansson, A. Fahlman, C. Nordling, K. Siegbahn and B. Lindberg, 'Structure Studies of Sulphur Compounds by ESCA', *Chemical Physics Letters*, 1 (11), (1968), 557–559.

[64] S. Hagstrom, C. Nordling and K. Siegbahn, 'Electron Spectroscopy for Chemical Analysis', *Physics Letters*, 9 (3), (1964), 235–236.

[65] J. Nordgren, H. Agren, L. Selander, C. Nordling and K. Siegbahn, 'Electron-Spectroscopy and Ultra-Soft X-Ray-Emission of Free Molecules', *Physica Scripta*, 16 (5–6), (1977), 280–284; J. Nordgren, H. Agren, L. Pettersson, L. Selander, S. Griep, C. Nordling and K. Siegbahn, 'New 10m Grazing-Incidence Instrument for Molecular X-Ray Studies', *Physica Scripta*, 20 (5–6), (1979), 623–626; H. Agren, J. Nordgren, L. Selander, C. Nordling and K. Siegbahn, 'Valence Electron-Structure of the SF_6 and CS_2 Molecules, Studied in High-Resolution X-Ray-Emission', *Physica Scripta*, 18 (6), (1978), 499–505; J. Nordgren and N. Wassdahl, 'Soft-X-Ray Fluorescence Spectroscopy Using Tunable Synchrotron-Radiation', *Journal of Electron Spectroscopy and Related Phenomena*, 72 (1995), 273–280.

[66] Transparency Market Research Report. 'X-Ray Photoelectron Spectroscopy Market – Global Industry Analysis, Size, Share, Growth, Trends and Forecast 2015–2023' (2015). A single user license for this report costs $5795 but a summary is available online at http://www.transparencymarketresearch.com/pressrelease/xray-photoelectron-spectroscopy-market.htm (accessed on 29 November 2017). Contact address: Transparenccy Market Research, State Tower, 90 State Street, Suite 700, Albany NY – 12207, United States

[67] G. C. Smith, *Surface Analysis by Electron Spectroscopy: Measurement and Interpretation* (Berlin: Springer, 1994).

[68] S. Hüfner, *Photoelectron Spectroscopy: Principles and Applications* (Berlin: Springer, 3rd edn, 2003).

[69] K. Siegbahn, 'Preface to Hard X-Ray Photo Electron Spectroscopy (HAXPES)', *Nuclear Instruments and Methods in Physical Research, Section A*, 547 (1), (2005), 1–7; G. Panaccione, G. Cautero, M. Cautero, A. Fondacaro, M. Grioni, C. Henriquet, G. Monaco, M. Mulazzi, F. Offi, L. Paolasini, G. Paolicelli, P. Pittana, M. Sacchi, G. Stefani and P. Torelli, 'Results and Perspectives in Hard X-Ray Photoemission Spectroscopy (HAXPES) from Solids', *Nuclear Instruments and Methods in Physical Research, Section B*, 246 (1), (2006), 106–111

CHAPTER TEN

ACCOUNTS OF MOSELEY AND VERSIONS OF HIS LAWS

JOHN L HEILBRON

Introduction

Henry Moseley's life and work lend themselves to the opposite genres of heroic romance and introductory physics. The first attempts to capture his life centered on his glorious death; the first historical accounts of his work, on its pedagogical utility. Part of his lesson is that stubborn hard work can overcome time and machines; another part, that even the most stubborn and dedicated do not get to the top without pulls from well-placed colleagues. The most important of these colleagues in Moseley's case were Rutherford, Darwin, and Bohr. Their pulls are our first consideration. Next comes an account of 'Moseley's Law'. It emerges that this 'law' is generally stated in physics books in ways that hide the great puzzle in atomic structure that it presented to Moseley and Bohr, and not infrequently in ways that make no physical sense. A third section offers a review of the scant early historiography on Moseley to suggest the process of his posthumous transition, not of hero to god, as in euhemeristic creations, but of hero to human being.

The Encounters

Moseley was twenty-five and Bohr twenty-seven when, during the spring and summer of 1912, they were research students together in Rutherford's laboratory in Manchester. Bohr was a double alien: a foreigner and a tyro in the laboratory's main if not sole research topic, radioactivity. Moseley was completely at home: a veteran student of radioactivity, known throughout the laboratory as a skillful and indefatigable experimenter.[1] They probably had no significant conversation during those months in 1912. They did not have a research interest in common, and they were not well matched temperamentally.

Moseley regarded foreigners from the heights of Eton and Oxford and had the stereotypical prejudices of people so nourished. He had only contempt for the students he had to teach and criticized Rutherford, who had a breezy down-under manner, for acting like a stage colonial. Moseley's general conservatism, which included respect for and probably belief in the teachings of the Church of England, would not have opened common ground with the freer-thinking Bohr, who inscribed himself as an 'agnostic' in the register of his university residence, Hulme Hall.[2] With his customary omniscience, Moseley informed his sister Margery that 'all agnostics are prone' to 'frequent mental worry'.[3] Bohr was not an agnostic but an atheist and his worry at the time was not his soul but the publication of his doctoral thesis.

There were countervailing factors that might have drawn the two together. Each was the son of a university professor who died young; and each had a devoted and supportive nuclear family, a sister and mother in Moseley's case, a brother, mother, and fiancée in Bohr's. Together with Charles Galton Darwin, like Moseley the grandson as well as the son of a scientist, and Georg von Hevesy, a descendant of Austrian barons on both sides of his family, Moseley and Bohr constituted the aristocracy of a laboratory led by the son of a flax farmer. Still they were not drawn together.

Bohr recalled in interviews conducted fifty years later that no one at Manchester paid any attention to him at first, except the cosmopolitan Hevesy and the colonial Rutherford. A social and inquisitive man, he found the English xenophobic and uncurious, 'polite and so on, but they wouldn't be interested in seeing anybody.… It takes half a year to get to know an Englishman'.[5] Bohr could not afford the time. So he stopped going to the laboratory after he had learned as much as he thought he needed to know. He then worked alone in his room at Hulme Hall.[6]

Neither Bohr nor Moseley was engaged in work of any importance during the first three months of their co-existence in Manchester. Then Bohr descended from Hulme Hall to demonstrate that some of Darwin's calculations about the electronic structure of the nuclear model were nonsense. Darwin amiably acknowledged his errors; the two became friends; and Bohr started to develop a new version of Rutherford's atom. By the time he left Manchester in late July 1912 for his wedding in Denmark on 1 August, Bohr had set down a preliminary foundation for what would become, some eight months later, the quantum atom. He had not finished his paper correcting Darwin before he left, however, and returned to Manchester on his honeymoon to complete it. And that is how we know that by August 1912 Bohr and Moseley had made some social contact; for, in the only surviving exchange of correspondence between them, dating from over a year later, Moseley asked to be remembered to Mrs Bohr.[7]

Two months after this fleeting meeting, Moseley learned about Laue's diffraction experiments. Manchester did not do X-rays so Moseley tried to reproduce Laue's results with γ-rays. His failure induced him to learn the relevant experimental technique from William Henry Bragg at Leeds and to team up with Darwin to interpret the meaning of the measurements. They discovered, as Moseley put it with characteristic cheek, that 'the [Germans] who did the work entirely failed to understand what it meant, and gave an explanation that was obviously wrong. After much hard work Darwin and I found out the real meaning'. They had hit on the same explanation given by William Lawrence Bragg a few days before they tumbled to it. That was a pity, as the explanation was worth a share in a Nobel Prize.[8] With Darwin's help, Moseley perfected his apparatus and identified sharp characteristic X-ray lines against the general background radiation. Their paper describing these 'high frequency spectra' appeared in July 1913, in the same issue of the *Philosophical Magazine* as Bohr's first paper on the quantum atom.[9]

Rutherford had tried hard to prevent Darwin and Moseley from taking up X-rays. He did not welcome experimental investigations in his laboratory in fields he did not control.[10] And he had dragged his feet when Bohr told him that the best evidence for the nuclear model was the principle of atomic number and the concept of isotopes, both of which Bohr considered obvious inferences from the distinction between the nuclear and electronic precincts of atoms.[11] Rutherford doubted both inferences. His model suggested that beta and gamma rays arose in the electronic structure. To account for the penetrating power of some gamma rays, he assumed the existence of an 'H' ring of electrons within the innermost ring in Bohr's theory (the K ring), where the hardest X-rays originated.[12] On this theory, beta and gamma rays produced one another inside the electronic structure by the same quantum processes, whatever they might be, that turned cathode rays into X-rays and X-rays into photoelectrons.

Rutherford had arrived at his theory of beta and gamma rays at just the time that Bohr revealed to him the concepts of atomic number and isotopy, and their natural fit with the nuclear atom. Since the one ruled out an H ring and the other appeared to require electrons within the nucleus, Rutherford did not show the enthusiasm for the new concepts that Bohr had expected. Instead, Rutherford elaborated his theory by trying to make a success of Moseley's failed experiment on γ-ray diffraction.[13] In the last version published before the war, a primary beta particle originating 'from or near the nucleus' stimulated hard K and soft L rays from the electronic structure (Moseley's high-frequency spectra) and also H rays (γ-rays harder than K-type X-rays).[14]

The Laws: Theoretical

The invention of X-ray spectroscopy made possible a neat test of quantum theory and the nuclear atom. The test made use of an empirical rule found in 1911 by a researcher at the Cavendish laboratory, Richard Whiddington: the threshold energy of a cathode ray just able to excite a K ray from an element of atomic weight A is proportional A^2. Since many experiments suggested that $A \approx 2n$, where n is the number of electrons in the atom, and since on Rutherford's model $n = Z$, where Ze is the nuclear charge, Bohr expected the threshold energy to go as Z^2. Invoking Planck's slogan $E = h\nu$, the frequency of the K quantum should also be proportional to Z^2. This test, whose execution would make Moseley famous, had not started, or, probably, been designed, when Bohr returned to Manchester in early April 1913 to defend his draft of the first installment of his three-part paper on atomic structure from Rutherford's editorial pen. Moseley was then tending his garden in the New Forest.[15]

Three months later Bohr was again in Manchester, this time to discuss the rest of his paper with Rutherford. It was then that he had some serious conversation with Moseley about the importance of K radiation for the demonstration of atomic number. Bohr was keenly interested in the question. He had tried to deduce the number of electrons in the innermost rings from X-ray data on the assumption that the mutual repulsion of the electrons in the K ring should diminish the effective nuclear charge on each of them; hence the frequency should be proportional not to Z^2, but to $(Z - \sigma_n)^2$, where σ_n represents the diminution and depends on the number of electrons n in the ring. But experiment gave him too little to go on. 'I was afraid that such an attempt would be too uncertain in the present state'.[16] Hence his enthusiasm over Moseley's work – not as a demonstration of atomic number, which Bohr regarded as a sure thing, but as an authoritative guide to the number of electrons in the innermost ring.

Just how much Bohr's enthusiasm and advice influenced the direction of Moseley's measurements of K and L spectra is not known. Moseley smudged the historical record by stating that he had undertaken his investigation to test 'Broek's hypothesis,' by which he meant an approximation to the idea of atomic number suggested by a Dutch lawyer. Moseley felt obliged to acknowledge Antonius van den Broek's hypothesis because he was the first to publish it. 'Broek' was scarcely heard from again; and his meteoric pass through the clouds of physics showed only that his idea was in the air.

It is suggestive that Moseley began to build his apparatus in August 1913, shortly after Bohr's July visit to Rutherford. At a minimum Bohr's contribution would have been to suggest that Moseley try to fit his data to an expression of the form:

$$\nu_K = \text{const.} \times R(Z - \sigma_n)^2,$$

where R is the universal Rydberg frequency of the Balmer formula.[17] The hint would help considerably in data reduction! A discussion with Bohr might lie behind Moseley's statement that his follow-up to the work with Darwin 'was undertaken for the express purpose of testing Broek's hypothesis, which Bohr has incorporated as a fundamental part of his theory of atomic structure'.[18] As we know, the test was whether the K frequency could be expressed as the square of a number that changed by one unit between neighboring elements in the periodic table. That is exactly what Moseley found. 'The results are exceedingly simple,' he wrote to Bohr in November 1913, 'and largely what you would expect'.[19]

The frequencies (for there turned out to be at least two K lines) came out to be proportional to $(Z-1)^2$. The frequency of the softer, Kα, had as its constant factor ¾R. That was close to the number whose calculation was the centerpiece of Bohr's theory of the hydrogen spectrum, and ¾, written as $(1/1^2 - 1/2^2)$, seemed to indicate a Balmer-like transition. To proceed further, to find the physics underlying both factors – the ¾ and the $(Z-1)^2$ – Moseley had available two misleading principles of atom-building invented by Bohr.

The first of these principles, for which Bohr gave no justification, required that every electron in an atom in its ground state have one and only one quantum of angular momentum no matter what its distance from the nucleus. The second principle, which Bohr claimed to base on ordinary mechanical considerations, predicted that heavier atoms would have more electrons in their innermost ring than light ones. From chemical evidence, Bohr placed lithium's electron outside the ring of the two equivalent electrons that, on Rutherford's model, must represent helium. He increased the innermost ring to four at nitrogen and to eight at neon. It remained there up to chromium ($Z = 24$), where Bohr temporarily stopped conjecturing about the internal arrangements of the atoms of the elements.

Ignoring the action of the outer electrons on the innermost ones, Bohr calculated the effective repulsion σ_n for various numbers of electrons. The value for a four-ring is 0.96, close enough to 1 to suggest that the innermost ring in the metals that Moseley had measured contained just four electrons, half of the 8 that Bohr had deduced on shaky mechanical grounds but enough to confirm the argument that the innermost ring enlarges its population with increase in Z.[20] The factor of ¾ implied that the entity giving rise to the Kα line dropped from a 2-quantum state to the 1-quantum ground state. Perhaps, Moseley suggested to Bohr, a fast cathode ray excites a ring of four electrons as a whole, giving each of them an additional quantum of angular momentum, and the excited ring disposes of its energy as a Kα quantum in returning to its ground state.[21]

Or, perhaps, the law permitted exceptions: 'Is it possible that really no inner ring exists and that it is one electron vibrating by itself?' Then the factor Z–1 would become Z and two elements somewhere – Moseley suggested the pair argon/potassium, where the chemical sequence reverses the order of atomic weight – would have the same Z. A young Swedish physicist, Ivar Malmer, proposed a worse complication: his measurements of the high-frequency spectra of antimony suggested that an element's Z value might not be the same for its K and L series.[22] Evidently, experimentalists were quite willing to entertain options that would ruin the nascent principle of atomic number! Ignoring these subversive suggestions, Bohr remained with the question what 'vibrating by itself' or vibrating at all might mean on a quantum theory of radiation. He answered that Kα might originate in the filling of a hole in the innermost ring created by the collision of a cathode ray with one of its electrons and that the simultaneous excitement of four electrons was energetically impossible.

Bohr's suggestion made the tantalizing Z–1 in Moseley's formula an approximation to $Z - Q$, where Q depends on the number of electrons in the ring and their screening before and after internal ionization. Bohr did not say what number satisfied $Q = 1$, but for the table he gave and the approximation he suggested, there was no good fit.[23] However, in a manuscript note Bohr was able to extract something from his model by following up the insight of Walther Kossel, a physicist close to Laue's group. Kossel suggested that Moseley's two K lines, α and β, represented the beginning of a spectrum, Kα arising by capture of an electron from the second (L) ring and Kβ from the third (M). The hardest possible K line would arise from the capture of an electron from outside the atom. Applying this consideration to element 83, Radium C, Bohr killed off Rutherford's β-γ theory: RaC's hardest possible K ray was far too soft to be its γ-ray.[24]

Moseley's last research centered on the L series of the heavier elements. In his final published paper, he gave measurements of Lα for 24 elements including seven rare earths. He was able to add three more elements at the heavy end of the earths after examining samples brought to him by Georges Urbain, who had extracted them by laborious fractional crystallization. What Urbain called the '*loi de Moseley*' unmasked the preparations that he had supposed to be the heaviest earth as a mixture of lighter ones. But whereas Urbain had needed a chemical laboratory and a decade to isolate and analyze his specimens, Moseley's desktop identity provider gave results in five minutes. They tried to jam the element Urbain thought he had found, which he called 'celtium', into the only two empty places that Moseley allowed the rare earths: 61, which does not exist naturally, and 72, which is not a rare earth. In the end, Moseley had to content himself with giving the wavelengths of Lα for Urbain's samples of numbers 70 and 71, and declare number 72 still at large.

The Laws: Empirical

The form $v_K = \text{const.} \times R(Z - \sigma_n)^2$ is not 'the law of Moseley' but a representation of the empirical data on the basis of Bohr's atom. What Moseley actually found he wrote as $v_K^{1/2} \propto N - a$, where N is a whole number that changes by one from element to element in the periodic table. The identification of N with atomic number and atomic number with Z fixes a value for a; without the Bohr theory Moseley would not have been able to choose $N - a$ from all possible values of $(N+x) - (x+a)$. He could have made room for any number of elements beneath sodium ($Z = 11$), the lightest element he examined, say one between hydrogen and helium, making $v_K^{1/2} \propto (Z+1) - 2$ and presenting the problem, on the Bohr theory, of explaining a screening constant of 2. The same problem existed for the L spectra, where Moseley's form was $v_L^{1/2} \propto N - 7.4$ with the assumption $N = Z$.

After the war, Rutherford asked James Chadwick to repeat the scattering experiments that had inspired the creation of the nuclear model. Since the theory of the experiments makes the scattering dependent directly on Z, Chadwick hoped to remove the ambiguity of Moseley's 'arbitrary choice of the number a', and so provide the missing 'experimental proof of Moseley's conclusion [$N = Z$]'. His results perhaps would not have been entirely persuasive to any one who had not accepted the conclusion already: Chadwick obtained, for platinum, silver, and copper, the three elements he studied, $Z = 77.4, 46.3$, and 29.3, respectively, which come within 1.5 percent of their atomic numbers 78, 47, and 29.[25]

Several other informed physicists commented on the arbitrariness of Moseley's identification of N with Z around the time that Chadwick undertook to remove it. For example, Frederick Soddy correctly stated Moseley's law as

$v^{½} = Q$, $\Delta Q = 1$, as did E.N. da C. Andrade.[26] Neither explicitly invoked Z. Arthur Erich Haas, whose precocious quantum atom had interested participants in the famous Solvay Council of 1911, observed that although the *Gesetz von Moseley* invited equating N with Z, it did not rule out more elements than ten below sodium. Mme. Curie, in a review of isotopy, and Arnold Sommerfeld, in the early editions of *Atombau und Spektrallinien*, took the same line: Moseley's $v^{½} \propto N - a$ supports only the supposition that $\Delta N = \Delta Z$, or, as Max Born put it in a text that summed up the old quantum theory, $Z - N = $ constant.[27]

Measurements of the X-ray spectra made by Manne Siegbahn, Ivar Malmer, and their associates in neutral Sweden during the war brought to light several new lines in the L series, a softer M series, and the doublet structure of Kα and Kβ, and confirmed indications of non-linearity obtained by Moseley in his $v^{½}$–plot of L lines. Although these findings did not impugn the claim $N = Z$, they loosened the connection between Moseley's law and Bohr's atom. Efforts to derive the one from the other did not succeed before the invention of electron spin and quantum mechanics destroyed the model that had guided Moseley. Reviewing the state of quantum theory in 1925, J.H. Van Vleck had no place for these efforts: 'we have now outgrown [them]'.[28] As early as 1921 French spectroscopists had published their suspicion that the mysterious a in Moseley's law was a misleading artefact. These included the main experts, Maurice de Broglie, Alexandre Dauvillier, and René Ledoux-Lebard. Dauvillier omitted Moseley entirely from a review of atomic structure and X-ray analysis.[29]

Since Moseley's law did not quite fit the facts, Ledoux-Lebard and Dauvillier proposed the empirical revision $v \propto N^{2.10}$ and, more generally, $v^{½} = a + bN + c/(d-N)$, the lower-case letters designating constants. Although they conceded, correctly, that such fits were 'illusory', others were devised. Malmer proposed $v/R = N^2 - NA_n + B_n$, where n differs from one series of rays to another. Paul Foote and F.L. Mohler, physicists at the U.S. National Bureau of Standards, gave, as their best recipe for Kα, $v/R = (¾)Z^2 - 1.65Z + 3.88$. The definitive second edition of A. H. Compton's *X-rays and electrons* preferred $(v/R)^{½} = 0.874(Z - 1.13)$.[30]

The first editions of the bible of the old-quantum theory, Sommerfeld's *Atombau*, acknowledge the fundamental importance of Moseley's law in ordering the periodic table, but relegate its connection to Bohr's quantum atom to the realm of 'historical interest', that is, of no scientific importance. Sommerfeld rephrased this judgment in later editions, writing in 1922 that 'it will be a marvel [*bewundernswert*] for all time that Moseley could take the first steps toward the theoretical explanation of the line spectra while he made his pioneering measurements'; a performance that, on further thought, Sommerfeld downgraded to 'remarkable [*bemerkenswert*]'.[31] His estimate may relate to the wartime call by Willy Wien and other right-wing German physicists with whom Sommerfeld sympathized to curtail acknowledgment of British science and scientists. Perhaps the same consideration applies to Walther Gerlach's text on the experimental basis of quantum theory, and Gustav Mie's on astrophysics, which do not mention Moseley when treating high-frequency line spectra.[32] Chauvinism might also be behind Oskar Klein's crediting his countryman Siegbahn with the 'profound study' of X-ray spectra, an opinion quickly neutralized by Maurice de Broglie's award of the profundity to Moseley.[33]

A further indication of the loss of connection between Moseley's law and the atomic theory that led him to it is the disappearance of the problematic constant *a* from statements of it. Robert Millikan was precocious in misquoting 'Moseley's discovery' as $\lambda_1/\lambda_2 = (N_1/N_2)^2$, λ standing for wavelength and the subscripts for different elements. Another Nobel Prize winner, Jean Perrin, offered $v_K \propto Z^2$. When Gerlach got around to citing the '*Moseley'schen Gesetz*', he wrote it as 'λZ^2 = konst'. In a second edition, he wrote the reverse, 'v_K = const. $Z^{1/2}$'.[34] Both misrepresentations are now found in standard reference works. Thus a current and often-reprinted text on atomic physics by an Oxford professor, and the latest edition of Van Nostrand's *Scientific Encyclopedia*, have $v^{1/2} \propto Z$; the *Encyclopedia Britannica* and the *Macmillan Encyclopedia of Physics*, $v \propto Z^2$. Most of these sources mention Moseley's success in ordering the elements and specifying missing ones, but by omitting the constant *a* and/or inverting the dependence of v on Z, they eviscerate the chief theoretical problem of atomic structure as Bohr and Moseley saw it.[35] As Sommerfeld would say, it is now only of historical interest.

Many older texts transcribe Moseley's law as he wrote it, particularly English books, which suggest that their authors read him before paraphrasing him. O. W. Richardson's *Electron Theory of Matter* (1916), G. W. C. Kaye's *X-rays* (1918), and J. A. Crowther's popular *Ions and Ionizing Radiations* (1924) are examples.[36] The American spectroscopist Harvey White gave a full and accurate review of Moseley's work in 1934, coincidentally with the publication of two German *Handbuch* articles, which together sufficiently exhausted the subject.[37] By the 1930s, with the advent of nuclear, cosmic-ray, and low-temperature physics, experimentalists had fresher subjects to pursue than spectroscopy and theorists had long since ceased to find inspiration in Moseley's Law.

The Man

In describing the development of physics during the war years, a Russian physicist, O. D. Chwolson, characterized Moseley's work as 'immortal' three times, although, with immortality, once would seem enough. He also observed that Moseley himself, having died during the war, was not immortal. That did not prevent George Sarton from enrolling Moseley among the 'immortals of science'. His early death only improved his immortality. 'He died in beauty. The sudden termination of this noble life on the battlefield was an additional glory. It bestowed on his personality a touch of romance and mystery, which completed its consecration'. This nonsense dates from 1927. During the war Moseley's death was taken as a symbol of the wastefulness of combat and the stupidity of generals. The upper classes that fell wastefully had been educated to do so. James Chadwick's comment on Moseley's death may indicate the thoroughness of the brainwashing. He wrote to Rutherford from the relative security of the prison camp in which he was held as an enemy alien – he had been working in Berlin when war broke out – that he was very sorry to learn about Moseley. 'Still he has [!] the satisfaction of knowing he has done his duty'.[38]

Some seized on Moseley's tragic end to argue that scientific men were too important to squander on the battlefield. The argument has been bent to suggest that the authorities had a policy of sending everyone to the front. It would be more correct to say that they did not have a policy of keeping back people whose special skills would be more useful at home. J. D. Bernal had it right: 'Moseley, who might have become the greatest experimental physicist of the century in England, was allowed to go out to Gallipoli and get killed'. Rutherford tried to have him transferred out of the Royal Engineers, but Moseley 'deliberately chose to share with others of his age the dangers of active service'.[39] The press reported Moseley's loss under captions like 'Sacrifice of Genius', which helped Rutherford and other senior scientists to argue the case of keeping their younger colleagues from exposure on the battlefield. In fact, by 1915, the need to develop strategic materials previously imported from the Central Powers, and increasing demands on new weapons, technologies, and countermeasures forced the Entente to put their scientists in the laboratory. The Germans also corrected their earlier profligacy with scientific manpower. Early in the war the physicists' news magazine, *Physikalische Zeitschrift*, regularly reported on colleagues killed in action; after 1916, no more names appeared.

Moseley's story – the story of the brilliant young man who worked with feverish haste to finish his fundamental investigation as if he had anticipated an early death – remained in the realm of romance until the discovery of his letters to his sister and mother in the early 1970s. The only notable items before then were Sarton's romantic article of 1927, inspired by his sympathetic concern that 'workers perhaps not too well known' receive the credit they deserve; Lister A. Redman's short summary of Moseley's work for clever sixth-form students of physics; and Bernard Jaffe's *Moseley and the Numbering of the Elements* (1971), an entry in the Science Studies Series that grew out of a postwar program to improve the teaching of physics. The series had as its aim inculcating a feeling for 'the most stirring and fundamental topics of science'.[40]

Sarton placed his factual review of Moseley's work within an entirely fanciful research environment. Thus he could write, '[Moseley] himself would hardly have dreamed of hitting [on] such a simple law', not knowing his announcement to Bohr, 'The results are exceedingly simple, and largely what you would expect'.[41] Redman published his brief account when senior physics master at the Kirkham Grammar School in Lancashire. Having spent some time with radar, he had a developed taste for apparatus and regarded Moseley's experimental technique as exemplary. As a good pedagogue he also sought to interest his students by describing Moseley's personality (very reserved), academic performance (poor at classics), and single-mindedness (no interest in music halls). He made use of the correspondence between Moseley and Rutherford, and looked for Moseley's family correspondence, but his plea for information about it apparently returned nothing useful. Had he found it and other documents to which it led, he would not have been able to characterize Moseley as a perfectly focused nerd. Although Moseley was reserved, he was not single-minded; he had a wide knowledge of gardening and natural history, and won school prizes in classics.[42]

Jaffe wrote his biography after retiring from teaching high-school physics and chemistry. His narrative, which was informed by his visits to sites connected with Moseley, used Moseley's correspondence with Bohr as well as with Rutherford, and aroused further interest in Moseley the man. But as Jaffe's technical presentation seldom rises above high-school level, it scarcely qualifies as history of science, and as the motivations he supplied are often jejune, it scarcely qualifies as biography. 'Moseley heard the weak whisper of Nature trying to yield to man another of its secrets'. He stopped to listen. 'The whisper became louder and clearer'. Thus Moseley's Law. When not amplifying nature's whispers, 'Moseley sometimes displayed strong, definite opinions, but never snobbishness, arrogance, or condescension'.[43] The family correspondence confirms his magnificent self-confidence and adds more than whispers of snobbishness and arrogance. It was his arrogance and stubbornness (to say nothing of his genius and the luck of time and place) that enabled Moseley to achieve his extraordinary results and win the commission that took him into battle.

The Life and Letters of an English Physicist (1974) was a sequel to papers describing Moseley's work at a more demanding level than Jaffe's and analyzing attempts to derive Moseley's Law from Bohr's quantum atom.[44] The book had no sequel, and its author's *Doktorvater* wondered why his student was wasting his time on a life that might have its dramatic interest but threw no new light on the structure of scientific revolutions. Moseley did not do normal science. When he could choose for himself, he insisted on entering a nascent field for which he was not prepared against the wishes and advice of the professor who was trying to normalize him. Had he survived the war, he would not have tried to overtake the Swedes in the exact measurement of X-ray frequencies.

It is worth considering what direction he would have taken, if for no other reason than to review some of the options of demobilizing physicists. We shall assume that with a Nobel Prize in Physics (that of 1917, shared with C. G. Barkla), a fundamental law in his name, and the even weightier qualification of being an Oxford man, born and bred, Moseley would have been preferred for the Clarendon professorship over the real-world recipient, Frederick Lindemann.[45] His only apparent default, his age, which might have mattered before the war, would not have figured immediately after it. Lindemann was only marginally older than Moseley, and William Lawrence Bragg, who replaced Rutherford at Manchester when Rutherford took over from J. J. Thomson at Cambridge, was three years younger.

Lawrence Bragg had served as a sound ranger during the war. He brought several of his comrades with him to his new post in Manchester. Together they took up research almost where he had left it, with an investigation of the intensity of reflected X-rays. Their immediate objective was to test calculations made by Darwin in 1914 that could provide minute details about the structure of crystals. Unfortunately, young Bragg was handicapped by inexperience in teaching and a diffidence exploited by a resentful staff used to the overbearing Rutherford and by the returning veterans among the undergraduates. Moseley would not have experienced this time-wasting torment. His predecessor Clifton was no Rutherford and left no resentful staff; Moseley was a tougher character than Bragg; and his having taught 'idiots elements' at Manchester might have prepared him better for boisterous undergraduates.[46]

Like Bragg, but for different reasons, Rutherford had teething problems in his new post. Wishing to establish the Cavendish as the preeminent laboratory for applied as well as purer physics, he drew up a budget for laboratories, personnel, and endowment that reached £200,000. No sum near that could be procured at Cambridge at the time. The laboratory remained overcrowded with mature undergraduates and research students, many of whom would have distinguished scientific careers elsewhere. They worked on a wide range of topics. Some followed up prewar problems in radioactivity and atomic structure. Others developed apparatus to detect isotopic composition and rays from radioactive substances, and machines to produce high voltages and powerful magnetic fields.

The most novel research followed up experiments that Rutherford had made during his few leisure hours towards the end of the war: sending α-particles from a natural radioactive source through air to probe the limits of Coulomb's law. He was surprised to discover an emergent particulate radiation more penetrating than the bombarding α-particles. He identified the penetrating radiation as a stream of hydrogen nuclei ('protons') knocked out of nitrogen nuclei by the alphas. He had in mind something like a game of marbles in which the incoming particle replaced the ejected one. He speculated that nuclei contained $N = Z$ protons and enough tightly held neutral pairs of protons and electrons to make up the rest of the atomic weight. Because these supposititious 'neutrons' would exhibit almost no external charge they would be hard to detect. Rutherford put Chadwick, who had returned from Germany, on the job. Chadwick could not put his full strength into it as his health was shaky and he had many other assignments. After he and Rutherford had disintegrated aluminum, phosphorus, and fluorine without finding neutrons, he proposed, in 1924, that they make 'a real search'.[47] They did not do so.

Lindemann had the advantage of a clean slate and empty laboratory when he arrived in Oxford after spending the war years at the Royal Aircraft Factory. He had grand plans to build a research school, but found money tighter than Rutherford did at Cambridge, and had to haggle for an annual budget of £1000. He tried to continue the low-temperature research he had done in Germany before the war. But the hydrogen liquefier he bought worked only in the presence of the German mechanic who built it. Lindemann gave it up, and research too. That was also in 1924.[48] Moseley would have faced the same financial and staffing problems at Oxford, but he would have known how to get his way and how to adapt his research to available resources. With his local contacts and international reputation, he would have been able to draw promising students who had never heard of Lindemann. What would they have done?

First, they would have submitted to the irritation of learning the language of the vanquished enemy since the quickest way to the research front in atomic physics was through Sommerfeld's *Atombau*.[49] They would have learned quickly that, despite considerable effort, there was no satisfactory quantitative derivation of their professor's 'law'. The qualitative understanding of the origin of the high-frequency spectra, however, seemed secure, and the frequency measurements had been carried to one or two places of decimals beyond Moseley's. Moreover, the 'horde of hungry Germans' Moseley had feared might scoop him in 1914, and other competitive Europeans, were at work on further calculations and measurements.[50] None of it would have seemed attractive to Moseley. He probably would have set up his old vacuum spectrometer to instruct research students in experimental technique and to look for the missing elements. Bohr might have told him that, on the latest Copenhagen principles of atomic structure, element 72 could not be a rare earth, and asked him to look for it. If so, or if Moseley had reached a similar conclusion on his own, he might well have found the missing element before Hevesy and Dirk Coster did in 1922, and 'hafnium' would be known as 'oxonium'. Thus encouraged, he would have put a research student or two on the trail of the three remaining missing elements below gold. Since two of them do not exist naturally, it would have been a frustrating enterprise.

For his own work he would have looked for a field that lent itself to the same sort of rapid survey technique that had proved so fertile in his X-ray work, a new field with only a little theory and the promise of a rich experimental harvest for the assiduous researcher. Can there be any doubt that he would have rushed into the salient Rutherford had opened into the nucleus? Moseley had mastered the handling of the radioactive materials used as sources. He would have run quickly through all the lighter elements suitable for targets looking for ejected protons and Rutherford's neutrons. Eventually he would have discovered the highly penetrating radiation beryllium emits when bombarded by α-particles. The Germans who discovered this effect in 1930 thought the penetrating rays were very energetic γ-rays. This identification became increasingly implausible as measurements of the velocities of protons ejected by the beryllium radiation accumulated: the suppositious gammas would have to have incredible energies to be able to give the protons their observed velocities. Everything fell into place, however, if the penetrating radiation was particulate. In 1932, Chadwick recognized that it consisted of neutrons.[52]

Let us assume that Moseley detected the beryllium radiation around 1925 with the same instrument – an improved Geiger counter – that its real discoverers used in 1930. With liberal applications of his stubbornness and resourcefulness, Moseley would have been able to acquire the necessary strong source of alpha particles.[53] He would have been as alert as Chadwick to Rutherford's conjectures about electron-proton pairs. Well, then, let Moseley identify the penetrating radiation as neutrons in the late 1920s. If nuclear physics proceeded after his counter-factual discovery, as in fact it did after Chadwick's, physicists would have had plenty of time to discover fission and begin to work on nuclear weapons before the outbreak of the Second World War. Would Germany or the Soviet Union have been able to develop them before its end? As it happens, they did not, perhaps because Moseley died at Gallipoli.

To some it may be easier to believe that a benign Providence, acting like Mark Twain's *Mysterious Stranger*, did away with Moseley for the good of humanity, than to credit that good physicists could have muddled Moseley's law into such aberrant forms as the triply wrong $v^2 \propto Z$. We may conclude that Moseley and his law still have work to do. The story of his achievements, if accompanied by an account of the errors that scientists have made in trying to tell it, might promote awareness of the injury anachronism does to understanding, and of the short change textbooks give to items 'of historical interest'.

Notes and References

[1] Moseley to his mother (Amabel Sollas), October 1910 and 5/6 July 1912, and to his sister (Margery), 7 April 1912 and 28 July 1912, Letters 44, 58 and 56 in J. L. Heilbron, *H. G. J. Moseley, The Life and Letters of an English Physicist, 1887–1915* (Berkeley: University of California Press, 1974), 176, 189–190 and 187–188 (hereafter Heilbron, *Moseley*).

[2] Finn Aaserud and J. L. Heilbron, *Love, Literature, and the Quantum Atom* (Oxford: Oxford University Press, 2013), 74–78 (hereafter Aaserud and Heilbron, *Quantum Atom*); Hulme Hall Register, s.v. 'Bohr'.

[3] Moseley to Margery, 1 May 1911, Letter 52 in Heilbron, *Moseley, op. cit.* 182–183.

[4] Aaserud and Heilbron, *Quantum Atom, op. cit.* 129–141, 159–161.

[5] Sources for History of Quantum Physics, 'Interview with Niels Bohr, 1962', (American Philosophical Society Library, Philadelphia), 26, 31, 32, 33 (quote), 38, 65 (hereafter SHQP).

[6] *Ibid*. 50.

[7] Moseley to Bohr, 16 November 1913, Letter 81 in Heilbron, *Moseley, op. cit.* 211.

[8] Moseley to his mother, 10 October 1912 (first mention of X-rays) and November 1912 (the explanation), Letter 61 and Letter 63 in Heilbron, *Moseley, op. cit.* 193 and 194–195.

[9] Moseley to his mother, 18 May 1913 (finishing with Darwin), Letter 74 in Heilbron, *Moseley, op. cit.* 204–205.

[10] Heilbron, *Moseley, op. cit.* 72; Aaserud and Heilbron, *Quantum Atom, op. cit.* 162–163; Niels Bohr, 'Reminiscences of the Founder of Nuclear Science and of Some Developments based on his Work', in Niels Bohr, *Collected Works* (12 vols., Amsterdam: North-Holland/Elsevier, published between 1972 and 2007), vol. 10, 385. This version includes footnotes not published in the original printing in *Proceedings of the Physical Society of London*, 78 (1961), 1083–1115.

[11] Aascrud and Heilbron, *Quantum Atom, op. cit.* 163.

[12] E. Rutherford, 'On the Energy of the Groups of Beta Rays from Radium', *Philosophical Magazine Series 6*, 24 (144), (1912), 893–894, in Ernest Rutherford, *Collected Papers*, 3 vols. (London: George Allen and Unwin, 1962–65), vol. 2, 292–293.

[13] E. Rutherford and H. Richardson, 'The Analysis of the β Rays from Radium B and Radium C', *Philosophical Magazine Series 6*, 25 (149), (1913), 722–734, 6 March 1913, and 'Analysis of the γ-rays of the Thorium and Actinium products', *Philosophical Magazine Series 6*, 26 (156), (1913), 937–948, (October 1913), in Rutherford, *Collected Papers*, vol. 2, 352, 420–421; and Rutherford and E. N. da C. Andrade, 'The Wavelengths of the Soft γ Rays from Radium B', *Philosophical Magazine Series 6*, 27 (161), 854–868, in Rutherford, *Collected Papers*, vol. 2, 440, 443.

[14] Rutherford waffled over the origin of the initial beta ray: in December 1913, being accused by Frederick Soddy of refusing to allow negative charges in the nucleus, he said that he had never pronounced on the matter; in February 1914, he granted that a nuclear source was reasonable, and pointed to Bohr; but in September 1914, he still allowed the possibility of an extra-nuclear origin. E. Rutherford, 'The Structure of the Atom', *Nature*, 92 (2302), (1913), 423, (6 December 1913), 423, in Rutherford, *Collected Papers*, vol. 2, 409; 'The Structure of the Atom', *Philosophical Magazine Series 6*, 27 (159), (1914), 488–498, in *Collected Papers*, vol. 2, 429; 'The Connexion between the β and γ Ray Spectra', *Philosophical Magazine Series 6*, 28 (165), (1914), 305–319, in *Collected Papers*, vol. 2, 475–478, 484.

[15] Heilbron, *Moseley, op. cit.* 84.

[16] Bohr to Rutherford, 10 June 1913, in Bohr, *Collected Works* (ref. 10), vol. 2, 586.

[17] Cf. N. Bohr, 'On the Constitution of Atoms and Molecules, Part II', *Philosophical Magazine Series 6*, 26 (153), (1913), 476–502 (hereafter Bohr II), on 499–500.

[18] Moseley, review of 'The Svedberg, *Die Existenz der Moleküle, Experimentelle Studien* (Leipzig, 1912)', *Nature*, 92 (2307), (1913/14), 554 (Oxford, 5 January 1914). E. Rutherford, 'Moseley's Work on X-Rays', *Nature*, 116 (2913), (1925), 316–317 (29 August 1925) confirms this statement.

[19] Moseley to Bohr, 16 November 1913, Letter 83 in Heilbron, *Moseley, op. cit.* 214.

[20] Bohr II, *op. cit.* (ref. 17), 482.

[21] Moseley to Bohr, 16 November 1913, Letter 81 in Heilbron, *Moseley, op.cit.* 213.

[22] I. Malmer, 'The High-Frequency Spectra of the Elements', *Philosophical Magazine Series 6*, 28 (168), (1914), 787–794, on 792 (hereafter Malmer, 'High-Frequency').

[23] Bohr to Moseley, 21 November 1913, Letter 83 in Heilbron, *Moseley, op. cit.* 215.

[24] SHQP, Bohr MSS, m/f 5, Bohr, 'Note', MS, late 1914 or early 1915.

[25] J. Chadwick, 'The Charge on the Atomic Nucleus and the Law of Force', *Philosophical Magazine Series 6*, 40 (240), (1920), 734–746, on 736, 742–743.

[26] Frederick Soddy, *The Interpretation of Radium and the Structure of the Atom* (London: J. Murray, 1920), 239–240; E. N. da C. Andrade, *The Structure of the Atom* (London: Bell, 1923), 99–100.

[27] Arthur Erich Haas, *Das Naturbild der neuen Physik* (Berlin: De Gruyter, 1920), 89-90, 107 n. 45; Arnold Sommerfeld, *Atombau und Spektrallinien* (Braunschweig: Vieweg, 1921), 174–175 (hereafter Sommerfeld, *Atombau*); Marie Curie, *L'isotopie et les éléments isotopes* (Paris: *La Société Journal de Physique*, 1924), 138–139; Max Born, *The Mechanics of the Atom* (London: Bell, 1927), 177 n.

[28] J. H. Van Vleck, 'Quantum Principles and Line Spectra', in National Research Council, *Bulletin*, 10 (4), (1926), 77–78.

[29] Maurice de Broglie, *Les rayons X* (Paris: *La Société Journal de Physique*e', 1922), 89; René Ledoux-Lebard and Alexandre Dauvillier, *La physique des rayons X* (Paris: Gauthier-Villars, 1921), 2, 175, 376, 384–386; Alexandre Dauvillier, *La technique des rayons X* (Paris: Paris: *La Société Journal de Physique,* 1924), 187.

[30] Malmer, 'High-Frequency', *op. cit.* (ref. 22) 787; Paul Foote and F. L. Mohler, *The Origin of Spectra* (New York: American Chemical Society, 1922), 48, 192, 204; A. H. Compton and S. K. Allison, *X-rays in Theory and Experiment* (New York: Van Nostrand, 1935), 32–37, 584–585, 590–592, followed by Compton's student J. D. Stranathan, *The 'Particles' of Modern Physics* (Philadelphia: Blakiston, 1942), 296–303.

[31] Sommerfeld, *Atombau, op. cit.* (ref. 27), 176; idem. (1922 3rd edn), 195; (1924, 4th edn), 263.

[32] Walther Gerlach, *Die experimentellen Grundlagen der Quantentheorie* (Braunschweig: Vieweg, 1921), 62–63, 82. Curiously, George Hevesy and Fritz Paneth do not mention Moseley in connection with X-ray spectra in the first edition of their *A Manual of Radioactivity* (London: Oxford University Press, 1926), but do in the second (1938), 63.

[33] Oskar Klein, *Entretiens sur les idées fondamentales de la physique moderne* (Paris: Hermann, 1938), 186–187, 237; Maurice de Broglie, *Atomes, radioactivité, transmutations* (Paris: Flammarion, 1939), 105–108.

[34] Robert Millikan, *The Electron* (Chicago: University of Chicago Press, 1917), 201; Jean Perrin, *Les atomes* (Paris: Alcan, 1927), 301; Walther Gerlach, *Materie, Elektrizität, Energie* (Dresden: T. Steinkopff, 1923), 127, and *Matter, Electricity, Energy* (New York: Van Nostrand, 1928, 2nd edn), 205.

[35] Christopher J. Foot, *Atomic Physics* (Oxford: Oxford University Press, 2005), 9–11; Van Nostrand *Scientific Encyclopedia*, 10th ed. (Hoboken: Wiley-International, 2008), s.v. 'Moseley's Law'; *Encyclopedia Britannica*, (Chicago: Encyclopedia Britannica, 15th edn, 1997), vol. 8, 350, vol. 14, 337; *Macmillan Encyclopedia of Science*, 4 vols. (New York: Simon and Schuster, 1991), vol.1, 73, improved at vol. 4, 1720.

[36] O. W. Richardson, *The Electron Theory of Matter* (Cambridge: Cambridge University Press, 1916), 514, 605–606; G. W. C. Kaye, *X-rays* (London: Longmans Green, 1918), 224–229; J. A. Crowther, *Ions, Electrons, and Ionizing Radiations* (New York: Longmans Green, 2nd edn, 1924), 182; (5th edn, 1929), 197; (7th edn, 1938), 175–176.

[37] H. E. White, *Introduction to Atomic Spectra* (New York: McGraw-Hill, 1934), 302–328, 311, 321, 325, 333, 335, 344, 348; Adolf Smekal, 'Quantentheorie', in *Handbuch der Physik*, 24:1 (Berlin: Springer, 1933), 33; H. Kuhn, 'Atomspektren', in *Hand- und Jahrbuch der chemischen Physik*, 9:1 (Leipzig: Akademische Verlagsgesellschaft, 1934), 93, 156–161.

[38] O. D. Chwolson, *Die Physik, 1914–1920* (Braunschweig: Vieweg, 1927), 54–55, 204, 208–213; George Sarton, 'Moseley, The Numbering of the Elements', *Isis*, 9 (1), (1927), 96–111, on 97 (hereafter Sarton, *Moseley*); Chadwick to Rutherford, 14 September 1915, in Andrew Brown, *The Neutron and the Bomb: A Biography of James Chadwick* (Oxford: Oxford University Press, 1997), 37 (hereafter Brown, *Chadwick*).

[39] J. D. Bernal, *The Social Function of Science* (London: Routledge, 1939), 171; E. R. Lankester, 'Henry Gwyn Jeffreys Moseley', *Philosophical Magazine Series 6*, 31 (182), (1916), 173–176.

[40] Dorothy Stimson, in her collection of Sarton's essays, *Sarton on the History of Science* (Cambridge, MA: Harvard University Press, 1962), ix; Lister A. Redman, *The Physics Teacher*, 3 (1965), 151–157; Bernard Jaffe, *Moseley and the Numbering of the Elements* (New York: Doubleday, 1971). Hereafter Jaffe, *Moseley*.

[41] Sarton, *Moseley, op. cit.* (ref. 38), 276; Moseley to Bohr, 16 November [1913], Letter 81 in Heilbron, *Moseley, op. cit.* 211–213.

[42] Lister A. Redman, *Physics Teacher, op. cit.* (ref. 39), 151–153, 157, and 'Physics in the English Schools', *The Physics Teacher*, 2 (1964), 118–125, on 118. Perhaps Redman knew Ivor B. N. Evans's brief romantic notice, 'H. G. J. Moseley', *Discovery*, 8 (1947), 341-344, which mentions but does not use the family correspondence.

[43] Jaffe, *Moseley, op. cit.* (ref. 40), 97, 28, resp.

[44] J. L. Heilbron, 'The Work of H.G.J. Moseley', *Isis*, 57 (3), (1966), 336–354, and 'The Kossel-Sommerfeld Theory and the Ring Atom', *Isis*, 58 (4), (1967), 451–485.

[45] The most powerful man on the Nobel Committee for Physics, Svante Arrhenius, nominated Moseley, and Moseley's results motivated the favorable consideration of Barkla's. Royal Swedish Academy of Sciences, Stockholm, Archives, Arrhenius's nomination, 30 January 1915, and Vilhelm Carlheim-Gyllensköld's report on Barkla, August 1918, pp. 11–12.

[46] David Phillips, 'William Lawrence Bragg, 31 March 1890–1 July 1971', *Biographical Memoirs of Fellows of the Royal Society [of London]*, 25 (1979), 75–143, on 96–99; John Jenkin, *William and Lawrence Bragg: Father and Son* (Oxford: Oxford University Press, 2008), 401–412; Moseley to his mother, September–October 1910, Letter 43, in Heilbron, *Moseley, op. cit.* 174. W. H. Bragg and W. L. Bragg, *X-rays and Crystal Structure* (London: Bell, 1924, 4th edn), 53–55, has a brief and cool account of Moseley's 'highly important rule'.

[47] J. G. Crowther, *The Cavendish Laboratory, 1874–1974* (New York: Science History, 1974), 183–191; Ernest Rutherford, 'Collisions of α Particles with Light Atoms', *Philosophical Magazine Series 6*, 37 (222), (1919), 537–561, April 1919, in *Collected Papers*, vol. 2, 565–567 (H atoms from alphas) and 'Nuclear Constitution of Atoms', *Proceedings*

of the Royal Society A, 97 (686), (1920), 374–400, received 3 June 1920, *Collected Papers*, vol. 3, 34 (the neutron).

[48] Jack Morrell, *Science at Oxford, 1914–1939: Transforming an Arts University* (Oxford: Oxford University Press, 1997), 386–394.

[49] Heilbron, *Moseley, op. cit.* 79–80.

[50] Moseley to his mother, 10 November 1913, Letter 80 in Heilbron, *Moseley, op. cit.* 211.

[51] *Ibid*. 135-139.

[52] Brown, *Chadwick, op. cit.* (ref. 37), 103–108; J. L. Heilbron and R. W. Seidel, *Lawrence and his Laboratory* (Berkeley: University of California Press, 1989), 144–147.

[53] Cf. Jules Six, *La découverte du neutron (1920–1936)* (Paris: CNRS, 1987), 24–42, 54–69.

CHAPTER ELEVEN

ARTEFACTS AND ARCHIVES: REINTERPRETING HARRY'S STORY IN A MUSEUM CONTEXT

ELIZABETH BRUTON, SILKE ACKERMANN AND STEPHEN JOHNSTON

Introducing an Exhibition Project

Henry 'Harry' Moseley is a name that matters in the history of science, certainly enough to justify the fresh view represented by this volume. But he is unfamiliar to a larger public. While Moseley's Law might be understood by chemists and physicists, neither his most important work nor even his name have a resonance with a general audience. Is it possible to engage such an audience with his life, science and legacy? That was the challenge taken up by the University of Oxford's Museum of the History of Science (MHS). With the generous support of the Heritage Lottery Fund (HLF) and a range of partners and lenders, the Museum ran a project which led to 'Dear Harry', a centenary exhibition in 2015 with an extensive accompanying public programme.

This chapter uses the lens of that exhibition project to provide a perspective on the themes and strands of the volume's other chapters. Emphasising the material and visual culture that are central to exhibition work, it documents the creation and development of the project, as well as its major outcomes, including public engagement with the history of science through the display of original experimental apparatus designed and used by Harry himself.

In keeping with the character of exhibition work, the chapter adopts a visually accessible 'photo essay' format, with a central narrative accompanied by a set of images and extended captions. Foregrounding material from the Museum's collection and its exhibition partners and lenders, it also draws on family papers and photographs recently rediscovered as part of the project's research.

The 'Dear Harry' exhibition marked the centenary of Harry's death on 10 August 1915 as well as the centenary of the Gallipoli campaign in which he tragically died, aged just twenty-seven. Part biography, part First World War centenary commemoration, and part history of science, the exhibition presented an intimate biographical portrait of Henry 'Harry' Moseley.

We first describe the origins and development of the exhibition, which depended on partnerships with the Royal Engineers Museum, Library and Archive; the Royal Signals Museum; the Department of Physics at the University of Oxford; and Trinity College, Oxford. The chapter then recapitulates the structure of the exhibition, drawing particular attention to some of the specific artefacts and archival items that gave the exhibition poignancy and impact.

The core of the exhibition consisted of three sections on Moseley as son, scientist and soldier, prefaced by an introduction providing the context of Gallipoli and its centenary commemoration, and concluding with a section on Harry's legacy. As a son, Harry appeared through his close relationship with his family, especially his mother and sister, and how this framed and supported his passion for science and the natural world around him. As a scientist, his university education in physics and his postgraduate researches were emphasised. And as a soldier, his training as a Royal Engineer and his service at Gallipoli were the focus.

These three central themes of son, scientist, and soldier were personal, but also representative of broader narratives. The exhibition explained not just the individual significance of Harry Moseley as one of the most exceptional scientists of his generation, but also explored his global contribution to the development of twentieth-century science. Moreover, Harry's life story highlights that of other sons, other scientists and other soldiers, through the loss that the scientific and international communities experienced in the face of war, the changing role of science and scientists in war, and the impact that the campaign at Gallipoli had on the Allied war effort.

In exploring the selection of objects and materials presented in the exhibition, the chapter shows how the decision to present a strongly human face to Harry's scientific research was implemented in practice, while also indicating the way in which Harry's life and legacy was set against the wider stage of international scientific discovery and the First World War.

In closely following the exhibition's structure and content, many other aspects of the larger project have had to be omitted here: the conservation of apparatus and archives in the Museum's collections; full details of the subsequent permanent redisplay of this important material with a much richer interpretation than previously realised; and the comprehensive programme of public events, education work, and digital resources which the project delivered. One element beyond the exhibition which does feature is the re-establishment of contact with the Ludlow-Hewitt family. Related to Harry through the marriage of his sister Margery, the family holds the Moseley family archives and several items discovered since the exhibition are included here. Singling out 'Dear Harry' for discussion does however reflect the depth of engagement which the exhibition provoked from museum visitors. It attracted 48,000 visitors and achieved an emotional response unprecedented in the Museum's temporary exhibition programme. In introducing Harry Moseley to a new and broad contemporary audience, the exhibition moved visitors to question what else was lost in war.

Fig. 1 Part of the 'Dear Harry' exhibition, showing his surviving scientific apparatus, a wall-mounted video touchscreen and a showcase of archival material from his schooling. (Source: Museum of the History of Science, University of Oxford).

The 'Dear Harry' Project

The exhibition 'Dear Harry: Henry Moseley, a scientist lost to war' ran at MHS between May 2015 and January 2016. Its origins however go back to 2011 when the upcoming centenary of Moseley's most important work suggested the value of re-examining the Museum's holdings. During the academic year 2011-2012 Kristen Frederick-Frost studied archival material and apparatus housed in MHS associated with Harry Moseley's X-ray spectroscopy, as well as related papers and objects held by the University of Oxford's Department of Physics. Her work reconnected the surviving material in the two organisations to create a novel and nuanced account of Harry's experimental process.[1] Frederick-Frost's research provided the groundwork for the 'Dear Harry' exhibition as well as Chapter 3 in this volume, but her focus on material culture also highlighted the more urgent need for conservation of vulnerable artefacts if they were to be preserved for future generations.

While encouraging the possibilities of new display, this academic work also made it clear that Harry's research could not be easily presented to a broad public. Almost nothing of the intellectual framework of early twentieth-century physics and chemistry could be assumed and the surviving experimental apparatus was fragmentary and obscure. To address the inaccessibility of the latter, the Museum built on an existing collaboration with the Hochschule für Technik und Wirtschaft in Dresden. Under the supervision of Professor Markus Wacker, a series of computer graphics students spent time animating objects from the MHS collection. In 2013 the Museum worked with Martin Wolff to create an animation of Harry's X-ray apparatus in action.[2]

These academic initiatives informed the more general project planning which began in 2013 with discussions between two of this chapter's authors, Dr Stephen Johnston at MHS and Dr Elizabeth Bruton, then postdoctoral researcher for 'Innovating in Combat: Telecommunications and Intellectual Property in the First World War', an AHRC-funded project on wartime telecommunications at the University of Leeds and MHS led by Professor Graeme Gooday.[3] As part of this project, Bruton had further researched Harry's role as a signals officer in the Royal Engineers and presented a lecture on this topic before the Royal Society, which formed a basis for her chapter elsewhere in this volume.[4] Through this lecture, contact was made with the Royal Signals Museum as well as fellow book editor and chapter author Professor Russell Egdell, then at Trinity College, Oxford, and co-author of a series of articles in the magazine *Stand To!* on the experiences of members of Ernest Rutherford's research group in the First World War.[5] Thus both the military dimension of the exhibition and the partnerships which would shape 'Dear Harry' began to develop. The crucial next step was to secure funding. Dr Silke Ackermann, newly-appointed as Director of MHS, led the team working on a HLF 'Our Heritage' grant application. The key decision was to make this project a dramatic departure for MHS.

The Museum's collection is of international standing and particularly celebrated for its unparalleled early Islamic and European mathematical and astronomical instruments. Objects rather than interpretation had traditionally been given pre-eminence in its permanent displays. But it was already clear that an exclusive focus on the highly technical nature of Harry's science would not engage the Museum's general audience. Instead, a larger vision of his life and death was needed. Surprisingly, the Museum had never previously created a major biographical exhibition, but it was this approach, crystallised around the three central themes of 'son, scientist, and soldier' that determined the physical display, the project partners and exhibition lenders, and the successful outcome of the 2014 application to HLF.

Through biography, the aim was to create an exhibition which was genuinely personal. Much of the correspondence edited by John Heilbron has the intimate familiarity of family letters through which Harry's character emerges strongly. In the exhibition, carefully selected quotations from throughout his life appeared in cases and on large graphic panels. These were complemented by a second continuous strand, the pocket diaries of his mother Amabel Moseley. While compiled principally as social diaries they nevertheless contributed a poignant and emotional impact, charting the life and legacy of her only son.

Each of the exhibition's three main sections benefited enormously from the contributions of the project's four partners: the Department of Physics and Trinity College, both of the University of Oxford; the Royal Signals Museum; and the Royal Engineers Museum, Library, and Archive. In addition, further loans came from Eton College, the Imperial War Museum, and Summer Fields School, Oxford, as well as private lender Dr Simon Vaughan Hunt. Images from the Bodleian Library and the Royal Society also figured prominently. The combination of personal and scientific material enabled us to tell an accessible and rounded story of Harry.

Life and Death at Gallipoli

The 2015 centenary of the conflict at Gallipoli, where Harry was killed, emerged as the most appropriate date to mark his life, and summer 2015 was earmarked for the exhibition, to exactly match the period of the campaign. Despite some increased awareness of the campaign through wider World War One centenary activities, formative feedback suggested that many of our visitors were unfamiliar with the campaign or even where Gallipoli was located. The exhibition therefore opened with an overview of the Gallipoli campaign, setting the scene with a short film featuring original British Pathé news footage of the campaign. The impact was contextualised with images and loan objects from the battlefield itself. As well as providing the rationale for the exhibition's staging in 2015, we were keen to emphasise the international context of Gallipoli – a First World War campaign which is far less visible in the British public consciousness than the trenches of the Western Front and more often associated with the Australian and New Zealand Army Corps (ANZAC).

As described in more detail in Chapter 5, the Gallipoli Campaign (*Çanakkale Savaşı*) in which Harry fought and died was one of the most futile episodes of the First World War, lasting from February 1915 to January 1916. British soldiers, together with troops from Australia and New Zealand (jointly known as ANZAC), Canada, India, and France fought against soldiers from the Ottoman Empire, Germany's ally. The Gallipoli peninsula was inhospitable; the ground was dry and barren; and water supply was a major problem for both sides. Allied forces were unprepared for the difficult rugged terrain as well as the depth and strength of resistance from Ottoman troops. This was illustrated in the exhibition through the map and evocative photographs shown in Fig. 2.

Extensive defences including fortifications, landmines, and barbed wire manned by experienced troops proved a strong opposition to Allied offenses with wave after wave of men on both sides being sent to their almost certain death. In the opening stage of our exhibition, we included examples of recovered Ottoman landmines and barbed wire lent by the Imperial War Museum, illustrating the defences which the Allied forces faced. As a result, when the invading armies were finally withdrawn in January 1916, there were more than 100,000 dead and more than 400,000 wounded.

Largely overshadowed then and now by the trench warfare of the Western Front – particularly in British histories of the war – the Gallipoli campaign was nonetheless a defining moment in the history of the Turkish, Australian and New Zealand nations.

In Turkey, the campaign is ranked as one of the greatest victories during the war, a final surge just before the Ottoman Empire crumbled prior to the establishment of the modern Turkish state in the early 1920s. In Australia and New Zealand, despite the defeat and large loss of life resulting from the campaign, it is considered a pivotal point in the establishment of a new national identity, marking the transition from nations of farmers to nations of heroic soldiers. In all three countries, the Gallipoli campaign commemorations are the most significant First World War events. As an introduction to a biographical exhibition, the setting of Gallipoli was intended to remind visitors

Fig. 2 Composite image created for the exhibition of a selection of photographs taken at Gallipoli by a Royal Engineer in the period June to August 1915 (Royal Engineers Museum Library and Archive, object 9803.4.6) together with a 1915 map of part of the Gallipoli peninsula (Bodleian Library shelfmark D30:24 (14)). The image was part of the exhibition's introduction, selecting twelve vivid postcard-size photographs from a group of sixteen taken by a Royal Engineers Signaller at Gallipoli during the summer period when Harry also served. These were combined with an artillery map used by the British Army during the Gallipoli campaign. Maps were an important part of the 'Dear Harry' exhibition as formative feedback had revealed that many MHS visitors were unfamiliar with the Gallipoli campaign as well as where Gallipoli was located. The British Army artillery grid map from the Bodleian Library collections supplied exhibition visitors with a sense of the geography and terrain of the region. It is one of a series of 1:20,000 scale maps produced by the Survey of Egypt in July and August 1915. Based on captured Turkish maps, they replaced the inadequate and inaccurate 1:40,000 maps used in the early stages of the campaign. These improved maps reproduced Ottoman place names but also added new Allied terms such as Anzac Bay, and were overlaid with the standard 600 yard square grids of an artillery map. This particular map was further adapted and personalised, evidently during the campaign, with manuscript additions by the unknown hand of a British Army officer, showing prominent features of the August 1915 offensive in which Harry took part. These deeply personal artefacts survived the brutal conditions of the campaign and represented the broader Allied experience, augmenting Harry's personal correspondence. (Source: MHS exhibition team by courtesy of the Royal Engineers Museum, Library and Archive and the Bodleian Library, University of Oxford).

that, although we offered an individual approach, our focus on Henry Moseley picked out just one among so many other sons, scientists and soldiers, and that his death allowed us to ask bigger questions about what else was lost in war.

Son

Born in Dorset in 1887 to a distinguished scientific family, Harry moved at a young age to Oxford where his father Henry Nottidge Moseley held the prestigious role of Linacre Professor of Human and Comparative Anatomy. The family had a comfortable life and travelled often but Harry's father – who was previously healthy and physically robust – was struck down by severe physical illness and died just two weeks before his son turned four and in the same year that Harry's sister Amabel also died.

As we have seen in Chapter 1, two of Harry's sisters died when Harry was still young. Elizabeth (known as Betty) was born in 1883 and died of TB aged sixteen in May 1899; her death certificate notes that her mother was present at her death.[6] A third sister, Amabel, was discovered through census records

Fig. 3 Three Moseley family photographs showing a young Harry aged nearly five on 1 November 1892, just over a year after his father's untimely death; an undated photograph of Margery Moseley as a young woman, wearing a fashionable cap; and an undated photograph of their mother Amabel. After Henry Nottidge Moseley's death aged only forty-seven in 1891, Harry's mother moved the family to Chilworth near Guildford in Surrey. With sole responsibility for the upbringing of her three surviving children, Amabel Moseley actively and determinedly encouraged their intellectual and wider interests. It was during the first year in Guildford that the photograph of Harry beside his little chair was taken. (Source: Ludlow-Hewitt family archive, by courtesy of William le Fleming).

only after the 'Dear Harry' Exhibition had opened and so only two of Harry's sisters, Margery and Betty, were included in the displays.[7] The younger Amabel Moseley, born in 1886, was named after her mother and died of tubercular meningitis aged five in 1891.[8]

The remaining Moseley family members – Harry, his mother Amabel, and his older sister Margery – were close-knit and corresponded frequently and warmly. They were also regularly photographed, with examples shown in Fig. 3 as well as earlier in Chapter 1. Harry and Margery were especially close and had a shared love for natural history. Harry first went to school at Summer Fields, in north Oxford as a boarder in 1897 where his early education was classical in focus with no formal science teaching.[9] Harry was initially inconsistent in his studies, though bright and curious about the natural world, in particular collecting birds' eggs – see Fig. 4.

> My dear Mar,
>
> Parsons found the 4+1 (Tree Pipit!) eggs, in a tiny whole in a willow tree, only just big and enuff for Parsons' hand, too small for mine. In-getting out the eggs he broke one, so off course we took all. I wonder if the one at Marston Ferry had eggs in it.

Fig. 4 Left and above: A letter from Harry to his sister Margery, c.1900 and a Summer Fields school photograph of the cast of the school play 'A Regular Fix', 1900. Despite a certain shyness, Harry made friends at Summer Fields through a shared passion for collecting birds' eggs. Around 1900 and while at Summer Fields, Harry wrote to his sister of an expedition (transcribed as Letter 4 in Heilbron, *Moseley, op. cit.* 43). If not simply the spelling mistake of a boy, 'anuff' in the fifth line of the letter may be a family inside joke. At Summer Fields, Harry took part in activities beyond the formal curriculum. In the Fifth Form in 1900, he acted in a play 'A Regular Fix', described as a 'most amusing farce' by the school magazine. The moustachioed and costumed Harry stands in the middle of the back row of the cast photograph, but he may not have been entirely convincing in his role: the magazine reported that 'Moseley had the difficult part of the young lover, which is always a hard one to act'. It is worth noting that as an undergraduate he sported a very similar – now real – moustache. (Source: Ludlow-Hewitt family archive by courtesy of William le Fleming; and Summer Fields School, Oxford).

As described in more detail in Chapter 1, at age thirteen Harry won a scholarship to Eton College where he won prizes in classics, chemistry and physics and took up rowing. Harry's academic performance at Eton was noted and rewarded. In our exhibition, we displayed two key artefacts demonstrating this: a volume of work judged particularly notable by teachers and therefore 'sent up' to the school's headmaster as examples of excellence, including exercises by Harry in science and mathematics; and a copy of Charles Darwin's

Fig. 5 Trinity College 2nd Torpid Crew of 1910. Harry took part in traditional student activities such as rowing. He appeared in five crew photographs from the College's Boat Club including the 'Torpid' races of 1910. Of the nine young men in this photograph, two were German and four including Harry died in the First World War. The two German crew members were Albrecht von Bernstorff (far left, seated), a German Rhodes Scholar, who served as a diplomat in London during the inter-war years and was a vocal opponent of the Nazis during their rise to power, eventually resulting in his execution in Lehrter Strasse prison in Berlin in 1945; and L. R. E. Schmidt (second from right, seated), the son of one of the directors of Krupps, Essen. The international mixture and fate of these nine men was broadly reflective of that of many other young men and sports teams in Oxford from this era. (Source: Archive of Trinity College, Oxford, by courtesy of the President and Fellows of Trinity College, Oxford).

The Descent of Man (1899) given to Harry as a prize in Spring 1904. The book is now held by Trinity War Memorial Library, to whom it was presented by his mother in her son's memory.

Scientist

After finishing at Eton, Harry returned to Oxford in October 1906 where he studied for a four-year natural sciences degree at Trinity College as a Millard Scholar. Archival materials and photographs from the college, including Harry's entry in the Trinity College admissions register, written in his own hand, featured in the exhibition. After Mathematical Moderations, he chose to specialise in physics, while also taking part in traditional student activities such as rowing as well as joining the University of Oxford's Officer Training Corps.

Beyond his required studies, Harry was also an active member of Oxford's informal scientific societies, including the University's Alembic Club where

Fig. 6 A gelatine print photograph of H. G. J. Moseley in the Balliol-Trinity Laboratories, Oxford, 1910 (MHS MS Museum 118; inv. 18874). Before he left Oxford for his first academic post at Manchester University, Harry posed for this iconic photographic portrait in the shared Balliol-Trinity college laboratory, located just across the road from MHS on Broad Street, Oxford. The processed and slightly brightened version of the image shown here was placed at the division between the 'Son' and 'Scientist' sections in the exhibition and was designed to catch visitors' attention as they moved through the first section of the exhibition. (Source: MHS exhibition team, from the Museum of the History of Science, University of Oxford).

small groups of undergraduates would meet to review and discuss recent developments in chemistry, with support from a few more senior university scientists. Alembic Club term cards and dinner programmes featured in the exhibition to demonstrate Harry's active membership of the society as well as his wider scientific interests beyond the University curriculum.[10]

In the summer of 1910, Harry graduated with a second-class degree in Physics. Only one student out of a total of fourteen gained a First Class degree in Physics.[11] Later in the summer of 1910, and as described in more detail in Chapter 2, Harry left Oxford for his first academic post, working under Sir Ernest Rutherford (1871–1937), the 'father of nuclear physics' at Manchester University. Harry's most important research began in Manchester and was continued when he moved back to Oxford in late 1913 for the last year of his scientific career.

The significance of Harry's results on the X-ray spectra of the elements was quickly recognised as a 'great discovery': his original apparatus along with related graphs and documents were gathered together after Harry's death and preserved by Professor John T. S. Townsend, a mathematical physicist and the first Wykeham Professor of Physics at Oxford (1900–1941). The materials were available for preservation in part because Harry's research was

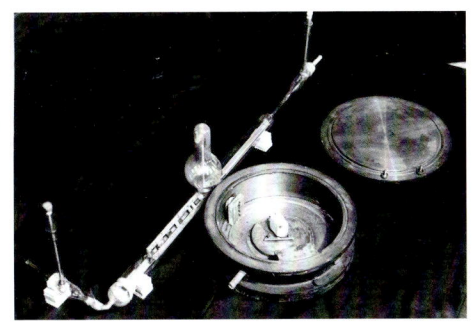

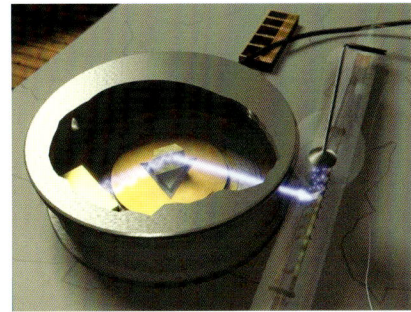

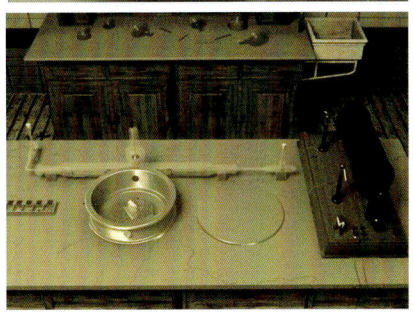

Fig. 7 Photograph of Harry's reassembled experimental apparatus taken after its arrival at MHS, together with two stills from an animation of Harry's X-ray apparatus in action. The panel on the opposite page shows a photograph of the new permanent re-display of the apparatus. The animation was produced for the exhibition by computer graphics student Martin Wolff from the Hochschule für Technik und Wirtschaft in Dresden, under the supervision of Professor Markus Wacker. It was intended to engage visitors with Harry's experimental apparatus and show how it would have worked in practice. Though available online (see note 3), its primary purpose was to be placed alongside the static display of the apparatus in the 'Dear Harry' exhibition; it now plays the same role for the new permanent display about Harry and his experiments in MHS. The full video demonstrates how Harry fired electrons at samples of chemical elements to generate characteristic X-rays and analysed their wavelengths using a crystal of potassium ferrocyanide as his monochromator. Through this apparatus of his own design, Harry established experimentally the physical reality of the atomic number of the elements. This was one of the most fundamental discoveries ever made about the physical nature of matter, but the point was not easily explained to visitors using simply the surviving experimental apparatus. Two of the main outcomes of the HLF project were to conserve the apparatus so that it would be available for future display and research (as Chapter 3 shows) and to re-display and re-interpret the apparatus in a way that engaged and educated general and specialist audiences alike. By joining a rich biographical context to the didactic benefits of animation, the exhibition sought to motivate visitors to understand Harry's achievement. The apparatus had previously been displayed as a rather unloved assemblage in the bottom of a showcase in the MHS basement gallery, lacking obvious visual and aesthetic appeal or interpretation to enable intellectual access. Taking seriously both history and science provided a way of re-invigorating not only the display of original apparatus in the Museum but the long tradition of display and interpretation in science museums. (Source: Museum of the History of Science, University of Oxford; Martin Wolff/MHS).

incomplete: it was the potential continuation of Harry's scientific research should he return from war that led to its preservation rather than immediate re-use or cannibalisation. However, while originally retained for practical reasons, the apparatus subsequently became an object of commemoration: scientific material often survives for complex reasons.[12]

Much of the material was gifted to MHS by the Physics Department in 1935 although the large-scale graph of Moseley's Law drawn by Harry himself as well as some other material related to his work at Oxford remained with the Department of Physics. The archival holdings at MHS include material transferred from the Inorganic Chemistry Laboratory via E. J. Bowen in 1964 as well as more personal material deposited by Professor John L. Heilbron in the mid-1970s on behalf of (the late) Alfred Ludlow-Hewitt, Harry's brother-in-law.

Despite a research career of only forty months, by 1914 Harry was already contemplating his next move and applying for professorships in physics. Testimonials from the MHS collections were included in the 'Dear Harry' exhibition.[13] Although only in his mid-twenties, Harry could gather support from the most distinguished physicists of the period. He was sufficiently confident of his research to put it into abeyance for three months so as to travel with his mother to Australia via Canada to attend the meeting of the British Association for the Advancement of Science (BAAS).

Soldier

As described in more detail in Chapter 5, after Britain declared war on Germany in early August 1914, Harry rushed back to England from Australia from the BAAS meeting. Like other young men on both sides of the conflict, he was eager to serve his country as a soldier.

Harry abandoned his immediate plans for scientific research, instead intending to enlist in Kitchener's New Army as a Royal Engineers signals officer, as indicated by the 'flag-wagging' he had practised on the way back to Britain from Australia.[14] Lacking surviving materials directly related to Harry's own military service, we chose examples of signalling apparatus from the historic collections of our project partner the Royal Signals Museum, including signalling flags, a heliograph and tripod, a telescope as well as a field telephone and telegraph wire to demonstrate the types of apparatus which Harry would have used during his training and practice as a Royal Engineers' Signals Officer. Harry began his training at Aldershot in October 1914, for which we showed his mother Amabel's diary which noted that she had just missed Harry in Oxford as Amabel had taken a longer journey back to the UK from Australia.[15] Throughout his training and military service, Harry kept his mother and sister updated with his progress and activities through regular correspondence with some telling quotations printed in large font on the exhibition wall.

In February 1915, Harry's unit was attached to the 13th Division of Kitchener's 'New Army' and Harry was allocated responsibility for the communications of the 38th Brigade. Harry and his unit assumed that they would be sent to France and the Western Front. But Gallipoli became their destination as the British Government sought to break the gridlock which set in after the first land offensive on the peninsula.

In June 1915, Harry and his division arrived in Alexandria where Harry wrote his will, leaving his estate to the Royal Society – again, a section of this remarkable text was printed on the exhibition wall in large text. By July 1915, these inexperienced soldiers of Kitchener's 'New Army' had arrived at Helles, at the southernmost tip of the Gallipoli peninsula. After initial combat experience at Helles, Harry's division then landed near ANZAC Cove in early August 1915 and began to attack the strategic high ground held by Ottoman forces at Chunuk Bair (*Conk Bayırı*). As a Signals Officer, Harry would have been kept busy maintaining communications lines and reporting battlefield developments – with an example of exactly contemporary activities shown in Fig. 8 below. In the murderous four days that followed, thousands of men on both sides died. One of these men was Harry, killed during an Ottoman counter-attack on 10 August 1915, aged just twenty-seven.

His body, like that of so many others, was never found, but is most likely among the over 600 unidentified soldiers from both sides buried in the small cemetery at 'the Farm' near to where he was killed. Harry's name is recorded on the Helles Memorial, the Commonwealth battle memorial for the whole Gallipoli campaign located at the foot of the Gallipoli peninsula – as shown in the torn map in Fig. 9.[16]

Legacy

Two days after the news of Harry's death had reached his family, it was reported publicly in *The Times* on 1 September under the sub-headline 'A Brilliant Physicist' with later British newspapers reporting his death under headings such as 'Sacrifice of a Genius' and 'Too Valuable to Die'.[17] News of Harry's death spread quickly and internationally; even German newspapers commented on his loss, though he was now formally an enemy.[18]

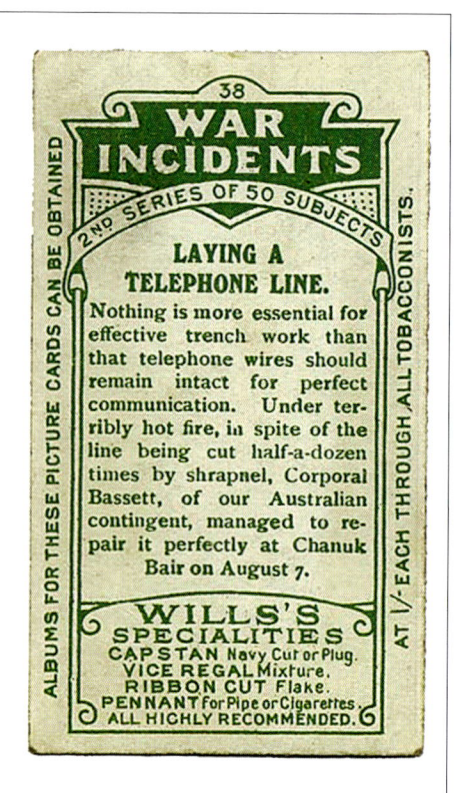

Fig. 8 Reproduction of a Wills's cigarette card of Gallipoli signals work at the battle of Chunuk Bair featuring New Zealander Corporal Cyril Bassett. This card from 1915 shows Corporal Bassett of Divisional Signals, New Zealand Engineers, New Zealand Expeditionary Force, laying telephone lines at Chunuk Bair on 7 August 1915, the action for which he was awarded the Victoria Cross (VC), making him the first and only New Zealander to be awarded a VC in the Gallipoli Campaign. The terrain shown does not match the reality of the Gallipoli peninsula nor Chunuk Bair. Bassett's responsibilities as a signaller closely replicate Harry's in the same action. Unlike Harry, Bassett survived the war, going on to have a successful banking career and a return to military service in the Second World War. As Brigade signals officer, Harry's role was to provide a communications system using a mixture of telephone and telegraph cables, messenger runners and visual signalling. Signallers were issued the standard British Army field telephone which could be used on poor-quality lines damaged by bombardment and stretched by limited supplies of wire. This telephone model was extremely successful in the flat, muddy terrain and trenches of the Western Front as well as the arid hills and gullies of Gallipoli. (Source: Images available in the public domain via the Australian Government's Department of Veterans' Affairs 'Gallipoli and the Anzacs' website http://www.gallipoli.gov.au/bravery-awards-at-gallipoli/cyril-bassett-vc.php – accessed 1 December 2017).

Fig. 9 Torn-off section of map of Gallipoli 1: 250,000 from the REMLA archive (7501.4.3). Harry was one of the many thousands of soldiers killed or injured on both sides during the brief Gallipoli land campaign which ran from 25 April 1915 through to early January 1916. In our 'Dear Harry' exhibition, the limited scope and ultimate failure of the Allied land campaign was illustrated by this torn-off section of a map provided to the British Army: due to the limited movements of the Allied forces, only one half was needed during the campaign. (Source: Royal Engineers Museum, Library and Archive; Museum of the History of Science, University of Oxford).

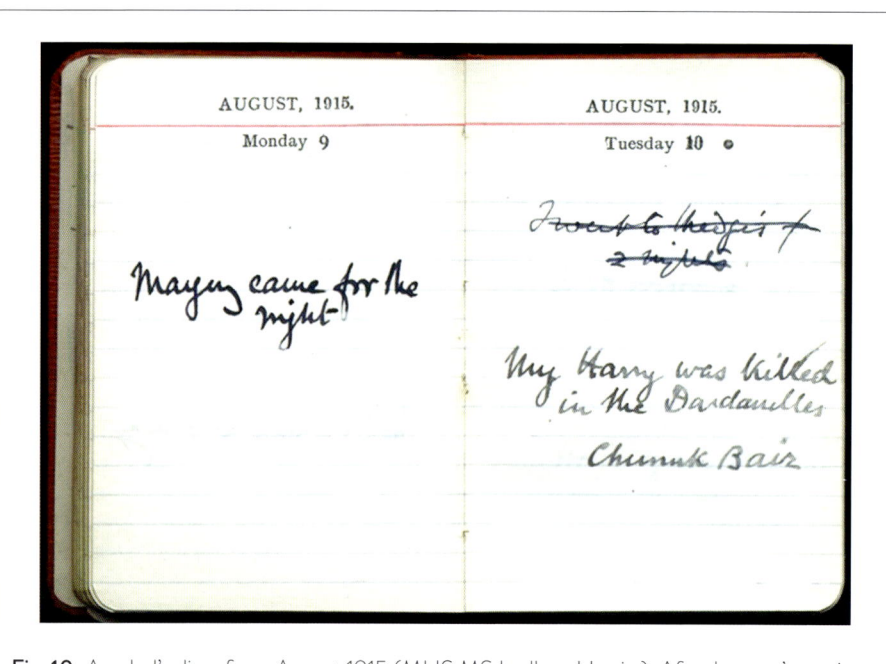

Fig. 10 Amabel's diary from August 1915 (MHS MS Ludlow-Hewitt). After her son's tragic death at Gallipoli, Harry's mother Amabel used her social diary to record her final moments with her son as well as the details of his death. Due to the confusion of the Battle of Chunuk Bair, news of Harry's death did not reach his family until Monday 30 August, almost three weeks after he was killed. The War Office telegraph arrived around 2.15pm, according to her diary, and Amabel immediately sent for Margery, her daughter and Harry's sister. Amabel went back in her diary to 10 August – the day her son died – and wrote in somewhat shaky handwriting a poignant and simple note: 'My Harry was killed in the Dardanelles Chunuk Bair'. It was around this time, with the same pen and hand as she noted the details of her son's death, that Amabel went through her social diary, striking through the entries unrelated to her son and annotating and updating some of the entries relating to Harry. Amabel's social diary was thus transformed into a deeply personal memorial to her son, recording her final and rare moments with Harry as he completed his Royal Engineers training early in 1915. To the entry for Sunday 6 June 1915, which originally read 'Harry went 5.30. I walked with him to the top of Romsey Hill', she added a later note after his death: 'The last time I saw him'. (Source: Ludlow-Hewitt family archive by courtesy of William le Fleming; diary image from Museum of the History of Science, University of Oxford).

Across the political and military divide, the international scientific community was shocked by Harry's death. His former supervisor at Manchester, Sir Ernest Rutherford, especially condemned the waste. Rutherford led the public and academic outcry at the death of Harry and other young scientists – partly from sheer distress, partly in anger that so great a genius could be sacrificed to so little purpose. Upon receiving the news of Harry's death, Rutherford wrote a heart-felt letter and moving obituary of Harry in the scientific magazine *Nature* published in September 1915, a month after Harry's death:

> Scientific men of this country have viewed with mingled feelings of pride and apprehension the enlistment in the new armies of our promising young men of science – with pride for their ready and ungrudging response to their country's call, and with apprehension of irreparable losses to science.[19]

Rutherford concluded with a plea that Harry's death not be in vain:

> It is a national tragedy that our military organisation at the start of the war was so inelastic as to be unable, with a few exceptions, to utilise the scientific services of our men, except as combatants in the firing line. Our regret for the untimely death of Moseley is all the more poignant.[20]

In the final exhibition case, we included letters which illustrated the shock-waves generated internationally by Harry's death: several letters exchanged between French chemist Georges Urbain (who had visited Harry in Oxford) and Rutherford share their grief and profound dismay at their young colleague's death, with Urbain writing movingly that news of Harry's death '*me plonge dans la stupeur*' (has made me utterly speechless).[21] The letters were later given by Rutherford to his friend, the Oxford chemist Nevil Sidgwick, who was keen to keep some record of Harry in Oxford. Today they are preserved in the MHS archives alongside other records – publications and notes – and recollections of Harry's work gathered together by Rutherford and Harry's grief-stricken mother Amabel in order to establish and preserve Harry's scientific and personal legacy.[22]

As discussed in Chapter 6, Harry was nominated for the 1915 Nobel Prizes in both Physics and Chemistry – both nominations were put on hold for that year, with the assumption that he could be considered for later years. His untimely death meant he was not eligible for the Prizes in 1916 and beyond. Rutherford, who had tried unsuccessfully to get Harry out of frontline service, used Harry's death as a rallying call for scientists to be put to better use by the British military. Rutherford deplored the death of such a scientific talent and campaigned for scientists' specialist skills to be more effectively harnessed in wartime rather than being sent to fight and die on the battlefield. By the end

Fig. 11 Above and opposite page: Undated group photograph of British nursing assistants in France including Amabel Sollas (née Moseley), 1915–1918; Buckingham Palace telegram, postmarked 3 September 1915; and Amabel's 1918 diary (MHS MS Ludlow-Hewitt). Harry's mother Amabel chose to remember her son in a personal and practical manner. Her diaries show that between 1915 and 1918 she served irregularly in hospital canteens in France, providing food and refreshments for wounded soldiers as well as nursing assistance. In the undated group photograph (unknown at the time of the 'Dear Harry' exhibition), Amabel is shown in a nursing uniform; other photographs from this period show her in a more utilitarian uniform suitable for working in a canteen. Another artefact only recently uncovered in the Ludlow-Hewitt family archives and therefore again unknown at the time of the 'Dear Harry' exhibition is a telegram from their Majesties to Amabel, postmarked 3 September 1915: 'The King and Queen deeply regret the loss you and the Army have sustained by the death of your son in the service of his Country. Their Majesties truly sympathise'. Although standard in form, the survival of this artefact demonstrates its importance and emotional resonance with those left behind. On 10 August 1918, with the war still underway, Amabel marked the anniversary of Harry's death in her personal diary, writing plainly: 'My anniversary 3 years'.

> **POST OFFICE TELEGRAPHS.**
>
> Buckingham Palace
> OHMS
>
> TO Mrs A. A. Moseley 43 Woodstock Rd Oxford
>
> The King and Queen deeply regret the loss you and the army have sustained by the death of your son in the service of his Country their Majesties truly sympathise

> Week. AUGUST, 1918.
>
> Thur. 8
>
> Fri. 9
>
> Sat. 10 My anniversary 3 years

of the war, Rutherford's goal had been achieved in part due to the impact of Harry's death: talented scientists were thereafter directed into research or intelligence work wherever possible. Thus, Harry's death indirectly contributed to a new vision of the role of science and scientists during the war.

Perhaps the most immediately accessible aspect of Harry's scientific research and achievements can be seen in the Periodic Table of Elements, which today retains Harry's underlying structure and framework, with each element placed by its atomic number. For the exhibition, we showed a modern Periodic Table of Elements, highlighting the four 'missing' elements predicted in Harry's final version of his 'X-ray plot', all of which were subsequently discovered. More generally, Harry's scientific research and ideas signposted the direction of travel physics would take in the twentieth century and they remain of influence upon science and scientific research today. Nearly sixty years after his death, Isaac Asimov acknowledged Harry's work and scientific potential:

> In view of what he [Harry] might still have accomplished … his death might well have been the most costly single death of the War to mankind generally.[23]

The impact of Harry's death on science was costly, but perhaps the deepest and most strongly felt impact was the personal one felt by the remaining, grief-stricken members of his family: his mother Amabel and elder sister Margery. At the heart of the 'Dear Harry' exhibition is the personal and moving story of a family, one family of many representing all those who sent their husbands, sons, and fathers off to war never again to return. And it was also through the 'Dear Harry' exhibition project that we reconnected with relatives of the Ludlow-Hewitt family that Harry's sister Margery had married into, and who generously provided the new family material featured throughout this volume. Furthermore, the family kindly shared details of the personal impact felt by Harry's mother Amabel and his sister Margery and how they created their own legacy in the aftermath of Harry's death.

Amabel died in 1928, a few years before the first visit by Allied and ANZAC mothers to Gallipoli. The exhibition closed with some simple but powerful and resonant words. These much-quoted conciliatory remarks on Turkey's former enemies are attributed to Kemal Atatürk, then President of Turkey, on the occasion of ANZAC day 1934:

> *Those heroes who shed their blood and*
> *lost their lives … you are now lying*

in the soil of a friendly country.
Therefore, rest in peace.

There is no difference between the Johnnies
and Mehmets to us, where they lie side by side
here in this country of ours …
You, the mothers, who sent their sons from
far-away countries, wipe away your tears.

Your sons are now lying in our bosom and
are in peace. After having lost their lives on
this land they have become our sons as well.

Notes and References

[1] Kristen Frederick-Frost, 'An Artifact-Based Study of Henry Moseley's X-Ray Spectroscopy', (Masters Thesis, University of Oxford, 2012). Frederick-Frost won the University's Jane Willis Kirkaldy Senior Prize in the History of Science.

[2] See 'Henry Moseley's X-ray Spectrometer' on the MHS Oxford YouTube channel at https://www.youtube.com/watch?v=UTp9jAQpf7c (accessed on 1 December 2017; with over 8,500 views). The work was documented in Martin Wolff's undergraduate dissertation and featured in an article in the Museum Association's professional journal *Museum Practice*, available online at https://www.museumsassociation.org/museum-practice/museum-films/15012014-museum-of-the-history-of-science (accessed on 1 December 2017).

[3] See http://blogs.mhs.ox.ac.uk/innovatingincombat/ (accessed on 4 December 2017) for further details.

[4] Elizabeth Bruton, '"Sacrifice of a Genius": Henry Moseley's role as a Signals Officer in World War One', a lunchtime lecture at the Royal Society, London, October 2013. See a video and PowerPoint slides from the talk at https://royalsociety.org/science-events-and-lectures/2013/henry-moseley/ (accessed on 1 December 2017).

⁵ John Richardson, Russell Egdell, Nick and Harold Hankins, 'Rutherford, Geiger, Chadwick, Moseley, Cockcroft and their role in the Great War – Part 1', *Stand To!*, *The Journal of the Western Front Association*, 86 (2009), 28-34; 'Part 2 – John Cockroft', 87, (2009/2010), 42–45; 'Part 3a – Henry Moseley', 89 (2010), 46–50; 'Part 3b – Henry Moseley', 90 (2010/2011), 29–35.

⁶ Death Certificate FE 409587: 1899 Deaths in Sub-District of Cranleigh, Surrey entry number 278 Elizabeth Moseley, died 24 May 1899. Details kindly passed on by Clare Hopkins.

⁷ The younger Amabel Moseley was included in 1891 census records not available to John L. Heilbron when he wrote *H. G. J. Moseley, The Life and Letters of an English Physicist 1887–1915* (Berkeley: University of California Press, 1974), hereafter Heilbron, *Moseley*: census information is closed for 100 years. She was unknown to the surviving family while the 'Dear Harry' exhibition was still running, although photographs of her were subsequently discovered within the Ludlow-Hewitt collection.

⁸ H. C. R. Ludlow-Hewitt, *The Ludlows, the Hewitts and the Moseleys* (unpublished family history), 9.

⁹ For further detail of Harry's time at Summer Fields, see two 'Dear Harry' guest blog posts by the late Gavin Hannah, former history master at Summer Fields and editor of 'Summer Fields, the First 150 Years' (London: Third Millenium, 2014): 'Harry Moseley at Summer Fields, 1897–1901' at http://www.mhs.ox.ac.uk/moseley/2015/09/02/harry-moseley-at-summer-fields-1897-1901/ (accessed 4 December 2017); and 'New Boys 1897: Moseley's Year-group at War, 1914–18' at http://www.mhs.ox.ac.uk/moseley/2015/11/09/new-boys-1897-moseleys-year-group-at-war-1914-18/ (accessed 4 December 2017).

¹⁰ See MHS MS Museum 151: term cards for the programme of the Alembic Club and MHS MS Museum 152, a box of Alembic Club Annual Dinner programmes and related material.

¹¹ *Oxford University Calendar 1911* (Oxford: Clarendon Press, 1911), 173 identifies the numbers of Honours Degrees in Natural Science (Physics) as follows: I (1), II (8), III (5), IV (0). Heilbron, *Moseley, op. cit.* 173 mentions the eight Class II degrees but states incorrectly that there were no Class I degrees. In a letter to his mother written immediately after his exams, Harry anticipated that there would be 'one first and many seconds'. See Moseley to his mother, Trinity College, Oxford, [21 June 1910], Letter 41 in Heilbron, *Moseley, op. cit.* 171–173.

¹² S. J. M. M. Alberti, 'Why Collect Science?', *Journal of Conservation and Museum Studies*, 15 (1), (2017), 1–43. Available online at https://www.jcms-journal.com/articles/10.5334/jcms.150/ (accessed 14 December 2017).

[13] See MHS MS Ludlow-Hewitt 1, Harry's letter with testimonials and published papers, gathered together in a paper binding to form his application for the post of Professor of Physics at Birmingham University, 1914.

[14] In a letter to his sister Margery written on board Cunard RMS *Lusitania* while travelling back from Australia to the UK, Harry wrote: 'I have been reading up a smattering from War Office manuals, and practising flag wagging Morse and semaphore while crossing the Pacific'. Moseley to Margery, 28 September [1914]. Letter 106 in Heilbron, *Moseley, op. cit.,* 250–251.

[15] MHS MS Ludlow-Hewitt, Moseley's mother's diary, 17 October 1914.

[16] For details of MHS director Dr Silke Ackermann's visit to Gallipoli following in Harry's footsteps including her visit to the Helles war memorial, see http://www.mhs.ox.ac.uk/moseley/2015/06/08/death-at-the-farm-re-visiting-harry-on-gallipoli/ (accessed on 1 December 2017).

[17] 'Fallen Officers. *The Times* List of Casualties'. *The Times*, 1 September 1915 (issue 40948), 11. 'Sacrifice of a Genius' and 'Too Valuable to Die' appeared in the *Evening News* on 23 October 1919. Closer to the time of his death, obituaries of Harry appeared in the British press in *The Times* on 1 September 1915 (cited above); the *Glasgow Herald* on 2 September 1915; *The Oxford Times* on 4 September 1915; the *Manchester Guardian* on 8 September 1915 and elsewhere; all cited in Heilbron, *Moseley, op. cit.*, 124–125.

[18] Harry's death was reported in Germany in K. Fajans, *Naturwissenschaften* 4 (1916), 381–382 and elsewhere. See Heilbron, *Moseley, op. cit.* 124–125.

[19] E. Rutherford, 'Henry Gwyn Jeffreys Moseley', *Nature,* 96 (2393), (1915), 33–34.

[20] *Ibid*.

[21] MHS MS Museum 118, autograph letter from G. Urbain, 26 September 1915.

[22] MHS MS Museum 118, autograph letters from Amabel Sollas to Sir Ernest Rutherford, 27 September 1915 and 5 October 1915.

[23] Isaac Asimov, *Asimov's Biographical Encyclopedia of Science and Technology* (New York: Doubleday and Company, 1972), 921.

APPENDIX I

THE ELECTRONIC STRUCTURE OF ATOMS AND THE LABELLING OF X-RAY SPECTRA

RUSSELL G EGDELL

At the time that Moseley conducted his experiments, and in the years immediately thereafter, the currently-accepted theoretical framework that is needed to understand X-ray spectra in detail did not exist: a full appreciation of the processes involved in X-ray emission became possible only after the emergence of quantum mechanical descriptions of the electronic structure of atoms in the interwar years. Nonetheless a labelling system for X-rays was quickly developed, starting with letters near the middle of the alphabet (K, L and M), with subsequent refinement by the addition of Greek letters and numerical subscripts, as for example in the symbol $L\alpha_1$. This appendix outlines the now-standard text book way of describing the electronic structure of atoms[1] and then explains how the often-confusing and unsystematic nomenclature for X-ray spectra – which is still used today – relates to our current models.

In wave-mechanical models of electronic structure, electrons cannot be regarded as particles with an identifiable position and momentum: instead, everything that can be known about an electron is encapsulated in its 'wave-function'. Wave-functions are calculated by solution of a second-order three-dimensional differential equation – the Schrödinger equation – that takes into account the sources of potential acting on the electron. Once the wave-function is known, the probability of finding an electron at a given point in space is given by the square of the wave-function, while other properties such as energy and momentum can be deduced by applying mathematical 'operators' to the wave-function. Quantisation does not arise for free electrons, but where electrons are confined by a potential of some sort (as in an atom) quantisation arises quite naturally from boundary conditions and the requirements that the wave-function must be single-valued and continuous.

Electrons in atoms occupy a series of 3-dimensional 'orbitals', which gives rise to the idea of series of 'shells' of electrons surrounding the nucleus. The orbitals do not constrain the electrons to a circle (as in the Bohr Model) or even a three-dimensional shell of fixed radius, but instead involve a 'smeared-out' distribution with different probabilities of finding the electron at different points in space. This probability distribution does, however, maximise at a well-defined distance from the nucleus and, remarkably, the maximum corresponds to the so-called 'Bohr radius' for the single electron in hydrogen. Each orbital is characterised by a set of three quantum numbers n, l, m_l: the need for three is a consequence of the three-dimensional nature of space.

The principal quantum number n must be a positive integer: in hydrogen n uniquely determines the energy of the orbital. For each n value, l may take on integer values from $n-1$ all the way down to zero. The quantum number l specifies the angular momentum of the orbital, or equivalently, the angular form of the wave-function. Conventionally l values of 0, 1, 2 and 3 are represented by the symbols s, p, d and f. Taken together, n and l determine the radial form of the orbital wave-function and the energy of the orbital. Next m_l specifies the component of the total angular momentum about a given axis, and may take on values $l, l-1,$ and so on down to $-l$. For each l value there are therefore $2l+1$ possible m_l values. So we find a single s orbital for each value of n, three p orbitals when $n \geq 2$, five d orbitals when $n \geq 3$ and seven f orbitals when $n \geq 4$. As mentioned above, in hydrogen all the orbitals with a given n value have the same energy. However, in atoms with more than one electron, orbitals with the same n but different l values have different energies because they experience differing amounts of repulsion from the other electrons in the atom: in general the order of energies for a fixed value of n is $s<p<d<f$.

When time is introduced as a fourth dimension in *relativistic* quantum mechanics, a fourth quantum number is needed. All fundamental particles can be characterised by an intrinsic 'spin' angular momentum: for electrons the spin s is 1/2 (in units of $h/2\pi$), and the fourth quantum number m_s can take on values $\pm\frac{1}{2}$. According to the Pauli Exclusion Principle no two electrons in an atom can have the same values for all four quantum numbers. This means that an s orbital can accommodate at most two electrons, each set of p orbitals six electrons and so on.

As we progress from one atom to the next in the periodic table, orbitals fill up according to what is known as the Aufbau Principle, which in its simplest form states that successive electrons are fed into the lowest available orbital. The regular progression in the filling orbitals envisaged by this Principle is complicated by the fact that the energy of an orbital depends on the repulsion from other electrons in the system as well as on the electron-nuclear attraction, and the energy of an orbital depends on the occupancy of the orbital.

Nonetheless, these ideas allow one to specify an 'electron configuration' for a given atom. The electron configuration of an atom identifies each orbital by its integer value of n and the value by s, p, d etc., and also gives the occupancy as a superscript. So lithium with a nuclear charge $Z=3$ and three electrons has an electron configuration $1s^2 2s^1$, sodium with $Z=11$ is $1s^2 2s^2 2p^6 3s^1$, and so on.

The chemistry of an element is largely determined by its outermost electrons, which may be transferred to, or shared with, other atoms. The way in which outer electrons interact with electrons in neighbouring atoms gives rise to chemical bonds of various sorts. Thus atoms with similar outer electron configurations have similar chemical properties, as in the case of lithium and sodium. This is the basis of the chemical periodicity that is encapsulated in the modern periodic table.

Although the Bohr model is now mainly of historical interest, it still provides the basis of modern teaching in chemistry, and undergraduates are encouraged to discuss atomic energy levels in terms of the formula $E_n = -R \times Z_{eff}^2 / n^2$. Here the full nuclear charge Z is replaced by an effective nuclear charge given by $Z_{eff} = Z - \sigma$, where σ is what is called the 'shielding constant'. The task is then to develop an understanding of how the radial form of the wave-functions for different orbitals influences the extent of the shielding and the energies of the orbitals.

This modern description of electronic structure gives a basis for understanding X-ray emission spectra as measured by Moseley in his pioneering experiments. Consider, for example, copper, one of the elements described in his paper on the 'High-Frequency Spectra of the Elements'.[2] A free copper atom has the electron configuration $1s^2 2s^2 2p^6 3s^2 3p^6 3d^{10} 4s^1$, although in solid copper metal the outer $4s$ electrons are delocalised into the 'sea' of electrons that are responsible for the metallic behaviour and for holding the atoms together. Bombardment of copper by fast electrons, as in Moseley's X-ray tube, can eject an electron from the innermost $1s$ orbital leaving an electron configuration $1s^1 2s^2 2p^6 3s^2 3p^6 3d^{10} 4s^1$. This is said to have a 'hole' in the $1s$ orbital. An electron from the higher energy $2p$ orbital can drop down to fill this hole, with emission of a characteristic Kα X-ray photon; Kβ photons arise from a $3p$ to $1s$ transition. Softer (that is lower-energy and less penetrating) L-type X-rays are produced when a hole in the $2s$ or $2p$ shell is filled by an electron from the $3p$ or $3d$ shell 'falling down' in a similar way. The K, L notation, with letters close to the middle of the alphabet, was first introduced by Charles Barkla, who wanted to leave room in his labelling system for X rays both harder and softer than those he had discovered.[3] Although the Bohr model left no room for J-type rays (or anything even harder), Barkla spent about half of his Nobel Lecture delivered in 1920 describing his 'evidence' for J rays.[4] An alternative to Moseley's approach to X-ray emission spectroscopy pioneered by Maurice de Broglie involves irradiation of a sample by higher energy X-rays.[5] One advantage of this approach is that the continuous background of so-called Bremstrahlung radiation, that proved problematic to Moseley, is absent from the spectrum. The Bremstrahlung photons are emitted by deceleration of fast electrons as they enter the target, and obviously are absent when spectra are excited by X-rays.

A further refinement in relativistic quantum mechanics is that there is coupling between the electron-spin angular momentum and the orbital angular momentum. This splits atomic sub-shells with $l \geq 1$ to give two distinct states with total angular momenta equal to $j = l \pm 1/2$. The j value of these so-called spin-orbitals is identified by a subscript and the component with the lower j value is always the more stable. In discussing X-ray spectra, the successive spin-orbitals are identified in a labelling system using the letters K, L, M and N for values of the principal quantum number n equal to 1, 2, 3 and 4. A subscript is added to indicate the ordering within a shell of fixed n. Thus we have in order of increasing energy:

$K_1 \equiv 1s_{1/2}$

$L_1 \equiv 2s_{1/2}$, $L_2 \equiv 2p_{1/2}$, $L_3 \equiv 2p_{3/2}$

$M_1 \equiv 3s_{1/2}$, $M_2 \equiv 3p_{1/2}$, $M_3 \equiv 3p_{3/2}$, $M_4 \equiv 3d_{3/2}$, $M_5 \equiv 3d_{5/2}$

$N_1 \equiv 4s_{1/2}$, $N_2 \equiv 4p_{1/2}$, $N_3 \equiv 4p_{3/2}$, $N_4 \equiv 4d_{3/2}$, $N_5 \equiv 4d_{5/2}$, $N_6 \equiv 4f_{5/2}$, $N_7 \equiv 4f_{7/2}$

X-ray lines discussed in the primary references in this book are all labelled according to the so-called Siegbahn notation.[6] They may be assigned to the following transitions.[7]

$K\alpha_1$	$L_3 \rightarrow K_1$	$K\alpha_2$	$L_2 \rightarrow K_1$		
$K\beta_1$	$M_3 \rightarrow K_1$	$K\beta_2$	$N_3, N_2 \rightarrow K_1$	$K\beta_3$	$M_2 \rightarrow K_1$
$L\alpha_1$	$M_5 \rightarrow L_3$	$L\alpha_2$	$M_4 \rightarrow L_3$		
$L\beta_1$	$M_4 \rightarrow L_2$	$L\beta_2$	$N_5 \rightarrow L_3$	$L\gamma_1$	$N_4 \rightarrow L_2$
$M\alpha_1$	$N_7 \rightarrow M_5$				

The magnitude of the so-called spin-orbit splitting increases very rapidly with atomic number and is proportional to Z^4. The effects of spin-orbit coupling were not evident in Moseley's K-shell spectra of relatively light elements, where he was unable to separate the $K\alpha_1$ and $K\alpha_2$ components. However, Moseley's L-shell spectra of the heavier elements showed the presence of four lines, with with $L\alpha_2$ and $L\beta_1$ clearly resolved, thus implying a splitting of the $2p$ level.[8] In retrospect, the splitting of the $2p$ shell into L_2 and L_3 sub-shells by spin-orbit coupling may be considered to be the first observation of the effects of electron spin on core atomic energy levels.

Notes and References

[1] For accessible accounts of the quantum theory of atoms, see P. W. Atkins, *Molecular Quantum Mechanics, Parts I and II* (Oxford: Oxford University Press, 1970); and F. L. Pilar, *Elementary Quantum Theory* (New York: McGraw-Hill International, 1990).

[2] H. G. J. Moseley, 'The High-Frequency Spectra of the Elements', *Philosophical Magazine Series 6*, 26 (156), (1913), 1024–1034.

[3] C. G. Barkla, 'The Spectra of the Fluorescent Röntgen Radiations', *Philosophical Magazine Series 6*, 22 (129), (1911), 396–412.

[4] http://www.nobelprize.org/nobel_prizes/physics/laureates/1922/bohr-lecture.pdf (accessed on 26 November 2017).

[5] M. de Broglie, 'Sur la Spectroscopie de Rayons Secondaire Emis hors des Tubes à Rayons de Rontgen, et les Spectres d'Absorption', *Comptes rendus de l'Académie des sciences,* 158 (1914), 1493–1495.

[6] M. Siegbahn, *Spektroskopie der Röntgenstrahlen* (Berlin: Springer-Verlag, 1931).

[7] R. Jenkins, R. Manne, R. Robin and C. Senemaud, 'Nomenclature, Symbols, Units and their Usage in Spectrochemical Analysis. Part 8. Nomenclature System for X-Ray Spectroscopy', *Pure and Applied Chemistry*, 63 (5), (1991), 735-746; Lawrence Berkeley National Laboratory, *X-Ray Data Booklet* (Berkeley: Center for X-Ray Optics, 2009).

[8] H.G.J. Moseley, 'The High-Frequency Spectra of the Elements. Part II', *Philosophical Magazine Series 6,* 27 (160), (1914), 703–713.

APPENDIX II

PRIMARY ACCOUNTS OF MOSELEY'S DEATH

The four primary accounts of Moseley's death reproduced in this Appendix were preserved by the Ludlow-Hewitt family, in whose keeping the original letters remain. These letters were microfilmed by John Heilbron, but were not reproduced in his biography in 1974. The microfilms may be accessed via the Archive for History of Quantum Physics and through the University of California, Berkeley.

1. Letter from Moseley's Commanding Officer to Moseley's family, dated 13 August 1915. The letter is addressed to 'Dear Sir' and was intended for Moseley's father.

13th Signal Co
13th Division
BME Force
13th August 15

Dear Sir,

I very much regret to inform you that Lieut Moseley, commanding one of my sections, was killed in action during the recent fighting. He was shot through the head and died instantly. I am glad to know he was spared any suffering from wounds. He was a most promising officer – extraordinarily hard working and took an immense amount of interest in his work. I am very sorry indeed to lose him, and I beg to offer you my very sincere sympathy. This sorrow at his loss is felt by all the officers and men of my Company equally with myself.
Believe me,
Yours truly.

H Crocker Captain
Cmdg 13th Divisional ?…

2. Letter from G.E. Chadwick to Mrs. Sollas, dated 16 August 1915

On Active Service
HQ 38th Inf Bd
13th Div
Aug 16th 1915

Dear Mrs Sollas,

I feel I must write you a line & sympathise with you most sincerely for the loss of your son. I was the machine gun officer on the staff to which your son belonged & we were good friends. If you wish I can give you an account of those two (?) fearful days, but will not do so till I hear from you for fear of weighing on your already heavy burden. Let it suffice to say that your son died the death of a hero, sticking to his post to the last. He was shot clean through the head & death must have been instantaneous. In him the Brigade has lost a remarkably capable Signalling Officer & a good friend.

To him his work always came first & he never let the smallest detail concerning it go by him unnoticed. When he was shot the Turks had got round our right flank within 200 yards of us firing into our right rear. I was close to your son & he was calmly telephoning a message to the Division. I am the only one of the Bde unhurt & how I got away will always remain a mystery to me. My only wish is that your son could have shared my luck.

I am afraid this is a brutal (?) letter Mrs Sollas and you must forgive my clumsy phrasing for I am no epistle writer. What, however, I mean to convey is my deepest sympathy & sorrow for you & your family. The remainder of the Bde (very few in numbers I am afraid) join me in this. If there is anything I can do for you please let me know,
I remain yours very sincerely,

G Emilé (?) Chadwick

3. Letter from G.E. Chadwick to Mrs. Sollas dated 27 September 1915

Hd Qts
38th Inf Bde 13th Div
B.M.E.F.
Sept 27th 1915

Dear Mrs Sollas,

I must first apologise for taking so long to answer your letter which I received about 10 days ago, but we have had a very busy time here lately & I have not been able to find a single minute to sit down till now.

I cannot bring myself, Mrs Sollas, to pen again a description of those ghastly days, but if you write to Mother, she will send you a copy of my description of the actual fighting. Her address is:

> Mrs Spencer (?) Chadwick
> 39 Hanover Gate Mansions
> Regents Park
> London, N.W.

Forgive me for giving you this trouble, but I lost so many friends during those four days that I am doing my best to forget it.

No, Mrs Sollas, I found nothing of importance on your gallant sons [sic] body. His Sam Browne belt & personal effects I had packed up carefully & sent home to you. I sincerely hope you will get them alright. Your informants, the wounded soldiers of the 10th and 11th Divisions were not quite right. The 38th Brigade did not make a reconnaissance for the new landing, but we reconnoitred for the attack on the 6th and 7th of August. Your son did not do this, it was not his job, but he ran wires to places near us so that we got our reports back with great speed.

Poor Capt Baker reconnoitred the Right flank & I the left & a very unpleasant job it was as we had to go into all sorts of horrible places. Your son gave the utmost assistance throughout & got our messages through at the utmost speed.

I could tell you of hundreds of instances of his good work. Perhaps, if God grants me life, when I return, you will allow me to come down & call on you and if you can stand the strain, I can go through all that part of the campaign which concerns your poor son.

For the moment, Mrs Sollas, you must content yourself with the knowledge that your son always put his work before everything & no obstacle was allowed to obscure his object in view. No sacrifice was too great for him to make so long as that which he was given to do was done. Ever cool, calm & collected in dangers, & there are many here, he had besides the brains of a genius the courage of a lion & a determination such as is seldom found in the young man of today.

As a signalling officer he was perfection itself. I speak technically as I am a Signal Service man myself, though not officially as such out here. Well do I understand your feelings at the loss of such a son, Mrs Sollas, he was a son any Mother would be proud of.

All registered letters I have returned to you, but I have availed myself of your kind permission and have used the parcel for the men.

Yes the pyjamas you sent me are the joy of my life. Thank you so much for them. If there is anything further I can do please let me know. I have taken many photographs, your son appears in some of them. If you would care for them as soon as I can get them developed I will let you have them. Hoping you are recovering somewhat from your terrible blow.
I remain,

Yours very sincerely,
G. Emilé (?) Chadwick

4. Transcript of Chadwick's account of the Battle of Chunuk Bair, dated 14 August 1915. The handwriting is different from that in Chadwick's letters to Amabel Sollas and the account was probably transcribed by Chadwick's mother after the previous letter. The transcript gives the incorrect date for the advance.

Aug 14th/15
On the evening of the 7th [*sic*; 8th] we left our bivouacs to move into a position from which we were going to attack. We were to take a hill, from a position just above a farm. The farm itself is on a hill, or rather just behind the <u>crest</u> of a hill, and there is a small dip between the 'farm hill' and the one we were going to attack. The distance between the two hills is about 300 yards, and the ground is open except on the right where there was a wooded

nullah, or deep ditch running down to the sea about 2 mls back. This is as good a description as I can give without mentioning names. When I come back with the help of a map I shall be able to explain it better.

However to proceed. We left our bivouacs intending to move up a nullah to the right of the 'farm hill' & get into position <u>also</u> on the right of the 'farm hill' and so attack our objective (which we will call X) from the right. Guides were sent to lead us, and we started off on the evening of the seventh [*sic*; eighth].

When we got about half way up the nullah we were blocked by a party of mules carrying ammunition and rations. There was a fearful confusion. Imagine for a moment a ravine with precipitous sides rising up 600ft on either side, a small path 3ft wide, and a whole brigade advancing up it and 100 mules coming down it!

There was nothing left for us to do but to turn round and come back, for the mules with heavy loads could not possibly turn in so narrow a space. Then the most fearful night march I have ever tackled started. It was imperative for us to attack at 5 a.m. and here we were prevented from getting to our starting point. The general decided to go back and attack hill X from the left. It was his only way out of it but to attack hill X from the left was murder, and we all knew it!

Then as I say we started. It was a <u>cross country</u> march over high cliffs, deep nullahs, broken river beds and thickly wooded country. At last however we just got to our position on the stroke of time, but of course only the head of the column was up to time. The attack was carried out by tired and jaded men, who had marched over ghastly country in the pitch dark falling headlong down (?) holes, and climbing up steep and slippery inclines. Of course the attack was disjointed and failed. The good old East Lancs led the show & they were practically wiped out. They started 800 strong and came out 227. Colonel Cole-Hamilton, Capt Lutyens, Lieuts Debenham & Trimmer were killed. Capt. Gayer, Major McCormick, Lieuts Wood, Watson and Smith were wounded.

The Hampshires were as badly cut about as we were. Bull was magnificent. He rallied what was left of the Battalion under fire and held on an advanced position all day. At dusk he brought them all back to Farm Hill in an orderly manner bringing in what wounded were left also. All through the day he lay out in an advanced position within 200 yds of the enemy coolly giving his orders and returning now and again to report. I saw it all as I was supporting the attack on the right with 8 guns. I am afraid they were not much use but probably saved the Batt. from being wiped out entirely. I have sent in a strong recommendation for Bull and sincerely hope it will go forward. I spoke to General Shaw personally about it and feel confident it will, tell his mother and brother that the old Graf is a fellow to be proud of! No man could have done more.

Well after the attack had failed we hung on to our farm hill position all night. At about 3 a.m. we were attacked heavily from the front right. You never saw such a sight or heard such fire. My old machine guns did great business. As line after line of Turks came on we mowed them down. I got so excited I pushed a gunner away from a gun and fired it myself. We kept them back with fearful loss but to my horror as day dawned we turned and saw our troops on our right retiring helter skelter. The hill on our right was higher than ours and extended further back, there was a deep nullah between the two hills. The Turks manned the hill and pumped lead into us from our right rear.

This was my first experience of enemy fire without cover, and it was very unpleasant, most like a hailstorm only very much worse. They were not 300 yds from us. The General fell shot through the heart. Our signalling officer Moseley was killed and Baker (Brigade Major) wounded in the arm. I saw the position was pretty hopeless and at Baker's request rushed down…

••••

At this point, the narrative ends. Chadwick's mother was asked to transcribe Chadwick's account of the Battle of Chunuk Bair describing the circumstances that led to Moseley's death by Moseley's mother. The transcription ends with Moseley's death.… Editors.

APPENDIX III
MEMORIALS TO MOSELEY AND ONLINE LECTURES

PLAQUES, BUILDINGS AND LECTURE THEATRES NAMED IN HONOUR OF HENRY MOSELEY

Commemorative Tablet, Science School, Eton College, Windsor, Berkshire SL4 6DW

'This tablet is placed here to remind Etonians that H. G. J. Moseley (K. S. 1901–1906) began his study of physics in this laboratory. He was killed in the Gallipoli Peninsula on 10 August 1915. In his short life he gave to science the discovery of the atomic numbers of the elements'.

> *The rays whose path here first he saw*
> *Were his to range in ordered law*
> *A nobler law made straight the ray*
> *That leads him 'neath a nobler way'*

There is an annual Moseley Physics Prize.

Royal Society of Chemistry, National Chemical Landmark Plaque, Clarendon Laboratory, Parks Road, Oxford OX1 3PU

'Clarendon Laboratory where H. G. J. Moseley (1887–1915) completed his pioneering studies on the frequencies of X-rays emitted from the elements. His work established the concept of atomic number and helped reveal the structure of the atom. He predicted several new elements and laid the ground for a major tool in chemical analysis'. The plaque was unveiled by the Vice-Chancellor of the University of Oxford, John Hood, on 24 September 2007.

Blue Plaque, 48 Woodstock Road, now part of St. Anne's College, Oxford OX2 6HS

'The family home of Henry Gwyn Jeffrey's Moseley (1887–1915), who established the atomic numbers of the chemical elements. Killed in action at Gallipoli'.

Moseley Lecture Theatre, Schuster Building, University of Manchester, Manchester M13 9TX

The lecture theatre houses a plaque mounted on the back wall 'In memory of Henry Gwyn Jeffreys Moseley M.A. Oxon. Lecturer in Physics and John Harling Fellow in the University of Manchester. Lieutenant in the Royal Engineers killed in action on Gallipoli August 10 1915 aged twenty-eight years. Discoverer of the law defining the order and number of the elements'.

Henry Moseley X-Ray Imaging Facility, Alan Turing Building, The University of Manchester, Manchester M13 9PY

'Instrumentation purchased through funding from the Higher Education Funding Council for England, Engineering and Physical Research Council, and multiple industry partnerships enable 2D, 3D and 3D over time (so-called 4D) analysis of materials to be combined with elemental, atomic and isotropic characterisation techniques'.

Moseley Room, Second Floor, Lindemann Building, Clarendon Laboratory, Parks Road Oxford OX1 3PU.

Until 2007, this room had a small display of Moseley memorabilia, including an X-ray tube and Moseley's original X-ray plot, but these are now on display outside the Martin Wood Lecture Theatre. The Oxford Physics Department has established a Henry Moseley Society open to benefactors to the Department who make a contribution of £1,000 or more.

WAR MEMORIALS

Helles Memorial, Gallipoli

Moseley's name appears with over 20,000 others, ordered by regiment and rank. The monument honours all British and Indian soldiers with no known grave killed in the Gallipoli campaign, along with Australians killed at Helles.

Summer Fields School, Oxford

There are 139 names on the Memorial Brasses outside the Chapel. The names are listed alphabetically.

Eton College

Moseley is one of the 1,157 Old Etonians killed in the Great War out of the 5,650 who enlisted.

University of Manchester Quad War Memorial
The plaque is located in the quadrangle of the original Owens College Building. Moseley is listed along with 510 other members of the University killed in action. Names are ordered by seniority of the service or regiment, with further ordering by rank.

Trinity College, Oxford, War Memorial Library
Moseley appears as one of 154 Trinity men killed out of 816 serving officers. There were fifteen deaths in the Gallipoli Campaign, including eight in August 1915. Names are ordered first by date of Matriculation at Trinity and then alphabetically. The names of five German and Austro-Hungarian alumni killed in action are recorded on a separate plaque in the library.

St Giles' Church, Oxford
Moseley is listed second in chronological order of date of death of seventeen names marked on the St Giles' Church, Oxford World War One memorial of parishioners killed in the war between 1914 and 1919. See http://www.st-giles-church.org/stgiles/new-notices/war-memorial-100th-anniversary/ (accessed 27 November 2017).

MOSELEYUM: A MEMORIAL CHEMICAL ELEMENT?
There have been several campaigns to name a chemical element after Moseley.
1. Writing in *Science* in 1925, Richard Hamer suggested that element 43 (whose discovery seemed imminent) should be named Moseleyum when it was eventually disscovered.[1] This suggestion was endorsed by the Editors of *Nature*[2] and *Science*.[3]
2. Following adoption of the name roentgenium for element 111 in 2005, the popular science writer and neurologist Oliver Sacks argued in *Chemistry International* that element 112 should be named after either Henry Moseley or Frederick Soddy.[4]
3. Following the preparation of elements 113, 115, 117 and 118 by nuclear fusion reactions, Eric Scerri argued in *American Scientist* in 2014 that 'Master of the Missing Elements' Henry Moseley should be recognised in the name of one of these new elements.[5]
4. Following ratification of elements 113, 115, 117 and 118 by the International Union of Pure and Applied Chemistry in December 2015, the *Dorset Echo* argued that a chemical element should be named after the 'incredible Weymouth physicist Henry Moseley'.[6]

5. The suggestion in the *Dorset Echo* received further support from E. Scerri, P. P. Edwards, FRS, and R. G. Egdell in a letter to *Chemistry World* in March 2016.[7]
6. The *Chemistry World* letter was followed by a letter to *The Times,* signed by Scerri, Edwards and Egdell, with further support from Sir John Meurig Thomas, FRS (former Director of the Royal Institution and former Master of Peterhouse Cambridge), Nobel Laureate in Chemistry Roald Hoffmann, and the distinguished German chemist Friedrich Hensel. This letter was published on 27 February 2016, and was accompanied by a short feature article by the Science Editor, Tom Whipple, outlining Moseley's remarkable career.[8] However the new elements were eventually named as nihonium (113), moscovium (115), tennessine (117), and oganesson (118): the first three names celebrate the discovery of the elements in Japan, Russia and the United States, while the fourth name recognises the important role of Professor Yuri Oganessian in establishing the existence of these elements.[9]

Notes and References

[1] R. Hamer, 'Moseleyum'·, *Science,* 61 (1573), (1925), 208–209.

[2] Editorial, 'Current Topics and Events'. *Nature,* 115 (2893), (1925), 545.

[3] Editorial, 'Moseleyum and the Names of Elements', *Science,* 61 (1583), (1925) 510.

[4] O. Sacks, 'Honoring a Hero', *Chemistry International,* 27 (4), (2005), 10.

[5] Eric R. Scerri, 'Master of the Missing Elements', *American Scientist* 102 (September-October), (2014), 358–365.

[6] *Dorset Echo,* 13 January 2016.

[7] E. Scerri, P.P. Edwards and R.G. Egdell, 'Make Room for Moseleyium', *Chemistry World,* 13 (22), (2016), 44.

[8] E. Scerri, R.G. Egdell, R. Hoffmann, F. Hensel, J. M. Thomas and P. P. Edwards, 'Element of Truth', *The Times,* 27 February 2016, 24; T. Whipple, 'War Hero Unlocked Science's Great Secret', *The Times,* 27 February 2016, 19.

[9] https://iupac.org/iupac-announces-the-names-of-the-elements-113-115-117-and-118/ (accessed 27 November 2017).

FILMS, LECTURES AND BROADCASTS AVAILABLE ON-LINE

'"Sacrifice of a genius": Henry Moseley's role as a Signals Officer in World War One', E. Bruton, lecture delivered on 11 October 2013 at the Royal Society.
https://royalsociety.org/science-events-and-lectures/2013/henry-moseley/

'Bohr and Moseley in Manchester from a Manchester Perspective', N. Todd, lecture delivered on 14 August 2014 as part of meeting organised by the Institute of Physics History Group at the University of Manchester to mark the centenary of the Bohr Atom.
https://www.youtube.com/watch?v=jOoQQ9xVK2U

'H. G. J. Moseley in Interaction with Niels Bohr and Ernest Rutherford', J. L. Heilbron, lecture delivered on 14 August 2014 as part of meeting organised by the Institute of Physics History Group at the University of Manchester to mark the centenary of the Bohr Atom.
https://www.youtube.com/watch?v=EYIhx5YFNg0

'Moseley's Law', J. Wark, lecture delivered to alumni of Trinity College, Oxford, 16 December 2014.
https://www.youtube.com/watch?v=_KNZiQsUoA8

'Death of a Physicist', BBC World Service Feature Broadcast, BBC World Service Feature Broadcast, 4 August 2015.
http://www.bbc.co.uk/programmes/p02y4hfy

'The Mystery of Matter: Search for the Elements'.
'Episode Three: Into the Atom'.
Episodes were premiered on 19 August 2015 and are available on DVD from the United States Public Broadcasting Service. Supplementary Material but not the three main episodes may be viewed online in the UK.
http://www.pbs.org/program/mystery-matter/#about-series

'Henry Moseley (1887–1915): a scientist lost to war', R. G. Egdell, presented in the Talking Science series of public lectures at the Rutherford Appleton Laboratory on 18 September 2015.
https://vimeo.com/141668109

'Henry Moseley: Science's Lost Hero', Diamond Light Source Film, released by Diamond Light Source Communication Team, 17 May 2016.
http://www.diamond.ac.uk/Public/Multimedia/Films/2016/Moseley.html

BIBLIOGRAPHY

I. PRIMARY SOURCES

1. Archives

American Philosophical Society, Sir Charles Galton Darwin Papers

Accademia Nazionale delle Scienze detta dei XL, Rome

Centre de Ressources Historiques de l'ESPCI ParisTech, Moseley Letters

Eton College
  Eton College Archives
  Eton College Chronicle
  Eton College Debating Society
  Eton College Scientific Society

Moseley Family Archives, Amabel Moseley and Margery Ludlow-Hewitt (née Moseley) Papers

Niels Bohr Institute, Copenhagen, Hevesy Papers

Oxford University
  Oxford Magazine, 1908-1909
  Oxford Museum of the History of Science, Ludlow-Hewitt Papers

Royal Swedish Academy of Sciences, Stockholm, Center for History of Science, Svante Arrhenius Papers

Royal Swedish Academy, Nobel Foundation Committee for Chemistry

Trinity College, Oxford
  Trinity College Archives
  Trinity College Boat Club

University of California (Berkeley), Archives of the History of Quantum Physics, microfilms of Moseley letters

University of Cambridge Library, Rutherford Papers

University of Manchester Archives, Sir Arthur Schuster Papers

2. Published Works

a. Publications of H. G. J. Moseley

Moseley, H. G. J. and K. Fajans, 'Radioactive Products of Short Half Life', *Philosophical Magazine Series 6,* 22 (130), (1911), 629–638.

Moseley, H. G. J., 'The Number of β-Particles Emitted in the Transformations of Radium', *Proceedings of the Royal Society of London Series A*, 87(1912), 230–255.

Moseley, H. G. J. and W. Makower, 'γ Radiation from Radium B', *Philosophical Magazine Series 6,* 23 (134), (1912), 302–310.

Moseley, H. G. J. and C .G. Darwin, 'The Reflection of X-rays', *Philosophical Magazine Series 6,* 26 (151), (1913), 210-232.

Moseley, H. G. J. and C. G. Darwin, 'The Reflection of X-rays', *Nature,* 90 (2257), (1913), 43–57.

Moseley, H. G. J., 'The Attainment of High Potentials by the Use of Radium', *Proceedings of the Royal Society of London Series A,* 88 (605), (1913) 471–476.

Moseley, H. G. J., 'The High-Frequency Spectra of the Elements', *Philosophical Magazine Series 6,* 26 (156), (1913), 1025–1034.

Moseley, H. G. J., 'The High-Frequency Spectra of the Elements. Part II', *Philosophical Magazine Series 6,* 27 (160), (1914), 703–713.

Moseley, H. G. J. and H. Robinson, 'The Number of Ions Produced by the β and γ Radiations from Radium', *Philosophical Magazine Series 6,* 28 (165), (1914), 327–337.

'Moseley, H.G.J., 'Atomic Models and X-Ray Spectra'*, Nature,* 92 (2307), (1914), 554.

b. Obituaries and Biographical Commentaries

Anon., 'The Waste of Ability', *The British Medical Journal,* 2 (2855), (18 September 1915), 449.

Anon., 'Classics of Science: Moseley's Atomic Numbers', *The Science News-Letter,* 15 (416), (30 March 1929), 203–204.

Anon., 'Nuclear Secrets', *Time,* 18 (25), (21 December 1931), 1–2.

Anon., 'Moseley, Henry Gwyn Jeffreys', in *The Columbia Encyclopedia*, 6th edition (New York: Columbia University Press, 2000).

Bohr, Niels, 'Henry Gwyn Jeffreys Moseley', *Philosophical Magazine Series 6*, 31 (182), (1916), 174–176.

Bourne, G.C., 'Memoir on Henry Nottage Moseley', in H. N. Moseley, *Notes by a Naturalist* (New York: G. P. Putnam's Sons, 1892), v–xvi.

Ferreira, R., 'Photographs of Moseley', *Isis*, 60(2), (1969), 233.

Gilbert, Bruce, 'Henry Moseley: Understanding Atomic Numbers', *Chemistry Review,* 23 (1), (2013), 30–33.

Gorin, George, 'Mendeleev and Moseley: The Principal Discoverers of the Periodic Law', *Journal of Chemical Education,* 73 (6), (June 1996), 490–493.

Hamer, Richard, 'Moseleyum', *Science*, 61 (1573), (20 February 1925), 208–209.

Heimann, P. M., 'Moseley and Celtium: The Search for a Missing Element', *Annals of Science*, 23 (4), (1967), 249–260.

Heimann, P. M., 'Moseley's Interpretation of X-ray Spectra', *Centaurus*, 12 (4), (1968), 261–274.

Heilbron, John L., 'The Work of H.G.J. Moseley', *Isis*, 57 (3), (1966), 336–364.

Heilbron, John L., *H. G. J. Moseley: The Life and Letters of an English Physicist, 1887–1915* (Berkeley and Los Angeles: University of California Press, 1974).

Heilbron, John L., 'Moseley, Henry Gwyn Jeffreys (1887-1915)', *Oxford Dictionary of National Biography* (2004) https://doi.org/10.1093/ref:odnb/35125 (accessed 7 February 2018).

Heilbron, John L., 'Moseley, Henry Gwyn Jeffreys', *Complete Dictionary of Scientific Biography* (New York: Charles Scribner's Sons, 2008), vol. 9, 542–545.

Huxley, Julian, *Memories* (London: Allen and Unwin, 1970).

Jaffe, Bernard, *Moseley and the Numbering of the Elements* (Garden City: Anchor Books, 1971; London: Heinemann, 1972).

Kopal, Zdenek, 'H. G. J. Moseley (1887–1915)', *Isis,* 58 (3), (1967), 405–407.

Lankester, E .R., 'Henry Gwyn Jeffreys Moseley', *Philosophical Magazine, Series 6,* 31 (182), (1916), 173–176.

Özel, Mehmet Emin, 'A Young and Famous Scientist at Dardanelles War Henry Gwyn Jeffreys Moseley (1887-915)', *NeuroQuantology,* 6 (1), (2008), 72–74.

Reynosa, Peter, 'An Ode to Henry Moseley (ca. 2004)', *The Huffington Post,* 7 January 2016, http://www.huffingtonpost.com/peter-reynosa/an-ode-to-henry-moseley_b_8924690.html?ir=Australia (accessed on 7 February 2018).

Rutherford, Ernest, 'Henry Gwyn Jeffreys Moseley', *Nature,* 96 (2393), (1915), 33–34.

Rutherford, Ernest, 'H. G. J. Moseley, 1887-1915', *Proceedings of the Royal Society A,* 93 (655), (9 October 1917), xxii–xxviii.

Rutherford, Ernest, 'Identification of a Missing Element', *Nature,* 109 (2746), (1922), 781 (translation of Georges Urbain, in *Comptes rendus de l'Académie des sciences*).

Rutherford, Ernest, 'Moseley's Work on X-Rays', *Nature,* 116 (2913), (1925), 316–317.

Sarton, George, 'Moseley – The Numbering of the Elements', *Isis,* 9 (1), (1927), 96–111.

Scerri, Eric R., 'Master of Missing Elements', *American Scientist,* 102 (5), (2014), 358–385.

Siegbahn, Manne, 'The Prize in Physics', in H. Schuck and R. Sohlman, *Nobel: The Man and his Prizes* (Amazon: Grizzell Press, 2004).

Woodward, B.B., 'Moseley, Henry (1801–1872)', *Oxford Dictionary of National Biography* (Oxford: Oxford University Press, 2004; online edn May 2007).

Woods, Gordon, 'Harry Moseley: Discoverer of Atomic Number Z and the Prince of Eton Physicists', *Chemistry Review,* 13 (3), (2004), 30–32.

c. Contemporary Sources

Hartog, P. J., *The Owens College, Manchester A Brief History of the College and Description of its Various Departments* (London: Waterlow and Sons, 1900).

Mendeleev, Dimitri Ivanovich, *Principles of Chemistry* (London: Longman's Green, 1905), 2 vols.

Ramsay, W., 'Ancient and Modern Views Regarding the Chemical Elements', *Smithsonian Institution Report* (Washington, DC: Smithsonian Institution Press, 1911), 183–197.

Schuster, A. and R. S. Hutton, *The Physical Laboratories of the University of Manchester: A Record of 25 Years' Work* (Manchester: University of Manchester Press, 1906).

Thomson, J., *The Owens College, its Foundation and Growth* (London: Waterlow and Sons, 1886).

II. SECONDARY SOURCES

1. Books and Articles

Andrade, E.N. da C., 'Rutherford at Manchester, 1913–14', in J. B. Birks, (ed.), *Rutherford at Manchester* (London: Heywood, 1962), 27–42.

Asimov, Isaac, *Asimov's Biographical Encyclopedia of Science and Technology: The Lives and Achievements of 1510 Great Scientists from Ancient Times to the Present Chronologically Arranged* (Garden City: Doubleday, 2nd edn., 1982).

Asimov, Isaac, 'Made, not Found', *The Magazine of Fantasy and Science Fiction,* 67 (December 1984), 123–133.

Aspinall-Oglander, C. F., *Military Operations: Gallipoli* (London: William Heinemann, 2 vols., 1928–1932).

Authier, André, *Early Days of X-ray Crystallography* (Oxford: Oxford University Press, 2015).

Badash, Lawrence, 'Radioactivity before the Curies', *American Journal of Physics*, 33 (2), (1965), 128–135.

Badash, Lawrence, 'British and American Views of the German Menace in World War I', *Notes and Records of the Royal Society* [of London], 34 (1), (1979), 91–121.

Broadbent T.E., *Electrical Engineering at Manchester University* (Manchester: Manchester School of Engineering, 1998).

Biagioli, Mario and Jessica Riskin (eds.), *Nature Engaged: Science in Practice from the Renaissance to the Present* (Basingstoke: Palgrave Macmillan, 2012).

Bowen, Edmund J., 'The Balliol-Trinity Laboratories at Oxford, 1853–1940', in *Notes and Records of the Royal Society of London*, 25 (2), (1970), 228–230.

Bowen, Edmund J., *Chemistry in Oxford: The Development of the University Laboratories*, by Sir Harold Hartley (Cambridge: Cambridge University Press, 1966).

Brock, William H., *The Fontana History of Chemistry* (London: Fontana Press, 1992).

Charlton, H.B., *Portrait of a University: To Commemorate the University of Manchester, 1851–1951* (Manchester: Manchester University Press, 1951).

Crawford, Elisabeth, *Arrhenius: From Ionic Theory to Greenhouse Effect* (Canton, MA: Science History, 1996).

Crawford, Elisabeth and Robert Marc Friedman, 'The Prizes in Physics and Chemistry in the Context of Swedish Science', in Carl-Gustaf Bernhard, Elisabeth Crawford, and Per Sörbom (eds.), *Science, Technology and Society in the Time of Alfred Nobel* (Oxford: Pergamon Press for the Nobel Foundation, 1982), 311–331.

Crawford, Elisabeth, 'Arrhenius, the Atomic Hypothesis, and the 1908 Nobel Prizes in Physics and Chemistry', *Isis,* 75 (3), (1984), 503–522.

Crawford, Elisabeth, *The Beginnings of the Nobel Institution: The Science Prizes, 1901-1915* (Cambridge: Cambridge University Press, 1984).

Darwin, C.G., 'The Discovery of Atomic Number', *Proceedings of the Royal Society A*, 236 (1206), (1956), 285-296.

Darwin, C. G., 'Moseley and the Atomic Number of the Elements', in J. B. Birks (ed.), *Rutherford at Manchester* (London: Heywood, 1962), 17–26.

Darwin, C. G., 'Moseley's Determination of Atomic Numbers', in P. P. Ewald (ed.), *Fifty Years of X-Ray Diffraction* (Utrecht: Oosthoek's Uitgeversmaatschappij, for the International Union of Crystallography, 1962), 550–563.

Duhem, Pierre, *The Aim and Structure of Physical Science* (trans P. Wiener), (Princeton: Princeton University Press, 1954).

Eckert, M., 'Max von Laue and the Discovery of X-ray Diffraction in 1912', *Annalen der Physik (Berlin)*, 524 (5), (2012), A83–A85.

Fiddes, E., *Chapters in the History of Owens College and of Manchester University, 1851–1914* (Manchester: Manchester University Press, 1937).

Eve, A. S. *Rutherford: Being the Life and Letters of the Rt. Hon. Lord Rutherford, O.M.* (Cambridge: Cambridge University Press, 1939).

Fontani, M., M. Costa, and M. V. Orna, *The Lost Elements: The Periodic Table's Shadow Side* (Oxford: Oxford University Press, 2015).

Forman, Paul, 'The Discovery of the Diffraction of X-rays by Crystals: A Critique of the Myths', *Archive for the History of Exact Sciences*, 6 (1), (1969), 38–71.

Forman, Paul, 'Charles Glover Barkla', *Dictionary of Scientific Biography* (New York: Scribner's, 1970), 456–459.

Fox, Robert, Graeme Gooday, and Tony Simcock, 'Physics in Oxford: Problems and Perspectives', in Robert Fox and Graeme Gooday (eds.), *Physics in Oxford, 1839-69: Laboratories, Learning and College Life* (Oxford: Oxford University Press, 2005), 5–18.

Fox, Robert, 'Context and Practices of Oxford Physics, 1839–79', in Robert Fox and Graeme Gooday (eds.), *Physics in Oxford, 1839–69: Laboratories, Learning and College Life* (Oxford: Oxford University Press, 2005), 24–79.

Frederick-Frost, K.M., 'For the Love of a Mother – Henry Moseley's Rare Earth Research', *Historical Studies in the Natural Sciences*, 47 (4), (2017), 529–567.

Friedman, Robert Marc, 'Nobel Physics Prize in Perspective', *Nature*, 292 (5826), (1981), 793–798.

Friedman, Robert Marc, *The Politics of Excellence: Behind the Nobel Prize in Science* (New York: Freeman & Times Books, Henry Holt & Co., 2001).

Gall, A., 'Rutherford's Glassblowers – Otto Baumbach and Felix Niedergesass', *Science Technology: The Journal of the Institute of Science Technology* (2004), 3–8.

Gardiner, J. H., 'The Origin, History & Development of the X-ray Tube', *Journal of the Röntgen Society*, 5 (20), (1909), 66–80.

Gooday, Graeme, 'Robert Bellamy Clifton and the "Depressing Inheritance" of the Clarendon Laboratory, 1877-1919', in Robert Fox and Graeme Gooday (eds.), *Physics in Oxford, 1839-69: Laboratories, Learning and College Life* (Oxford: Oxford University Press, 2005), 80–118.

Greenwood, N.N. and A. Earnshaw, *Chemistry of the Elements* (Oxford: Pergamon Press, 1989).

Gordin, Michael D., *A Well-Ordered Thing: Dimitrii Mendeleev and the Shadow of the Periodic Table* (New York: Basic Books, 2004).

Hartog, P. J., *The Owens College, Manchester: A Brief History of the College and Description of its Various Departments* (London: Waterlow and Sons, 1900).

Heilbron, John L., *Elements of Early Modern Physics* (Berkeley and Los Angeles: University of California Press, 1982).

Heilbron, John L., *The Dilemmas of an Upright Man: Max Planck as Spokesman for German Science* (Berkeley: University of California Press, 1986).

Heilbron, John L., *Ernest Rutherford and the Explosion of Atoms* (Oxford: Oxford University Press, 2003).

Hirosige, T., 'The van den Broek Hypothesis', *Japanese Studies in the History of Science*, 10 (1971), 143–162.

Hopkins, Clare, *Trinity: 450 Years of an Oxford College Community* (Oxford: Oxford University Press, 2005).

Howarth, Janet, '"Oxford for Arts": The Natural Sciences, 1880–1914', in M. G. Brock and M. C. Curtoys (eds.), *Nineteenth-Century Oxford, Part 2, The History of the University of Oxford* (Oxford: Oxford University Press, 2000), VII, 457–459.

Jenkin, John, *William and Lawrence Bragg, Father and Son* (Oxford: Oxford University Press, 2008).

Kragh, Helge and P. Robertson, 'On the Discovery of Element 72', *Journal of Chemical Education,* 56 (7), (1979), 456–458.

Kragh, Helge, 'Elements No. 70, 71, and 72: Discoveries and Controversies', in C.H. Evans (ed.), *Episodes from the History of the Rare Earth Elements* (Dordrecht: Kluwer Academic Publishers. 1996), 67–89.

Kragh, Helge, *Neils Bohr and the Quantum Structure* (Oxford: Oxford University Press, 2012).

Kragh, Helge, 'The Isotope Effect: Prediction, Discussion and Discovery', *Studies in History and Philosophy of Science, Part B,* 43 (3), (2012), 176–183.

Lagerkvist, Ulf, *The Periodic Table and a Missed Nobel Prize* (Singapore: World Scientific, 2012).

LeLong, B., 'Translating Ion Physics from Cambridge to Oxford', in Robert Fox and Graeme Gooday (ed.), *Physics in Oxford, 1839-1939: Laboratories, Learning, and College Life* (Oxford: Oxford University Press, 2005), 209–232.

Lettevall, Somsen and Sven Widmalm (eds.), *Neutrality in Twentieth-Century Europe: Intersections of Science, Culture, and Politics after the First World War* (New York and London: Routledge, 2012).

MacLeod, Roy and Russell Moseley, 'Breadth, Depth and Excellence: Sources and Problems in the History of University Science Education in England, 1850-1914', *Studies in Science Education*, 5 (1), (1978), 85–106.

MacLeod, Roy, 'Education – Scientific and Technical', in Gillian Sutherland (ed.), *Government and Society in Nineteenth Century Britain: Commentaries on Education: British Parliamentary Papers* (Dublin: Irish Academy Press, 1977), 196–225.

MacLeod, Roy, 'The Mobilisation of Minds and the Crisis in International Science: "The *Krieg der Geister* and the Manifesto of the 93"', *Journal of War and Culture Studies*, 10 (3), (2017), 1–21.

MacLeod, Roy, 'Science and the Professors' War, 1914–1919', in Marysa Demoor, Cedric van Dijck, and Sarah Posman (eds.), *Intellectuals and the Great War* (Eastbourne: Sussex Academic Press, 2017).

Marsden, Ernest, 'Rutherford at Manchester', in J.B. Birks (ed.), *Rutherford at Manchester* (London: Heywood, 1962), 1–16.

Pais, Abraham, *Inward Bound: Of Matter and Forces in the Physical World* (Oxford: Clarendon Press, 1976).

Pauli, W. *et al.*, *Niels Bohr and the Development of Physics* (London: Pergamon Press, 1955), 1–11.

Richardson, John, Russell Egdell, Nick and Harold Hankins, 'Rutherford, Geiger, Chadwick, Moseley, Cockcroft and their role in the Great War – Part 1', *Stand To!, The Journal of the Western Front Association,* 86 (2009), 28–34; 'Part 2 – John Cockroft', 87 (2009/2010), 42–45; 'Part 3a – Henry Moseley', 89 (2010), 46–50, Part 3b – Henry Moseley', 90 (2010/2011), 29–35.

Scerri, Eric R., *The Periodic Table: Its Story and Its Significance* (Oxford: Oxford University Press, 2007).

Scerri, Eric, *A Tale of Seven Elements* (Oxford: Oxford University Press, 2013).

Scerri, Eric, *A Tale of Seven Scientists and a New Philosophy of Science* (New York: Oxford University Press, 2016).

Simcock, Tony, 'Laboratories and Physics in Oxford Colleges, 1848–1947', in Robert Fox and Graeme Gooday (eds.), *Physics in Oxford, 1839-69: Laboratories, Learning, and College Life* (Oxford: Oxford University Press, 2005).

Scheckter, John, '"Modern in Every Respect": The 1914 Conference of the British Association for the Advancement of Science', *The Journal of the European Association for Studies of Australia,* 5 (1), (2014), 4–20.

Taylor, F. Sherwood, 'The Teaching of Science at Oxford in the Nineteenth Century', *Annals of Science,* 8 (1), (1952), 82–112.

Trenn, T. J., *The Self-Splitting of the Atom: The History of the Rutherford-Soddy Collaboration* (London: Taylor and Francis, 1977).

Turner, G. L. E., 'The Discovery of Atomic Number', *Bulletin of the Institute of Physics and the Physical Society,* 16 (2), (1965), 54–55.

Widmalm, Sven, 'Science and Neutrality: The Nobel Prizes of 1919 and Scientific Internationalism in Sweden', *Minerva,* 33(4), (1995), 339–360.

2. Scientific Works

Armstrong, Henry E., 'The Classification of the Elements', *Proceedings of the Royal Society of London*, 70 (459), (1902), 86–94.

Barkla, C. G. and C. A. Sadler, 'Homogeneous Secondary Röntgen Radiations', *Proceedings of the Physical Society of London*, 21 (1907), 336–373.

Barkla, C. G., 'The Spectra of the Fluorescent Röntgen Radiations', *Philosophical Magazine Series 6*, 22 (129), (1911), 396–412.

Barkla, C. G. and J. Nicol, 'Homogeneous Fluorescent X-radiations of a Second Series', *Proceedings of the Physical Society of London*, 24 (1911), 9–17.

Baskerville, Charles, 'The Elements Verified and Unverified', *Science*, 19 (472), (1904), 88–100.

Bohr, N., 'On the Constitution of Atoms and Molecules', *Philosophical Magazine Series 6*, 26 (151), (1913), 1–25; *idem,* 'On the Constitution of Atoms and Molecules. Part II – Systems Containing a Single Nucleus', *Philosophical Magazine Series 6*, 26 (153), (1913), 476–502; *idem*, 'On the Constitution of Atoms and Molecules. Part III – Systems Containing Several Nuclei', *Philosophical Magazine Series 6*, 26 (155), (1913), 857–875.

Bragg, W. H. and W. L. Bragg, 'The Reflection of X-rays by Crystals', *Proceedings of the Royal Society A*, 88 (605), (1913), 428–438.

Bragg, W. H. and W. L. Bragg, *X-rays and Crystal Structure* (London: G. Bell, 1915).

Broek, Antonius van den, 'Intra-Atomic Charge', *Nature*, 92 (2300), (1913), 372–373.

Broek, Antonius van den, 'Intra-Atomic Charge and the Structure of the Atom', *Nature*, 92 (2304), (1913), 476–478.

Coster, D. and G. Hevesy, 'On the Missing Element of Atomic Number 72', *Nature,* 111 (2777), (1923), 79.

Geiger, H. and E. Marsden, 'On a Diffuse Reflection of the α-Particles', *Proceedings of the Royal Society A*, 82 (557), (1909), 495–500.

Geiger, H. and E. Marsden, 'The Laws of Deflexion of α-Particles through Large Angles', *Philosophical Magazine*, 25 (148), (1913), 604–628.

Hansen, H. M and S. Werner, 'On Urbain's Celtium Lines', *Nature*, 111 (2788) (1923), 461.

Hevesy, G. von, *Chemical Analysis by X-Rays and its Application* (New York: McGraw Hill, 1932).

Rutherford, E., 'The Mass and Velocity of the α-particles expelled from Radium and Actinium', *Philosophical Magazine*, Series 6, 12 (70), (1906), 348–371.

Rutherford, E. and T. Royds, 'The Nature of the Alpha Particles from Radioactive Substances', *Philosophical Magazine Series 6,* 17 (98), (1909), 281–286.

Rutherford, E., 'The Scattering of Alpha and Beta Particles by Matter and the Structure of the Atom', *Philosophical Magazine Series 6,* 21 (125), (1911), 669–688.

Siegbahn, Manne, *The Spectroscopy of X-rays* (London: Oxford University Press, 1925).

Siegbahn, Manne, *Spektroskopie der Röntgenstrahlen* (Berlin: Springer-Verlag, 1931).

Siegbahn, K., C. Nordling, G. Johansson, J. Hedman, P. F. Heden, K. Hamrrin, U. Gelius, T. Bergmark, L.O. Werme, R. Maanne and Y. Baer, *ESCA Applied to Free Molecules* (Amsterdam: North-Holland, 1969).

Townsend, J., *The Theory of Ionization of Gases by Collision* (London: Constable & Co., 1910).

Townsend, J., *Electricity in Gases* (Oxford: Clarendon Press, 1915).

Urbain, Georges, 'Sur un nouvel element qui accompagne le lutécium et le scandium dans les terres de la gadolinite: le celtium', *Comptes rendus de l'Académie des sciences,* 152 (1911), 141–143.

Urbain, Georges, 'Les numéros atomiques du néo-ytterbium, du lutécium et du celtium', *Comptes rendus de l'Académie des sciences,* 174 (1922), 1349–1351.

Werner, A., 'Beitrag zum Ausbau des periodischen Systems'*, Berichte der deutschen chemischen Gesellschaft,* 38, (1905), 914–921.

CONTRIBUTORS

Silke Ackermann studied history, oriental languages and cultures, and the history of science at Frankfurt University, before moving to England in 1995 as curator of European and Islamic scientific instruments and other medieval and early modern collections at the British Museum. For the next sixteen years, she served in many managerial and curatorial roles at the BM, and with the BM's team developed the Zayed National Museum in Abu Dhabi. In 2012, she became Professor for Cultural Tourism and Management at the University of Applied Sciences in Schwerin (Germany), where she was later appointed president. In 2014, she returned to the UK as Director of the Museum of the History of Science in Oxford and as a professorial fellow of Linacre College. She is the first female director of any Oxford University museum.

Elizabeth Bruton is Curator of Engineering and Technology at the Science Museum, London. Previously, she was the Heritage Officer at Jodrell Bank Discovery Centre, University of Manchester; Co-curator of the Exhibition, 'Harry's Story: Henry Moseley, a scientist lost to war', Museum of the History of Science, Oxford; and postdoctoral researcher on a project dealing with 'Innovating in Combat: Telecommunications and intellectual property in the First World War', supported by the UK Arts and Humanities Research Council at the University of Leeds. Her research interests include communication history, military history, history of electrical engineering, Victorian technologies, and institutional innovation.

Russell G. Egdell is Emeritus Professor of Inorganic Chemistry in the University of Oxford and Emeritus Fellow of Trinity College, Oxford. Over a period of forty years, he has taught and lectured on most areas of inorganic chemistry. His research is at the borderline between solid-state chemistry and solid-state physics, with a focus in recent years on the combination of X-ray photoelectron spectroscopy with X-ray absorption and X-ray emission to probe the electronic properties of metal oxides and their surface: much of this work has exploited high intensity synchrotron radiation. He has published over 250 academic papers, review articles and book chapters.

Kristen M. Frederick-Frost is a Curator of Modern Science at the National Museum of American History. Before joining the Smithsonian Institution, she held curatorial positions at the National Institute of Standards and Technology and at the Chemical Heritage Foundation. Her research explores the development of laboratory instrumentation and scientific note taking. Her interest in both stems from working as a research scientist for sounding rocket and satellite missions. She holds a PhD in Physics from Dartmouth College and an MSc in the History of Science, Technology, and Medicine from Oxford University, where she was introduced to Moseley's story and experimental apparatus.

Robert Marc Friedman is professor of history of science at the University of Oslo and Tetelman Fellow, Jonathan Edwards College, Yale University. His publications on the history of modern physical and environmental sciences include, *Appropriating the Weather: Vilhelm Bjerknes and the Construction of a Modern Meteorology* and *The Politics of Excellence: Behind the Nobel Prize in Science*. His dramatizations of history of science have received professional staging in several countries. These include 'Remembering Miss Meitner', 'Amundsen vs Nansen', and 'Transcendence'. His current work includes a study of geochemist V. M. Goldschmidt and a play about Einstein and Planck.

John L. Heilbron is a Professor of History, Emeritus, at the University of California, Berkeley, and an Honorary Fellow of Worcester College, Oxford. He has written some two dozen books, of which *Love, Literature, and the Quantum Atom* (with Finn Aaserud, 2013), *Physics: A Short History from Quintessence to Quarks* (2015), and *The History of Physics: A Very Short Introduction* (2018), all from Oxford University Press, are the most recent. He is also the editor of the *Oxford Companion to the History of Modern Science*.

Clare Hopkins read Latin at the University of Sheffield. Although both her parents were chemists, she trained in archive administration at University College of Wales, Aberystwyth, and since 1985 has been Archivist of Trinity College, Oxford. Besides managing an extensive archive that ranges in date from the sixteenth to the twenty-first centuries, she has established one of the finest collections of undergraduate ephemera in Oxford University. Clare is the author of several entries in the *Oxford Dictionary of National Biography*. Her *Trinity: 450 Years of an Oxford College Community* was published by Oxford University Press in 2005.

Stephen Johnston received his BA and PhD from Cambridge University and was a curator at the Science Museum London from 1987 to 1995. He joined the Museum of the History of Science, University of Oxford in 1995 and is now Deputy Director. His research and publications have focused on practical mathematics from the sixteenth to the eighteenth centuries. Prior to working on Henry Moseley, he curated exhibitions on a broad range of themes within the history of science, from 'Al-Mizan: Sciences and Arts in the Islamic World' (2010) to 'Geek is Good' (2014).

Roy MacLeod is Emeritus Professor of History at Sydney University, where he has written extensively in the social history of science, technology and medicine. He studied the history of science at Harvard and Cambridge (PhD, LittD), and has taught at many universities and institutes in North America, Europe, and Australasia. He co-founded the journal *Social Studies of Science*, and was for many years Editor of *Minerva*. He has recently published in the history of chemistry and nuclear history, and in the history of war and science. He is currently working on the discovery and use of the Rare Earths, and on the role of strategic resources in global geopolitics.

Francesco Offi is an Associate Professor of Physics at the Roma Tre University in Rome. He was awarded a first degree in physics at the University of Turin (Italy) and a PhD in Physics at the University of Halle (Germany). His scientific research deals mainly with experimental investigation of the electronic and magnetic properties of solid systems by spectroscopic techniques, particularly those involving X-rays. He teaches courses on electron spectroscopies, surface science techniques and optics.

Giancarlo Panaccione received a PhD in materials science in 1995 and is now a senior researcher at the IOM Institute of the National Research Council of Italy (CNR). His research is mostly devoted to the exploitation of synchrotron radiation spectroscopies for the study of the electronic and magnetic properties of solids, with an emphasis on highly correlated systems, quantum materials, and complex oxides and their heterostructures. He is coordinator of two research areas at CNR-IOM: *Phase transition and magnetic properties in strongly correlated oxides, metal-semiconductor interfaces, and anomalous metals*; and *Spin polarization research instrumentation in the nanoscale and time domain*. He has more than 170 publications in peer reviewed journals.

Eric R. Scerri received all his degrees in the UK before going to the United States in 1995 as a Caltech postdoctoral fellow. For the past eighteen years, he has been a lecturer at UCLA where he teaches general chemistry as well as courses in the History and Philosophy of Science. He is also the founding editor of the journal *Foundations of Chemistry*, and the author of several books with Oxford University Press, including *The Periodic Table, Its Story and Its Significance* (2007), *A Very Short Introduction to the Periodic Table* (2011), *A Tale of Seven Elements* (2013) and *A Tale of Seven Scientists and A New Philosophy of Science* (2016).

Neil Todd holds a BSc in theoretical and mathematical physics and a PhD in psychology, both awarded by the University of Exeter. Between 1994 and 2015 he was Lecturer and then Honorary Lecturer in Neuroscience at the University of Manchester. He is now an Honorary Senior Lecturer in the Department of Psychology at the University of Exeter, as well as a freelance musician, composer, science writer, dive master and historian. His interest in Rutherford's group in Manchester originally arose from the 'archaeological' evidence of radioactive contamination in the building where he worked in Manchester.

INDEX

NOTE: Page numbers in *italic* refer to illustrations

Academy of the Forty *see* Società Italiana delle Scienze detta dei XL
Accademia Reale dei Lincei, 158, 187
Ackermann, Silke, 244, 247, 299
actinium, 49, 175, 183, 186
Aftonbladet (Swedish newspaper), 141
Alan Johnston, Lawrence and Moseley Research Fellowship, 124
Alington, Cyril A., 21–2, *22*, 25, 304
Alington, Hester (*née* Lyttelton), 21
Alpha Particle X–ray spectrometer (APXS), 206
alpha particles, 4, 49, 51–2, 91–2, *96*, 96, 103, *205*, 206, 236, 237–8
alphon, 92–3, 103
aluminium, 97, 98; X–ray emission spectra, 211
Andrade, E.N. da C., 231
Ångström, Knut, 145
anode (in X–ray tube), 203, 209, 214
Arrhenius, Svante, 136–41, 144, 146–9
artificial transmutation, 61
Asimov, Isaac, 266
Assche, Pieter van, 182
astatine, 176, 183, 186
Aston, A.S., 59
Aston, Francis, 167
Atatürk, Kemal, 266
atomic structure: atomic number, 2, 5, 7, 8, 9–54, 60, 71, 73, 90–1, 98–102, 137, 175, 227, 230; atomic nucleus, 6–7, 52–3, 161; electronic, 15, 270–4; Rutherford on, 52–3, 71, 97; atomic weight, 60, 69, 71, 94, 96, 97, 99, 101–2; van den Broek on, 97
atoms: charge, 94
Auger electrons, 202, 214
Auger, Pierre, 202
Australia: HM in, 107, 110, 258

Bacon, Roger, 15
Baeyer, Adolf von, 142
Baldwin, Brigadier General Anthony Hugh, 114, 118–19
Balliol College, Oxford, 29–31
Barkla, Charles Glover: investigates secondary X–rays, 4, 69; wins Nobel Prize, 8, 146–9, 168, 235; on characteristic X–rays, 58; 'fringes', 76; van den Broek studies, 94, 103; on X–rays and labelling, 201, 272–3
Bassett, Corporal Cyril, VC, 260
Baumbach, Otto, 51, 61, 79, 81
Beaumont, J.W., 47
Becquerel, Henri, 48

Berg, Otto, 179, 181, 183
Bernstorff, Abrecht von, *254*
beryllium, 237–8
beta particle or beta rays (secondary corpuscular rays), 49, 201, 226
Birmingham, University of: HM applies for chair, 110
black–body radiation, 6
Blaserna, Pietro, 160
Bohr, Niels: influence on HM, 5–7; HM's relations with, 9; quantum theory, 53, 225; visits Manchester, 55, 59–60, 71, 225–6, 227; appointed Reader in Mathematical Physics in Manchester, 61; and nature of atomic number, 100–1, 137; van den Broek influences, 103; and Siegbahn's Novel Prize, 151; awarded Matteucci Medal, 168; influenced by Siegbahn, 200; works with HM, 224, 227–9; agnosticism, 225; on atomic number, 226–7; on atom building, 228, 232, 237, 270–2; correspondence with HM, 234; *Collected Papers*, 100
Boltzmann, Ludwig, 50
Böök, Fredrik, 143
Born, Max, 231
Bosanquet, Claude H., 179
Boscovich, Ruggero, 155
Bowen, E.J., 258
Boyle, Robert, 15
Bragg, William Henry and William Lawrence (together): explain radiation lines, 57–8; on X-ray spectroscopy, 59, 69, 142; contribute to Moseley Prize for Physical Research, 62; and Matteucci Medal, 162–3
Bragg, William Henry: war work, 61; influence on HM, 137, 226; nominated for Nobel Prize, 138, 142, 144, 148–9
Bragg, William Lawrence: on Moseley Memorial committee, 61; considered for Nobel Prize, 142, 144, 148–9; war work, 202; age, 235; replaces Rutherford at Manchester, 235
Brauner, Bohuslav, 187
Bremstrahlung, 213
brevium, 176
British Association for the Advancement of Science: Salford meeting (1903), 50; Australia meeting (Sydney etc., 1914), 83, 107, 110
Broeck, Elisabeth, 91
Broek, Antonius van den: on order of elements in periodic table, 5, 8, 59–60; on atomic number, 71, 90, 97, 99, 102, 227–8; background, 91; discovers alpha particle, 91–2; illus., *91*; and periodic table, 92–3, *92*, *93*, 95, *95*, 97–9, *98*, *99*; and distinction between atomic number and weight, 94–6; achievements, 102–3
Broglie, Louis–Victor de, 167, 200
Broglie, Maurice de, 149, 199–200, 231–2, 273
Brunetti, Rita, 187
Bruton, Elizabeth, vii, 8, 107, 244, 299
Bulford Camp, Salisbury Plain, 113
Burkhart, W.F., 190

Bush, Sam W. (instrument maker), 81

caesium, 184
calcium, 72, 78
Cambridge, University of: science studies, 15–16, 29; Cavendish Laboratory, 61, 236
Carlheim–Gyllenskjöld, Vilhelm, 146–7
Carter, Richard Henry, 26
cathodes, 69–70, 75, 226
Cauchois, Yvette, 184–5
celtium (proposed element name), 82–3, 85, 177–8, 230
Chadwick, Captain G.E., 117, 121–3, 276–8, 280
Chadwick, James, 62, 230, 233, 236, 238
Chandra space probe, 207
chemical atom, 55–6, 60, 238
Chemical News, 178, 184
ChemMin experiment, 206
Chwolson, O.D., 233
Clevedon, Somerset, 19
Clifton, Robert Bellamy, 33–4, 37, 46, 235
cobalt, 5, 14, 20, 60, 216
columbite, 180
Compton, A.H., 231
Conroy, Sir John, 31
Cook, Charles, 80, *81*
Copley Medal *see* Royal Society of London
copper, 272
Corbino, Orso Mario, 159–62, *159*, 166, 167
Cork, James, 188–90
Corson, D.R., 186
Coryell, Charles and Grace Mary 190
cosmic X–ray sources, 207
cosmic–ray physics, 232
Cossor, A.C. (company): manufactures X–ray tubes, 73, *74*, 75–6, 78
Coster, Dirk, 177–8, *178*, 192, 199, 237
Coulomb's law, 236
Cox, General H.V., 118
Crocker, Captain H., 122, 275
Crookes, Sir William, 50, 82
Crowther, J.A.: *Ions and Ionizing Radiations*, 232
crystals, 142
Curie, Marie, 49, 140, 231
Curie, Pierre, 49
Curiosity (NASA's Mars Rover), 199, 206
curium, 206
cyclotron, 181

Dalton, John, 45–6, 50, 62, 225; *A New System of Chemical Philosophy*, 46
Dardanelles *see* Gallipoli

Darwin, Charles, 57, 254
Darwin, Charles Galton: collaborates with HM in Manchester, 57–60, 69–71, 137–8, 224, 226; commissioned in Royal Engineers, 60–1, 202; contributes to HM memorial, 62; letters from HM on spectrum work, 72, 77; and X–ray spectroscopy, 204; and Bohr, 225; Lawrence Bragg tests calculations, 235
Dauvillier, Alexandre, 83–5, 177, 199, 231
Davidson, Maurice, 32
'Dear Harry' exhibition, Oxford (2015–16), 3–4, 9, 244–67, *246*
Debierne, André–Louis, 49
de Havilland, Hugh, 25
Department of Scientific and Industrial Research (DSIR), 124
Deslandres, Henri, 139
Devons, Samuel, 62
Dewar, James, 47
Diamond Light Source, Harwell, 210, *210*
Dirac, P.A.M., 6
dor, 187
Dresden: Hochschule für Technik and Wirtschaft, 247
Druce, J.G.F., 183–4
dwi–manganese, 179–80, 188
Dyer, John Maximilian, 25–6

Eastern Telegraph Company Mediterranean, 116
Eddington, (Sir) Arthur Stanley, 46
Edwards, P.P., 284
Egdell, Russell G., vii, 1, 8–9, 155, 175, 199, 247, 284, 299
Egerton, (Sir) Alfred, 26
Eggar, William, 25–6
Einstein, Albert: wins Matteucci Medal, 8, 155, 168; nominated for Nobel Prize, 146; equation for photoelectric effect, 202, 215
eka–caesium, 183, 188
eka–iodine, 183, 188
eka–manganese, 178–80, 188
electron energy loss spectroscopy (EELS), 209, *212*, 213
electrons, 4–6, 45, 48, 69, 99, 202, 228–9, 270–2
elements: high–frequency spectra, 60, 72; and K–type emissions, 72–3; and X–ray spectroscopy, 175; primordial superheavy, 177, 190–3
emanation (radioactive), 50–1, 54, 94, 98
Eton Chronicle, 19, 22
Eton College: HM attends, 14, 16–17, 20–3, 253; 'Army Class' (Corps), 16, 36; King's Scholars, 16, 20–1, *22*, 24, 30; science at, 23–4, *24*; Scientific Society, 26; sport at, 27; HM leaves for Oxford, 30
Evans, E.J., 61

Fajans, Kasimir, 55, 176
Fermi, Enrico, 161
Fernandez, Lorenzo, 187
First World War: Nobel Prizes in, 142–5; Central Powers excluded from international

Darwin, Charles Galton *continued*
 science after Armistice, 145–6
Fletcher, Horace, 22
Fogg, H.C., 188–90
Foote, Paul, 231
francium, 176, 183, 186
Frederick–Frost, Kristen M., 7, 67, 246–7, 299
Fricke, H., 200
Friedman, Robert Marc, 8, 136, 300
Friedrich, Walter, 142

Gaia, 90
Gallipoli: Anzac Cove, 1, 117–18, 259; HM killed and buried at, 8, 45, 120–3, 248; Dardanelles campaign (1915), 107, 115, 118–21, 249, *250*, 260–1, *261*; HM sent to, 115–16, 259; Cape Hellas, 116; signalling, 116, *261*; battle of Sari Bair, 118–19, *119*, 122; Suvla Bay, 118; battle of Chunuk Bair, 128, 130, 141, 269–70, 272, 288, 290
Galvani, Luigi, 157
gamma ray, 54, 158, 226, 229
gamma–ray spots, 57
Garbasso, Antonio, 166–7, *167*
Garibaldi, Giuseppe, 159
Geiger Counter, 51, 238
Geiger, Hans Wilhelm, 4, 51, 53–4, 61, 96–7, 102
Geiger–Marsden experiment, 8, 167, 206
Gentry, R.V., 191
Gerlach, Walther, 232
Germany: scientists and Nobel Prizes in First World War, 143–4; scientists boycotted then recognised after Armistice, 145–8
Giacconi, Riccardo, 207, 216
Giesel, Friedrich Oscar, 50
Glazebrook, Sir Richard, 115
Glendenin, Lawrence, 190
Godley, Major General Alexander, 120
Gogh, Vincent van, 91
gold, 175
Gooday, Graeme, 247
Goodheart, Arthur, 21
Granqvist, Gustaf, 149
Gregory, Sir Richard, 179
Griffith, Idwal, 34, 37
Gullstrand, Allvar, 142, 149–51, 200

Haas, Arthur Erich, 231
Habashi, Fathi, 183
Haber, Fritz, 148
hafnium, 1, 84, 175, 177–8, 192
Hahn, Otto, 176
Haldane, Richard Burdon (*later* Viscount), 36, 108

Hale, George Ellery, 139
Hamer, Richard, 179, 283
Hamilton, General Ian, 108
Hammarskjöld, Hjalmar, 141, 144
Hammarsten, Olof, 141–2
Hanawalt, J.D., 201
hard X–ray photoelectron spectroscopy (HAXPES), 216
Harris, J. Allen, 188
Hartley, (Sir) Harold, 29–30
Havilland, Hugh de *see* de Havilland, Hugh
Heilbron, John, 9, 30, 52, 100, 224, 248, 258, 300; *H.G. Moseley: The Life and Letters of an English Physicist*, 3, 235
helium, 4, 6, 50–1, 92, 102
Helvetica Chemica Acta, 185
Henderson, George, 30
Hensel, Friedrich, 284
Hertz, G., 201
Hevesy, Georg von, 59–60, 61, 82, 84–5, 177–8, 192, 225, 237
Hirsh, F.R., 184, 187
Hoffmann, Roald, 284
Hooke, Robert, 15
Hopkins, B. Smith, 187–8
Hopkins, Clare, 7, 14, 108, 300
Hughes Medal *see* Royal Society of London
Hulubei, Horia, 184–7
Hunt, Simon Vaughan, 248
Huxley, Julian, *22*, 24–5, 27–30
Huxley, Thomas Henry, 23
hydrogen: spectrum, 59; Broek on atom, 92–3; orbitals, 271

illinium, 188
Innes, P.D., 201
insertion device, 209–10
Institut International de Physique Solvay, 79
instrument makers and suppliers, 73, 75–6, 78–81, 85
International Research Council, 148, 150
International Union of Pure and Applied Chemistry (IUPAC), 187, 190–1
iodine, 183, 186
isotopes, 176, 190; radioactive, 191

Jaffe, Bernard: *Moseley and the Numbering of the Elements*, 234
James, Charles, 187–90
Jeffreys, John Gwyn, 17
Jervis, Henry, 32, *32*, 109
John Harling Fellowship, 57
John Hopkinson Electro–Technical Laboratory, Manchester, 47
Johnston, Stephen, 244, 247, 301
Joule, James, 46

Journal of the American Chemical Society, 188
Jowett, Benjamin, 29
Junior Scientific (Oxford Club), 36
K–shell hole states, 210
Kay, William, 51, 61
Kaye, G.W.C.: *X–rays*, 232
Keeley, T.C., 179
Kelvin, William Thomson, 1st Baron, 46, 158
King's Scholar *see* Eton College
Kirchoff, Gustav, 158
Kirkby, Paul, 37
Kitchener, Horatio Herbert, 1st Earl, 110, 114, 259
Klein, Oskar, 232
Knipping, Paul, 142
Knoll, Max, 207
Kossel, Walther, 229
Kuroda, P.K., 183

Lamb, Horace, 46
Langworthy, E.R., 46
Lankester, Sir Edwin Ray, 19, 115
Lattey, Robert, 29, 35
Laue, Max von, 56–7, 139, 142, 144, 166, 225–6
Lawrence, Ernest, 181
Ledoux–Lebard, René, 231
Lemnos (island), 116
Lenard, Philipp, 14
lepidolite, 184
Linac Coherent Light Source (LCLS), Stanford, 211, *212*
Lindemann, Frederick, 235–6
Lister, A.: *Moseley and the Numbering of the Elements*, 234
lithium, 228
livermorium, 192
Lodge, Sir Oliver, 110
Lorgna, Antonio Maria, 155–6
Loring, F.H., 183–4
Lovelock, J.: Gaia concept, 90
low–temperature physics, 232
Ludlow–Hewitt family, 245, 266
Ludlow–Hewitt, Alfred, 258
Ludlow–Hewitt, Edgar, 113
Ludlow–Hewitt, Margery (*née* Moseley; HM's sister): letters from HM, 19, 23, 26–8, 36, 109–10, 117, 253; illus., *20, 251*; learns of HM's death, 121, 262; and HM's religious conservatism, 225; marriage, 246, 266; closeness to HM, 252
lutecium (now lutetium), 82–3, 85, 177

M lines (spectroscopy), 149
McGill University, Montreal, 49–50

Mackenzie, K.R., 186
MacLeod, Roy, vii, 1, 301
Makower, W., 55, 61
Malmer, Ivan, 231
Manchester Literary and Philosophical Society, 53
Manchester, University of: HM at, 7, 15, 45–6, 52–5, 256; radioactive contamination, 7, 55; Physical Laboratories, 45, 47–8, *48*, *49*, 50; and chemical atom, 46; Owens College, 46, 61; group photographs, *56*, *58*; laboratory staff serve in forces, 60–1; post–First War changes, 60–2; HM memorial plaque, 61, *61*; Lawrence Bragg at, 235
manganese, 178–9
manganite, 179
Manhattan Project, Oak Ridge, Tennessee, 190
Marconi, Guglielmo, 160
Marinsky, Jacob, 190
Marsden, Ernest, 4, 26, 61, 96–7, 102
Marsden, Hugh, 26
masurium *see* technetium
Matteucci, Carlo, 156–8, *157*
Matteucci Medal (*Medaglia Matteucci*): awarded to HM, 8, 155; established, 157–8; nomination process, 158–9; committee report on HM, 161–2; gold content, 163; illus., *165*; and Nobel Prize laureates, 166–7, 169–71; status and reputation, 167–8; list of winners, 169–71
Mediterranean Expeditionary Force (First World War), 115
Meitner, Lise, 176
Mendeleev, Dmitri Ivanovich, 2, 71, 93, 137–8, 178–9, 183
Menger, Carl, 91
mica, 191
microscopes, electron, 207–9, *208*
Mie, Gustav, 232
Millard, Thomas, 31
Millikan, Robert, 232
Minder, W., 185
Mohler, F.L., 231
molybdenum, 179, 181
Monash, General (Sir) John, 118
monazite, 191
Moseley, Amabel (HM's mother) *see* Sollas, Amabel
Moseley, Amabel (HM's sister): death in childhood, 19, 251–2
Moseley, Elizabeth (Betty; HM's sister), 19–20, *20*, 252
Moseley, Henry Gwyn Jeffreys: birth, 1, 251; career and achievements, 1–4; nominated for Nobel Prize, 1, 8, 136–9, 144–5, 166, 167, 263; as army Signals Officer, 8, 114–15; awarded Matteucci Medal posthumously, 8, 155–6, 168; killed at Gallipoli, 8, 45, 61, 107, 120–3, 233, 248, 251, 259, 262; attends Eton, 14–23, 253; illus., *14*, *17*, *20*, *21*, *22*, *28*, *32*, *35*, *38*, *56*, *109*, *114*, *251*, *253*, *254*, *255*; at Summer Fields school, 16, 253; mother's influence on, 17; family background, 18–19; academic achievements at Eton, 23; scientific studies at Eton, 23–5; bird–nesting and natural history, 27–9, 252; rowing, 27, 35, 108, 114, 253, 255; applies for Oxford scholarship, 29–30; life at Trinity College, 31–2; studies physics at Oxford, 31, 33–5, 255; activities and interests

at Oxford, 35–6, 108–9, 255; joins Officer Training Corps, 36; awarded Second Class Honours degree, 38, 256; works under Rutherford in Manchester, 45, 52–3; dislikes teaching, 53; cooperates with Darwin, 57–9; meets and works with Bohr, 59–60, 224–5, 227–9; on atomic number, 60, 90–1, 94, 98, 103, 137, 175; moves from Manchester to Oxford, 60; memorials, 61–2, *61*, 259, 281–4; diagram, 67, *68*, 85; knowledge of chemistry, 90–1; military career, 107–11, 258; and periodic table, 107, 179, 266; applies for Birmingham physics chair, 110; signalling training in army, 111–13; declined by Royal Flying Corps as too heavy, 113–14; secures better equipment for Gallipoli, 115; sent to Gallipoli, 115, 259; will, 115, 123, 259; active service and life at Gallipoli, 116–18; and Barkla's Nobel Prize, 146–9; and Siegbahn's Nobel Prize, 150–1; Mateucci Medal committee report on, 161–2; X–ray spectroscopy, 175–6, 179, 185, 192, 204, 227; correspondence with Charles James, 188; staircase (X–ray emission spectra), *212*; collaborators and mentors, 224–5; conservatism, 224–5; on gamma–ray diffraction, 226; as loss to science, 233–4; biographies amd assessments, 234; speculation on future research, 236–8; amateur acting, *253*; experimental apparatus, 256, *257*, 258; obituary tributes, 259, 263; primary accounts of death, 275–80; published works, 288; 'High–Frequency Spectra of the Elements', 8, 67, *68*, 82, 137, 187, 192, 216; 'The Number of Beta–Particles Emitted in the Transformation of Radium', 54

Moseley, Henry (HM's grandfather), 18–19

Moseley, Henry Nottidge (HM's father), 18–19, 251

Moseley, Margery (HM's sister) *see* Ludlow–Hewitt, Margery

Moseley Research Studentship (*later* Alan Johnston, Lawrence and Moseley Research Fellowship), 123–4

Moseley's ladder, 60

Moseley's law, 4, 7, 67, 192, 224, 230–2, 234, 237–8, 244, 258

moseleyum: proposed as name for element 43, 179, 283

Mudros (island of Lemnos, Greece), 116

Murphy, Clarence J., 187, 190

Nagel, David, 31–3, *33*, 37, 90

National Academy of the XL *see* Società Italiana

National Physical Laboratory, 115

Nature (magazine), 5, 84, 94, 96, 98, 188, 263

near–edge X–ray absorption fine structure (NEXAFS), 213

Nernst, Walther, 140

neutrons, 236

nickel, 5, 10, 60, 102, 204

Nobel, Alfred, 148

Nobel Prize: HM nominated for, 1, 8, 136–9, 144–5, 167, 263; committees' impartiality and considerations, 140–3, 147–50; early history, 140; in First World War, 142–4; and Matteucci Medal winners, 166–7, 169–71; list of laureates not awarded Matteucci Medal, 171

Recipients: Arrhenius (chemistry), 137; Aston (chemistry), 167; Barkla (physics), 146–9, 168, 216; Braggs (physics), 57, 167; Curies (physics), 167; Einstein (physics), 146, 166; Fermi (physics), 161; Giacconi (physics), 207; Marconi (physics), 160; Perrin (physics), 184, 232; Ramsay (chemistry), 167; Röntgen (physics), 166; Rutherford (chemistry), 50, 167; Kai Siegbahn (physics; half–share),

Moseley, Henry Gwyn Jeffreys *continued*
    214, 216; Manne Siegbahn (physics), 136, 168, 216; von Laue (physics), 56
statutes, 141
Noddack, Ida *see* Tacke(–Noddack), Ida
Noddack, Walter, 179–83
nuclear charge, 5
nuclear fission, 192
nuclear physics, 232

Oak Ridge, Tennessee, 190–1
Offi, Francesco, 8, 155, 301
Officer Training Corps: established, 36
oganneson (element 118), 2
Ogawa Masataka, 179
Organessian, Yuri, 284
Oseen, Carl Wilhelm, 149
Ostwald, Wilhelm, 140
Owens College *see* Manchester, University of
Owens, John, 46
Oxford (city): Museum of the History of Science, 3, 9, 82–3, 244, 247–8; HM's family in, 20; *see also* 'Dear Harry' exhibition
Oxford, University of: natural sciences at, 15, 31; HM attends, 29, 255; Clarendon Laboratory, 31, 33–4; Alembic Club, 36, 110, 255–6; Officer Training Corps (OTC), 36, 108, 255; Electrical Laboratory, 104; rowing, *114*; HM graduates with second–class degree, 256; *see also* Balliol College; Trinity College
oxidation, 214

Pais, Abraham, 94
Panaccione, Giancarlo, 8, 155, 301
Paternò, Emanuele, 166
Pathfinder mission (to Mars), 206
Pauli Exclusion Principle, 271
Pauli, Wolfgang, 158
Peed, W.F., 190
Perey, Marguerite, 186–7
periodic table of elements: HM predicts four missing elements, 1; based on atomic number, 2; and X–ray spectroscopy, 7; Bohr on ordering of, 60; Mendeleev's, 71, 183; Broek on, 92–3, 95, 98–9; and Moseley's law, 107, 231, 266; identifications and gaps, 175–83
Perrier, Carlo, 181
Perrin, Jean Baptiste, 150, 184, 186, 232
Philosophical Magazine, 5, 53, 58–60, 70, 97, 100, 226
photon, 202, 211, 213, 215–16, 272–3
Physical Review Letters, 191
Physikalische Zeitschrift, 95, 233
pitchblende, 49
Planck, Max, 6, 55, 73, 100, 139, 142, 146
plum pudding model, 4

plutonium, 206
Poincaré, Henry, 167
polonium, 175
Porter, Thomas Cunningham, 24–5, 29
potassium ferrocyanide, 33, 72, 257
praeseodymium, 99
primordial superheavy elements *see* superheavy elements
Proceedings of the Cambridge Philosophical Society, 185
Proceedings of the National Academy of Science, 188
Proceedings of the Royal Society, 58
promethium, 1, 175, 187–90
protactinium, 176
Prout hypothesis, 92
pyrolusite, 179

quantitative wave theory, 56
quantum atom, 62, 225, 231
quantum numbers, 271
quantum theory, 6, 146, 227, 229

radiation: study of, 70–1; Planck's law of, 100; synchrotron, 191
radioactive contamination (Manchester), 7, 55
radioactivity: discovery, 45, 50; in astatine and francium, 183; disintegration of elements, 185
radium, 50–1, 96, 175, 183
radium B, 185
radium battery, 56
Radium C, 96, 229
radon, 175, 183, 185–6
Ramsay, Sir William, 51, 144, 167, 179, 188
rare earths, 67, 73, 82–5
Rattazzi, Urbano, 157
Rayleigh, John William Strutt, 3rd Baron, 47
Redman, Lister A., 234
Regnault, Henri Victor, 158
relativity, 146
resonant inelastic X–ray scattering (RIXS), 211
Reynolds, Osborne, 46
rhenium, 1, 175–6, 180, 182, 192
Richards, T.W., 140, 141, 144–5, 148
Richardson, O.W.: *Electron Theory of Matter*, 232
Righi, Augusto, 159–62, *159*, 166
Robinson, Harold Roper, 61, 201–2, 214
Roiti, Antonio, 159, *159*, 161–2, 166
Rolla, Luigi, 187
Rolleston, George, 19
Röntgen, Wilhelm Konrad von, 48, 142, 166
Roscoe, Henry, 46

Rosetta mission (to comet 67P), 206
Royal Corps of Signals, *112*, 258
Royal Engineers: HM commissioned in, 111
Royal Society of London: receives bequest in HM's will, 115, 123, 202, 259; Copley Medal, 157; Hughes Medal, 200; Moseley Research Studentship, 202
Royal Swedish Academy of Sciences, 136, 139, 142, 144, 150, 168
Royds, Thomas, 51
Ruska, Ernst, 207
Russell, A.S., 59
Russell, John Wellesley, 31
Rutherford back–scattering, 206
Rutherford, Ernest (*later* Baron): HM works with in Manchester, 1, 7, 45, 52–4, 256; on atomic structure, 4, 52–3, 71, 97, 227, 228; HM first hears of, 26; advertises assistant's post on Manchester, 36; research on radiation, 48–51; Nobel prize for chemistry (1908), 50; returns to Manchester from McGill (Canada), 50; illus., *52, 56*; on danger of radioactivity contamination, 54; Bohr cooperates with, 55, 59; affected by HM's death, 61; appointed to Cavendish Chair in Cambridge, 61, 236; wartime anti–submarine research, 61; contributes to HM memorial, 62; pays tribute to Manchester University, 62; letter from HM on X-ray tubes, 75; and HM's complaint about work at Oxford, 78; succeeds Schuster at Manchester, 79; and HM's results for celtium, 82; and rare earths research, 83; van den Broek studies researches, 94; on atomic weight, 96, 101–2; van den Broek prefigures, 103; attempts to keep HM from active service, 115, 233, 263; obituary of HM, 122, 263; and HM's work on atomic number, 137; Arrhenius promotes, 140; nominates Barkla for Nobel Prize, 146, 148; and presentation of HM's Matteucci Medal, 163; awarded Matteucci Medal, 167; and decay of radon, 185; and X-ray spectroscopy, 185, 206; promotes and supports HM, 224; on beta and gamma rays, 226; unwillingness to allow HM and Darwin to take up X–rays, 226; asks Chadwick to repeat HM's scattering experiments, 230; correspondence with HM, 234; money limitations, 236; on electron–proton pairs, 238; research group in First World War, 247; condemns waste at HM's death, 262; *Uranium Radiation and the Electrical Conduction Produced by It*, 49
Rydberg constant, 6, 228

Sacks, Oliver, 283
samarium, 99
Sarton, George, 233–4
Scacchi, Arcangelo, 158
scanning Auger microscopy (SAM), 216
scanning transmission electron microscopy (STEM), *208*, 209
Scerri, Eric R., 7, 9, 90, 282, 302
Schmidt, L.R.E., *254*
Schrödinger, Erwin, 6
Schuchardt, Dr (chemical supply house), 82
Schück, Henrick, 148
Schuster, Sir Arthur, 45–7, 50–1, 54, 79
Science (journal), 188
Scoones, Paul, 25
Segrè, Emilio, 161, 181, 186

semaphore signalling, 111–12, *111*
Sidgwick, Nevil, 263
Siegbahn, Kai, 202, 214–15
Siegbahn, Manne, 8, 136, 149–51, 168, 175–7, *176*, 199–203, 231–2
Siegbahn notation, 273
Siemens (company), 207
Sitzungsberichte der Preussischen Akademie der Wissenschaften phys.–math, 181
Società Italiana delle Scienze detta dei XL: awards Matteucci Medal, 155; Fellows, 156; foundation and name changes, 156
Soddy, Frederick: collaborates with HM, 50; on radioactivity decay, 51, 59; on place of element in periodic table, 98; letters from HM in Gallipoli, 117; questions HM's laws, 230; proposed element name, 283
Sollas, Amabel (*née* Jeffreys; then Moseley; HM's mother): illus., *17, 251*; influence on HM, 17, 38; marriage, 18; letters from HM, 25, 27, 29–30, 36–7; accompanies HM to Australia, 110; remarries, 113; HM sees for last time, 115; informed of HM's death, 121–3, 262, 276–8; acknowledges posthumous award of Matteucci Medal to HM, 163, *164*; pocket diaries, 248, 258, *262, 265*; telegram of sympathy from Buckingham Palace on HM's death, 264, *265*; wartime nursing, 264, *264*; death (1928), 266
Sollas, William J., 113
Solvay Institute *see* Institut International de Physique Solvay
Sommerfeld, Arnold, 151, 167, 200, 231; *Atombau*, 231–2, 237
space: X–ray spectroscopy in, 205–7
Spallanzani, Lazzaro, 155
spectrometers, 69–70, 78, *79*, 80, 83, *84*
Spektroskopie der Röntgenstrahlen (ed. Siegbahn), 201
spherical halos, 191
Spitzer, E.J., 190
Spring 8 facility (Japan), 210
Stark, Johannes, 146–8
stars: X–ray spectra, 206
Stewart, Balfour, 46
Summer Fields (prep school), 16, *253*; war memorial, *124*
superheavy elements, 177, 190–3
Surdo, Antonino lo, 167
Svedberg, Theodor, 150
Svenska Dagbladet (Swedish newspaper), 143
Sweden: neutrality, 141, 143, 146–7
synchrotrons, 209, 216

Tacke(–Noddack), Ida, 179–83
Tavani, Marco, 158
technetium ('masurium'), 1, 175–6, 178–83
Telephone D Mark III (DIII), 117, 119
13th Division: HM serves in, 114, 116–19, 122, 259
Thomas, Sir John Meurig, 284
Thomson, (Sir) John Joseph, 4, 46–8, 59
thorium, 50, 183

Tizard, (Sir) Henry, 36, 110, 124
Todd, Neil, 7, 45, 302
Townsend, John Sealy E., 34, 37, 81, 256
transmission electron microscopy (TEM), 208–9
Trinity College, Oxford: HM attends, 14–15, 30, 108; Millard Scholarship, 29, 90, 104, 108, 255; character and activities, 31–3, 36, 108; Torpids (rowing), 114, *254*
tungsten, 75
Twain, Mark: *Mysterious Stranger*, 238

uranium, 48–9, 183
uranium–X, 50
Urbain, Georges, 82–5, 98, 177–8, 230, 263

Valdares, Manuel, 185
Versailles Treaty (1919), 148
Victoria University of Manchester *see* Manchester, University of
Vittorio Emanuele II, King of Italy, 158
Vleck, J.H. Van, 231
Volta, Alessandro, 155–6
Volterra, Vito, 162

Waals, J.D. van der, 145
Wacker, Markus, 247, 257
Walters, Frank (Francis Paul), 30
Warburg, Emil, 142
wave mechanics, 6
wavelength dispersive X–ray fluorescence (WDXRF), 203
Weiner, C., 181
Welsbach, Carl Auer von, 82, 98
Werner, Alfred, 140
Whiddington, Richard, 100, 227
Whipple, Tom, 284
White, Harvey, 232
Wien, Willy, 232
Wilkins, John, 15
Williams, Charles Eccles Edmund, 16
Willis, Thomas, 15
Willstätter, Richard, 141, 144–5
Wolff, Martin, 247, 257
Wood, Robert, 167
Wooster, W.A., 185
Wynne, W.P., 178

X–ray absorption spectroscopy (XAS), 200–1, 211, 213–14
X–ray crystallography, 62
X–ray diffraction, 56–7, 192–3
X–ray emission spectroscopy (XES), 200–1, 203–7; in space, 204–6, *205*; synchrotron based, 209–10; and electronic structure, 272

X–ray fluoresence (XFS), 203
X–ray photoelectron spectra and spectroscopy (XPS), 201–3, 214–16, *215*
X–ray spectra: illus., *178, 180*; labelling, 270–4
X–ray spectroscopy: HM's work on, 7–8, 58–9, 69, 107, 149; development, 9, 150, 199–203, 216, 227; Braggs' research on, 69; Arrhenius on advances in, 137; Siegbahn develops, 150, 175; and discovery of new elements, 175–9; Kuroda on, 183; Yvette Cauchois designs equipment, 184; Rutherford uses, 185; portable, 203–4, *204*; in space, 204–7; in electron microscope, 207–8
X–rays: characteristic, 4, 58, 185, 201, 204, 206, 208, 213, 257; frequencies, 4, 7; K–type, 4, 13, 69, 71–3, 76–8, 82, 180, 199, 201, 207, 217, 227, 237, 270, 272–3; secondary emissions, 4, 69; tubes, 7, 69, 71–3, *74*, 75–9, *76, 77*; hard, 13, 79, 82–3, 86–7, 209, 213–14, 219, 223–4, 226, 236, 239, 282–3; soft, 13, 77–8, 82–3, 88, 201, 209, 223–5, 236, 238–9, 241, 250, 282; discovery, 45, 48; reflection, 58; high–energy, 67; L–type, 69, 71–3, 78, 185, 191, 195, 226–7, 229, 270, 272–3, 282; M–type, 199, 209; absorption edges and emission lines, 201; emission spectrum, 211; Rutherford's view of, 226
XMM–Newton mission probe (space), 206
xu unit, 200

Yntema, L.F., 188
Young, Thomas, 158
Young–Matteucci, Robinia, 158
ytterbium (*formerly* neoytterbium), 82–3

Zingales, Roberto, 183
zirconium, 177, 192

Zsigmondy, Richard, 150

Chemistry Review 2-4, i.e. 2   2004   2⁺ (.1-8
                                           i.e. Novent

Jan 5. Best man to JEWood
     9 Mother to CM Wood
    11   "   in law to Blake Bryon-
    22 p19 middle H not bom-ley (oxford)
   Top r/ 31 p59/Thomp sr
   Will se see 25a
    offense = offensive
       not - ence
   Estab + Pub Hels = £10h